Bike, Scooter, and Chopper Projects for the Evil Genius

Evil Genius Series

Bike, Scooter, and Chopper Projects for the Evil Genius

BRAD GRAHAM
KATHY McGOWAN

New York   Chicago   San Francisco   Lisbon   London   Madrid
Mexico City   Milan   New Delhi   San Juan   Seoul
Singapore   Sydney   Toronto

Library of Congress Cataloging-in-Publication Data

Graham, Brad.
    Bike, scooter, and chopper projects for the evil genius / Brad Graham, Kathy McGowan.
        p. cm. – (Evil genius series)
    Includes index.
    ISBN 978-0-07-154526-6 (alk. paper)
    1. Bicycles–Design and construction–Amateurs' manuals. 2. Tricycles–Design and
construction–Amateurs' manuals. 3. Pedicabs–Design and construction–Amateurs' manuals.
4. Motor scooters–Design and construction–Amateurs' manuals. I. McGowan, Kathy. II. Title.
    TL400.G689 2008
    629.227–dc22   2008008030

McGraw-Hill books are available at special quantity discounts to use as premiums and sales
promotions, or for use in corporate training programs. To contact a representative please visit
the Contact Us pages at www.mhprofessional.com.

Bike, Scooter, and Chopper Projects for the Evil Genius

1 2 3 4 5 6 7 8 9 0 QPD/QPD 0 1 3 2 1 0 9 8

ISBN 978-0-07-154526-6
MHID 0-07-154526-3

This book is printed on acid-free paper.

Sponsoring Editor	**Indexer**
Judy Bass	Warren Burggren
Editing Supervisor	**Production Supervisor**
David E. Fogarty	Pamela A. Pelton
Project Manager	**Composition**
Imran Mirza	Keyword Group Ltd.
Copy Editor	**Art Director, Cover**
Sharon Cawood	Jeff Weeks
Proofreader	
John Bremer	

About the Authors

Brad Graham is an inventor, robotics hobbyist, founder and host of the atomiczombie.com website. He is the coauthor of *51 High Tech Practical Jokes for the Evil Genius, 101 Spy Gadgets for the Evil Genius, Atomic Zombie's Bicycle Builder's Bonanza*—perhaps the most creative bicycle-building guide ever written—and *Build Your Own All-Terrain Robot*, from McGraw-Hill Professional.

Kathy McGowan is Mr. Graham's other "Evil Genius" half, providing administrative, logistical and marketing support for Atomic Zombie's many robotics, bicycle, technical, and publishing projects. She also manages the daily operations of their high-tech firm and several websites, including www.atomiczombie.com

Contents

Greetings! Glad that you decided to take an interest in the best hobby you will ever know! Maybe you are already a seasoned garage hacker and have decided to build a few of the projects from this book, adding your own special modifications, of course. Or, maybe you have never thought about taking an angle grinder to a working bicycle, chopping it into 50 pieces so you can twist and contort it into something unique and wonderful. Well, dude, let me tell you, once you start building your own custom vehicles, you will never stop. Some of your neighbors will love what you do, visiting you often to see what you are working on. Others will fear you like some type of mad Evil Genius, and others will simply give you that empty look that the dog often gives you when you try to make him understand complex math. Now, seriously, are you ready to be the "freak on the block," the dude who rides a pedal-powered chopper that looks sicker than the one for sale at the Harley store? Are you ready to beat your neighbors' Hummer from one set of lights to the next on a bicycle that weighs less than his spare tire? Well, if this all sounds good, then you are like me and are going to enjoy what you are about to read!

Nothing I build is impossible for anyone with a little motivation. I own only an angle grinder, basic AC welder, and the usual small hand tools. I have no lathe, no drill press, no chop saw, and often have no heat in my garage during the winter. I get most of my parts from the city landfill site, and I have no degree on my wall that says I am qualified for the engineering feats that I often achieve. What I do have is the desire to hack, weld, and ride! I have seen young garage hackers build great bikes after only a week of owning a lunchbox-sized welding machine. I have seen "old dogs" learn these "new tricks."

In fact, I have never seen anyone fail at this hobby if they put some effort and imagination to the grindstone. You can certainly build everything presented in this book and more by simply becoming motivated enough to get your hands dirty, sweat a little, and maybe even curse a bit. If it was easy to build a great invention, then every clown on the block would be doing it, and what would it be worth then? Cool rides do *not* come with serial numbers. A custom is a one of a kind! Enough ranting. Let's get down to business and start building.

Acknowledgments

Our Evil Genius collaborator, Judy Bass at McGraw-Hill, has always been our biggest fan, and we can't thank her enough for believing in us every step of the way. Thanks to Judy and everyone at McGraw-Hill for helping to make this project a reality. Completing two books in one is insane, but we did it! You're simply the best, Judy!

Thanks also to all of you who contact us, especially members of the "Atomic Zombie Krew," our international family of Evil Geniuses, bike builders and robotics junkies. We sincerely appreciate your support, friendship and feedback. You're the best creative "krew" in the world.

We also appreciate the help and enthusiasm of many friends, family and neighbors. Thanks for posing for pictures and accepting us for who we are. We're having an awesome time!

There are many projects, blogs, videos, a builders gallery and support at **www.atomiczombie.com**. We always look forward to seeing what other Evil Geniuses are up to. Hope to see you there!

Cool stuff, cool people!

WWW.ATOMICZOMBIE.COM

Getting Started

Project 1: A Complete Bicycle Autopsy

Building custom bicycles is a great hobby that can be learned by anyone with a desire to create. The skills needed to dismantle, alter and repair bicycle components can be easily learned, and the parts and tools you will need are quite inexpensive. Discarded or worn out bicycles offer many good parts, and can often be found at local scrapyards, city dumps, or yard sales for a few

dollars. Even if you plan to build a custom creation using all new parts, this hobby will seem inexpensive compared to many, as you can purchase a brand new bicycle at a store for less than a hundred dollars. The great thing about hacking and welding bicycles is that you will be working with all steel components, which are much stronger, more common, and much less expensive

Figure 1-1 *A typical all-steel suspension mountain bicycle*

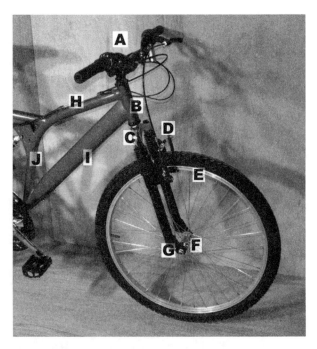

Figure 1-2 *Front component details*

than high grade aluminum or carbon fiber bicycle parts. If you have never torn a bicycle apart before, then this basic introduction will show you all you need in order to complete a total bicycle autopsy in minutes, stripping an entire cycle down to the individual parts using only a few basic hand tools. There will be some very useful tips and tricks presented that may save you a lot of frustration, especially if you are just starting out, so read through this entire section before embarking on any of the upcoming projects.

Figure 1-1 shows the most commonly available and inexpensive mountain bike available today, the all-steel frame suspension mountain bike from the local hardware store. This cycle cost me $120, and was used to make the StreetFox trike, as well as the LongRanger electric scooter presented in this book. The components are medium quality, and include aluminum rims, cantilever brakes, and suspension on both the front forks and rear triangle. Because the frame is made of steel, it can easily be cut and welded using any welder at all. Often, a bicycle like this can be found at a yard sale for a few dollars, although there may be a bit of rust on the frame, worn out tires, and the odd seized brake cable—nothing that we can't easily fix or replace. OK, now grab your toolbox, and let's tear this bicycle down to the individual parts.

Starting with the front of the bicycle, Figure 1-2 shows the parts that you should get to know by name. As per the letters, the components are:

A.   **Handlebars, Gooseneck, Brake Levers and Shifters.** Handlebars are held in place by the clamp on the gooseneck and are available in many widths and heights. Often, a mountain bicycle will

have straight or slightly curved handlebars such as these ones, whereas a road bike will have "curly" handlebars which allow the rider to hold on in two positions: a relaxed upper position, and a more aerodynamic "tuck" position. The gooseneck fits into the forks stem, and is held there by a wedge, which will be shown in greater detail later on. Goosenecks are available with two common stem diameters, so make sure you don't put the smaller sized gooseneck into the larger sized fork stem, or it will not be completely secured.

B.   **Head Tube and Fork Stem.** The head tube is the part of the frame that the fork stem is inserted into. The two cups on the top and bottom of the head tube carry the fork bearings, and will be shown in greater detail later. Head tubes are available in two common diameters, which means that there are also two common sizes of head tube cups and bearings. Again, always ensure that the parts are the same size, or there will be excessive friction in the steering system. There is no common standard for the length of the head tube, or the length of the fork stem, so you should keep matching parts together as a set as you collect them.

C.   **Front Forks.** Front forks come in a vast array of sizes, shapes and styles, ranging from the most basic straight leg style to the ultra heavy duty, triple-tree motocross style forks used for downhill mountain bikes. The front forks will fit only one size of front wheel properly, and the most common sizes for the bicycles you will be working with are: 26 inch, 24 inch, and 20 inch. Most modern front forks will also include the front brake mounting hardware, such as the one shown in Figure 1-2.

D. Front Brakes. The front brakes are the most important brakes on most bicycles, as they do the most work. Modern bicycles have cantilever brakes installed on the front forks, but you may also find some brakes that connect to the front forks using a single bolt through the crown of the fork. The type of brakes that connect to the fork using a single bolt are caliper style brakes, which are much less effective than the cantilever style shown in Figure 1-2, due to the fact that they do not exert as much friction on the front rim.

E. Front Wheel. Bicycle rims are available in many sizes and styles, but the 26-inch rim with 36 spokes is by far the most common wheel for an adult sized bicycle. Extremely cheap rims are made of steel, do not have stainless steel spokes and should be avoided due to poor braking characteristics and strength. Note that 20-inch diameter wheels are often used for children's bicycles and freestyle BMX bikes, and they can have as few as 26 spokes and as many as 48. BMX wheels with 48 spokes are extremely strong, which is why they are often chosen for trikes or load carrying cycles.

F. Front Hub. The front hub will have spoke hole drillings to match the rim, with 26 holes being the most common number of spokes for an adult bicycle. Decent quality hubs are usually made of aluminum, but you will most likely find both steel and aluminum hubs in your scrap pile. The hubs contain a pair of ball bearings to allow the hub to spin with minimal friction around the axle.

G. Front Dropouts. The front dropouts are slotted tabs on the front forks that allow the front axle to drop out of the forks once the nuts are loosened. Unlike the rear dropouts, the hole is not slotted, so it is not used to adjust the wheels position in the forks. There is usually a small hole above the axle slot where a special tabbed washer can help lock the front wheel in place in case one of the axle nuts comes loose.

H. Top Tube. The top tube runs from the head tube to the seat tube and is normally under compressive load on a bicycle frame. The top tube is usually the second largest diameter tube in a bicycle frame.

I. Down Tube. This tube runs from the head tube to the bottom bracket and is under tensile stress in a bicycle frame. This is normally the largest tube in a bicycle frame, and one of the most important in the strength of the frame.

J. Seat Tube. This tube is normally the same diameter on all bicycle frames as it has to carry the seat post, which fits snug inside the tube. The top of this tube will also have some type of clamp which will tighten around the seat post, allowing it to lock in place at the desired height. On a suspension bike frame, this tube may or may not have the duty of carrying the seat post. On the frame shown in Figure 1-2, it does not.

The most common components you will find at the rear of a bicycle are shown in Figure 1-3, and are labeled as follows:

A. Suspension Gusset. This part may differ depending on the style of suspension, but its basic purpose is to transmit the forces from the suspension spring into the frame in a way that does not induce damage on the frame. Typically, this part will be made from two steel plates with a thickness of 3/32 inches. The top of the rear

Figure 1-3 *Rear component details*

suspension spring will be held between the plates by a hollow bolt.

B. Suspension Spring. The rear suspension spring includes a high tension coil spring as well as a gas-filled shock absorber, so bumps and vibration are not transferred from the wheel into the frame. The spring is typically rated for 500–800 pounds of compression, which is due to the mechanical advantage gained, thanks to the position of the fulcrum on the rear triangle. A high quality rear suspension spring may have 3 or 4 inches of travel and cost more than a thousand dollars. The ones you will typically find on inexpensive bikes will have less than 2 inches of travel and cost only a few dollars to replace. The top ring is usually adjustable to offer a minimal amount of control over spring tension.

C. Rear Triangle. The entire moving part of the rear suspension is called the rear triangle. On any bicycle frame, this assembly includes the seat tube, seat stays (M), and the chain stays (L). The three parts actually form a triangle designed to carry the rear wheel. This assembly is extremely strong.

D. Rear Brakes. Much like the front brakes, rear brakes are available as cantilever style brakes as shown in Figure 1-3, including the mounting studs directly on the seat stays, or as bolt-on caliper brakes of lesser quality.

E. Cantilever Brake Studs. These studs are welded directly to the seat stays and allow the brake arms to pivot, placing the pads against the rim. It is best not to remove the brake studs, as their alignment is somewhat critical to proper brake operation.

F. Front Derailleur. The front derailleur forces the chain to move between the two or three front chain rings by derailing it slightly at the top as it enters the chain ring. The two plates that are on each side of the chain rub directly on the chain to force it to move.

G. The Chain. A bicycle chain is available in several sizes, although the pitch remains the same. A single speed bicycle chain is the widest style of bicycle chain, and is quite rigid from side to side as it does not have to run through a derailleur. BMX bikes and those with coaster hubs have a single speed chain. A bicycle with a derailleur must have a thinner, more flexible chain, due to the fact that the chain does not always make a perfect parallel run from the front chain ring to the rear free hub. A derailleur compatible chain is quite flexible from side to side, and is offered in various widths depending on the number of gears on the rear free hub.

H. Front Chain Rings. The front chain rings have between 20 and 50 teeth, usually having two or three on a crank set for a full range of gears. The front derailleur will move the chain between the chain rings to switch gears as the rider pedals forward. The smaller gear makes you pedal faster but delivers more torque to the rear wheel (for climbing), whereas the large chain ring makes you pedal slower, but propels the bicycle at faster speeds.

I. Crank Arm. Normally made of aluminum, the crank arm connects the pedals to the front chain rings so the rider can pedal the bicycle. The crank arms must convert reciprocation motion to rotary motion, much like the piston rod in a petrol engine. Crank arms are available as a left and right unit that connect to an axle, like the ones shown in Figure 1-3, as well as a single-piece-style crank arm, which is shaped like a large S, having both arms connected as a single unit.

J. Pedals. Offered in many varying styles and shapes, the pedals thread directly into the crank arms and allow the rider to put force down on the crank arms. Pedals have a left and right side, with the right side (chain ring side) having standard clockwise threads, and the left side having reversed threads. Pedal threads are also available in two standard sizes: the smaller size is used on three-piece crank sets, and the larger size is used on single-piece crank sets.

K. Crank Set Axle. The crank set axle is only available on a three-piece crank set, and must fasten the two crank arms together. We will examine these parts in more detail later.

L. Chain Stays. These two tubes run from the bottom bracket to the rear dropouts on each side of the rear wheel. These tubes are part of the rear triangle.

M. Seat Stays. These two tubes run from the top of the seat tube to the rear dropouts on each side of the rear wheel. These tubes are part of the rear triangle, often including the rear brake studs.

N. Rear Freewheel. The rear freewheel is a collection of small chain rings built onto a one-way clutch. When the freewheel turns clockwise, the rear hub turns with it. When the freewheel turns counterclockwise, the hub does not turn with it, which is why a bicycle can coast along with the cranks not spinning constantly. The effect of gear sizes is exactly the opposite of the front chain ring, with larger gears offering more torque, and small gears offering faster speeds. Most freewheels have between five and seven chain rings. To calculate the number of total gears on a bicycle, multiply the number of chain rings on the front crank set by the number of chain rings on the rear freewheel.

O. Rear Dropouts. The two slotted plates at the junction of the chain stays and seat stays are designed to hold the rear axle in place, and offer a bit of adjustment for the rear wheel. Because the slot extends for an inch or more, the rear axle can be moved along the slot, allowing the rear wheel to be adjusted slightly. On a single-speed bicycle, this adjustment is used to pick up any chain slack. The rear dropouts also hold the rear derailleur in place.

P. Rear Axle. The rear axle is a threaded rod that contains the rear hub bearings, cones, and rear axle nuts. On some bicycles, the rear axle also clamps the rear derailleur to the frame by placing it between the right side dropout and the axle nut.

Q. Rear Derailleur. Much like the front derailleur, the rear derailleur must force the chain to move across all of the rear freewheel chain rings in order to switch gears. The rear derailleur must also pick up chain slack, which is why it has a long body containing two idler gears on a spring loaded axle. The chain has a lot of slack when it is sitting on the two smallest chain rings, due to the fact that it does not have to travel as far as it does on the larger rings.

The rear derailleur pulls the chain around the rear chain ring, as shown in Figure 1-4. Because the chain must keep tension no matter which chain ring it may be on, the derailleur body must pivot back and forth to pick up the chain slack. The chain must also come in contact with at least half of the teeth on the chain ring or it may

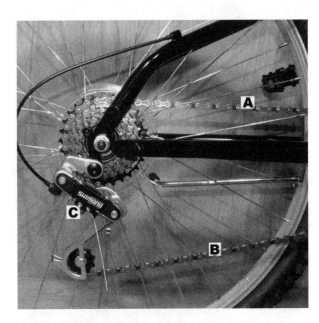

Figure 1-4 *Rear derailleur and chain details*

skip, which is why the upper guide wheel is directly under the rear axle. The point labeled (C) in Figure 1-4 shows the two small adjustment (limit) screws which control how far the derailleur can travel along the rear freewheel. If these screws are not set correctly, the chain may fall off the largest or smallest chain ring, or fail to reach them. The chain on the top of the chain ring is called the drive side chain (A), as it is always under tension when the cycle is being pedaled. The return side chain (B) is never under any tension as it simply returns back to the chain ring.

Remove the front wheel by loosening the two axle nuts so it can fall out for the fork dropouts. Of course, you must first release the front brake pads, or the wheel will become stuck between the brake pads and the front tire. Letting out all of the air in the front tire will also work, but simply releasing the front brake pads is much easier. As per Figure 1-5, press the brake arms together so the cable head can be removed from the brake arm slot (D). Also shown in Figure 1-5 is the cantilever studs (A), which have a built-in return spring, so keep the brake pads (C) away from the rim when not in use. The brake pads can be adjusted for different rim style by moving them along the brake arm and then locking them in place with the brake pad bolts (B). Properly adjusted brakes should not rub on the rim when idle, but sit as close as possible to the rim.

The rear wheel will come free from the rear dropouts once the rear axle nuts are loosened and the rear

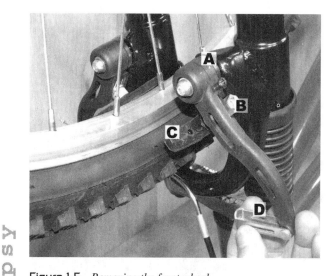

Figure 1-5 *Removing the front wheel*

Figure 1-7 *Removing the chain*

Figure 1-6 *Removing the rear wheel*

Figure 1-8 *Removing the rear derailleur*

derailleur is pulled back, as shown in Figure 1-6. Standing the bike upside down on the handlebars and seat makes maintenance a lot easier.

The chain can also be removed from the frame by taking out one of the links. A chain breaking tool (Figure 1-7) is highly recommended, as it is inexpensive and extremely useful, especially if you plan to build some of the projects in this book which require longer chains.

The rear derailleur is held to the frame either by the rear axle nut or by its own bolt threaded into a tab on the frame. Before it can be removed, the shifter cable must also be removed by loosening the locking nut, as shown in Figure 1-8. Also shown in Figure 1-8 is the cable end

cap, which can be pulled off the end of the cable using a pair of pliers.

Remove the pedals by placing a box wrench on the pedal stud, as shown in Figure 1-9. The pedal on the right side (chain ring side) has a normal thread, so it is removed by turning it counterclockwise. Sometimes pedals may be very difficult to remove, especially on older scrapped bicycles, so you may need to play dirty, using a hammer against the wrench, or even a blowtorch to heat up the crank arm. I have battled some stubborn pedals, but have only lost once, so chances are good that you will eventually get the pedals free if you don't give up.

Do not forget about the left-hand thread on the non chain ring side of the crank set, or you will be hammering on the wrench all night without any success! If you forget, simply look at the backside of the pedal

Figure 1-9 *Removing the pedals*

Figure 1-11 *Removing the crank arms*

Figure 1-10 *There are left and right pedals*

Figure 1-12 *Removing the crank axle bolt*

stud, where you will find an L or R stamped on to indicate which pedal you are working on. As shown in Figure 1-10, the L indicates a left side pedal, so you turn the wrench clockwise to loosen it.

The crank arms on a three-piece crank set are fastened to an axle by a bolt through the centers. As shown in Figure 1-11, the small plastic cap must by removed in order to get to this bolt. If the cap is slotted, it will unthread using a large flathead screwdriver. If it is just a cap, then pry it off with a screwdriver blade.

The crank axle will either have a nut or a bolt, but both threads are normal threads, so the wrench will turn counterclockwise to loosen the part. Figure 1-12 shows the socket wrench used to remove the bolt.

The crank arm may be another stubborn part to remove, especially if some corrosion has built up

between the axle and the crank arm. Use a long bolt, or some type of steel wedge to help bang off the crank arm, as shown in Figure 1-13. The wedge should always be placed in the crank arm and not the chain ring, or the thin chain ring will be bent. Again, a little heat via the blowtorch may make a really stuck crank arm budge.

Usually, handle grips can be removed by forcing them off from the inside edge. Of course, they may be glued on, or so stuck that you have to use the "ugly method" of removal, as shown in Figure 1-14. A single cut with a razor knife will get them off every time, but they are going straight in the trash can after that.

The handlebars are held in place by the clamp at the end of the gooseneck, and can be removed by loosening the nut as shown in the upper half of Figure 1-15. You will need the handle grip and levers removed from at least one side of the handlebars, so they can slide through the clamp before removing them completely. The actual

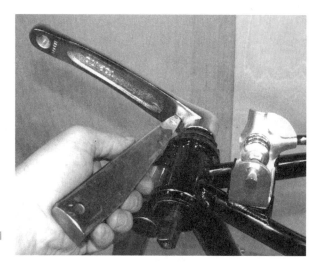

Figure 1-13   *Removing the crank arm*

Figure 1-14   *Removing old handle grips*

Figure 1-15   *Removing handlebars and gooseneck*

Figure 1-16   *Removing the gooseneck*

Figure 1-17   *The wedge-shaped gooseneck clamp*

gooseneck is held into the fork stem by a wedge shaped nut that is released by turning the bolt in the center of the gooseneck, as shown in the lower half of Figure 1-15.

A common "newbie" mistake is to loosen the gooseneck bolt as shown in the last photo, then start cranking the handlebars left and right, while trying to lift the gooseneck from the fork stem. Because of the way the wedge-shaped clamp locks in place, simply loosening the bolt in the center of the gooseneck will not always free it. You must tap the bolt down about ¼ inch after loosening it, as shown in Figure 1-16, in order to free the wedge.

As you can see in Figure 1-17, the wedge-shaped clamp will slide along the angled cut, creating a tremendous amount of friction in the fork stem, holding the gooseneck in place. Often, you can completely

Figure 1-18 *Removing the seat and seat post*

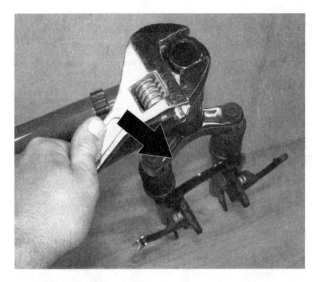

Figure 1-19 *Removing the front forks*

Figure 1-20 *The fork bearings*

remove the long bolt and still have the wedge stay securely fastened in the fork stem, which is why the tapping of the bolt head is necessary. Also, there are two common sizes of gooseneck stems, so do not put the smaller size in the larger sized fork stem or you may not get a good lock.

Removal of the seat and seat post is nothing special—just loosen the nut that holds the clamp around the seat tube to release the seat post. It is best to leave the seat installed when you take out the seat post—this way, you can use it as a place to hold onto, as you turn the post around while pulling upwards on the seat. Figure 1-18 shows the loosening on the nut around the seat post clamp.

The front forks have a threaded stem and are held in place by a threaded bearing race, a lock washer, and a top nut. The top nut can be removed by turning it counterclockwise, as shown in Figure 1-19. The lock washer and bearing race will usually come free by hand, as they are not supposed to be very tight. The lock washer has a tab, so it must be lifted straight off the fork stem, and the bearing race will unthread just like the top nut did.

The fork bearings will fall out of the head tube cups once the forks have been removed. As shown in Figure 1-20, the bearings have two sides, a ball side and a ring side. Always place the bearing in the cups' balls first, or they will not work properly. A stiff fork that does not spin freely is a clear sign of an improperly installed or wrong size bearing in the head tube cups. Yes, there are several sizes of bearing and cups, so keep the same parts together.

The fork hardware is shown in Figure 1-21. There will always be two bearings of equal size, a bearing race (larger ring), a lock washer (thin ring), and a top nut. Again, the bearings, cups and races must all match, so keep them all together as a set when salvaging bike parts.

Often, you will need to remove the head tube cups to weld or cut a head tube. Simply bang them out from the inside of the head tube, as shown in Figure 1-22, using some scrap steel rod or a long bolt. A few taps on each side of the cups should set them free with little effort.

The head tube cups shown in Figure 1-23 have two different heights. The larger cup is for the bottom of the head tube and the smaller one is for the top. If you put them in backwards, the fork hardware will not work

Figure 1-21   *The fork hardware*

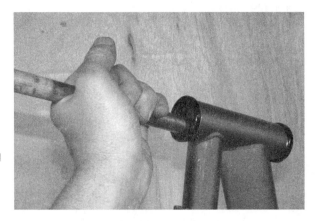

Figure 1-22   *Removing the head tube cups*

Figure 1-23   *Head tube cups have a top and bottom*

correctly, and the forks will seem very stiff when they are turned.

The rear suspension spring is held in place by the gussets on the frame and rear triangle by a pair of hollow bolts. Simply find the appropriate size hex keys, and remove the two bolts, as shown in Figure 1-24. Don't

Figure 1-24   *Removal of the rear suspension spring*

Figure 1-25   *Removal of the rear triangle*

worry, as the spring is not under any tension and will not go flying around the room once it is freed.

If your frame has rear suspension, then the entire rear triangle will come free from the frame by removing the bolt that holds it onto the pivot tube. This pivot is also fastened by a hollow bolt that requires a pair of larger hex keys to undo, as shown in Figure 1-25. You may have to tap one of the bolt halves out with a hammer once releasing the first bolt.

Figure 1-26 shows the rear suspension pivot parts once they are removed. The small tube that is welded to the frame has a pair of plastic plugs that act as a bearing surface for the pivot bolt, so don't forget to tap them out of your plan on welding or cutting this tube. Some higher quality suspension frames may actually have a bearing in place of the plastic plugs.

Figure 1-26   *The rear suspension pivot hardware*

Figure 1-28   *Removing the bottom bracket hardware*

Figure 1-27   *Taking apart the bottom bracket hardware*

Figure 1-29   *Scrounging for bicycle parts*

To remove the bottom bracket hardware, start by loosening the locking ring on the left side of the frame by turning it counterclockwise, as shown in Figure 1-27. This ring locks the bottom bracket bearing cup in place, and should be easy to remove by tapping on the slot with a hammer and chisel, or by using a pip wrench to grab it.

Once the locking ring is removed, the bottom bracket bearing cup will also unscrew from the left side of the bottom bracket shell in the counterclockwise direction. This cup will probably have a face that can be held with a wrench, and may require a bit of force to turn. When you free this cup, the two bearings and crank set axle will be freed, as shown in Figure 1-28. Just like the fork bearings, it's always balls into the cups, not the other way around. Also note that I never remove the right side cup from a bottom bracket, as this cup has a reversed thread and is a real pain to get out. There is usually no reason to remove it, and by leaving it in place, you never accidentally try to install a bottom bracket in reverse or thread the wrong cup into the wrong side.

Once you get bitten by the bicycle building bug, you will want to increase the size of your junk pile, so you can spend more time building and less time scrounging. Head to the local dumps and see if there is a "good neighbors' corner" or metal recycling area. Check your local scrapyards and flea markets, and hit yard sales. Tell all your friends and family that you take all metal scrap, especially bicycle parts, and before you know it, your garage will have a wonderful pile of junk for you to work with. I am lucky enough to have a metal scrap pile at the local city dump, where I can easily find 10 or more bikes every day I visit, but even if you have to dig around for junk, it doesn't take all that long to accumulate a huge collection of parts. Figure 1-29 shows a typical day at my favorite hardware store—the city dump scrap pile.

An angle grinder is a must for this hobby. This simple and inexpensive tool will be your most used weapon, and can reduce a frame into its individual components in minutes. You will also want three kinds of discs for your angle grinder: standard grinding discs, zip discs for

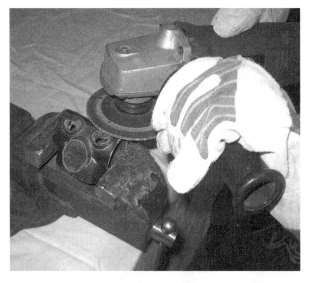

Figure 1-30 *The grinder is your best friend*

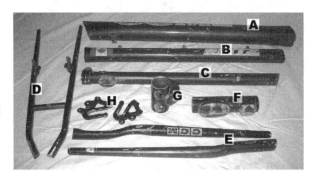

Figure 1-31 *Some more grinder handiwork*

cutting, and sanding (flap) discs for cleaning up welds, paint, and stickers. Figure 1-30 shows the trust angle grinder removing the excess material from a freshly severed bottom bracket.

If you plan to build your own custom frames, then you are going to need the basic building blocks. You could actually purchase new frame components, but I hear this is extremely expensive, and have never done so myself. It just seems silly to spend a hundred bucks on a few bottom brackets when I can find 10 of them for free at the dump, then hack them from the frame in a few minutes with an angle grinder. Figure 1-31 shows a completely hacked up mountain bike frame, with each part labeled as follows:

A. Down tube

B. Top tube

C. Seat tube and seat post clamp

D. Seat stays and rear brake studs

E. Chain stays

F. Head tube

G. Bottom bracket

H. Rear dropouts

With these basic building blocks and some spare metal tubing, you can build any project in this book, and many more that only your evil imagination can conjure up.

With a pile of scrap bicycle parts at your disposal, and an angle grinder ready to slice and dice, all you need is the magical box shown in Figure 1-32, and you will be a garage hacking guru. Seriously, for about $200, you can purchase a basic welder from a local hardware store and learn how to weld in less than a week. I have seen many new builders ask about purchasing a welder in our builder's support forum, and then that same person will send me a photo of a completed bike a week later—it's like magic. You can make all of the projects in this book using the most basic welder money can buy, and I actually use a department store AC welder for all of my work.

Every single project in this book and on our website was made with nothing more than a $200 department store welder, an angle grinder, and piles of bike parts. I only weld with a 6013 rod using an AC stick welder, and do not even own a drill press, so there is no reason

Figure 1-32 *This device will turn you into a building guru!*

why anyone can't build all of the projects presented here. If you really want to become a master of the arts, then you can take a welding course, which is usually only a few weeks long, and will teach you the finer points of welding. So, if you are just starting out, head down to the welding supply house and tell them you want a decent beginners' welding rig that will plug into whatever power source you have in your garage and allow you to weld thin walled tubing.

In addition to that basic welder, you will need a box of rod or wire (depending on your choice of welder), a pair of decent work gloves, a chipping hammer, welding helmet and, of course, some type of eye protection. Again, let me say that I only own a very basic AC welder, an angle grinder, and the gear shown in Figure 1-33, yet there is probably no vehicle I cannot build with these tools. I have no drill press, lathe, tube bender, or chop saw, so if I can make these cool machines, then so can you. With a little practice, you will be able to create

Figure 1-33   *Here is your battle gear!*

a showroom custom, and have the ability to turn your wildest imagination free, creating one-of-a-kind custom vehicles that rival any of the mass produced department store bikes available. So get that welder plugged in, collect a pile of junk and get to work!

Attitude and Style

The Gladiator Chopper is an over-the-top custom made from a fusion of bicycle parts, car parts and commonly available steel tubing. What surprises most people who ask me questions when I am out cruising on the Gladiator is that the entire chopper was made from junk parts using only a basic AC welder, an angle grinder and a hand drill. Yes, no other tools were used, and not one single part was machined or hard to find. This bike is proof that anyone with the drive to turn their ideas into a rolling work of art can do so without dropping hundreds of dollars at the machine shop, or requiring many years of mechanical or welding experience. I'm not going to lie to you—there is a lot of physical labor involved in making a bike like this, but the results are well worth the effort, since you will end up with a custom bike that will outshow any department store chopper that has been mass produced on an assembly line. Yes, I have a problem with a mass produced chopper—that is obvious in my writings! I mean, how can a bike with a six-digit serial number be considered a custom at all? A chopper, as its name implies, is a "chopped" stock bike, so how can one be mass produced and still be called a chopper? Dude, this just isn't right, which is why I built such a radical bike using parts and tools that anyone with a little love of the hobby could get their hands on.

Another thing that may surprise you when reading through this plan is that there is no confusing CAD drawing, jam packed with angles and measurements. The entire project is done step-by-step in such a way that you can substitute parts you have on hand, and make

modifications to suit your own style as each step progresses. Yes, you can build your chopper exactly like the Gladiator, as I have used the most common parts I could find, but I challenge you to go out on a limb and make slight modifications to the frame and parts used, even if this is your very first attempt at building a bike. I will give all measurements and details on the parts I used, but do not be afraid to experiment, since your own modifications will add a personal touch to your chopper, making it a true custom.

It's a good idea to read over the entire plan once before hacking up any bikes or heading to the scrapyard on a search for parts. By reading ahead, you will see why a certain part was chosen, and get some ideas on how to modify the design to suit your needs. With even a slight change in a few of the main frame tubes, you could make a completely different looking chopper, even though the core design has been followed. Don't be afraid to try new things, and remember that hard work is what makes this hobby successful, not spending money on mass produced products. Long live the real custom!

We will be building the rear of the chopper first, since it requires the most work and forms the building foundation of the bike. Once the rear of the Gladiator is complete, you have a lot more room to unleash your creativity. Looking back on when I first began the Gladiator concept, it's hard to believe that I started with nothing more than this old rusty pair of 15" x 7" car rims (Figure 2-1), but I really wanted to turn the most commonly available scrap into a rolling work of art, so

Figure 2-1 *A pair of old 15" x 7" steel car rims*

these rims were perfect. You can use practically any type of car rim you like, as long as it meets two requirements: it must be a 15-inch rim, and the center should be welded to the rim so you can cut it out. All rims will have a stamp on the inside showing the size, with the first number indicating the diameter, and the second indicating the width of tire that will be needed. 15" x 7" is a very common rim size, but you are free to use any width from 15" x 5" right up to 15" x 15", if you can find a set of monster tires to fit such a beastly rim. A 7-inch wide tire may not seem like such a monster on a car, but remember that those fat department store chopper rear tires are only 4.25 inches in width, so any car tire is going to dwarf them in comparison.

As for the rim diameter, 15 inches was chosen because it is a perfect size to spoke up with standard 20-inch bicycle spokes. Any other rim size will require you to find custom length spokes, which is indeed possible, but best left for the advanced wheel builder. I would recommend using a 15-inch rim to make things easier, unless you are certain you can find spokes to fit another size rim. Read ahead a bit and you will understand why the common stock steel rims are the ones you want, not some fancy aluminum aftermarket types.

As shown in Figure 2-2, almost all steel rims are held together by the welds around the edge of the center on the backside of the rim. There will most likely be five or six welds holding the center in place, each with a length of approximately 3 inches. Some rims may have an aluminum center that can be simply smashed out, like the one I chose for my OverKill chopper, but those require a bit more searching to locate, and since we need two identical rims for the Gladiator, I chose the most common style available. Of course, if you have access to another style rim that will allow the center to be removed with less work than grinding a weld, then feel free to use it.

Figure 2-2 shows the welds circled in black marker, along the edge of the rusty steel center. Since the rims will need to be painted after the weld removal surgery, a little rust is not such a big deal, but try to avoid any rims that have obvious bends or nicks around the edge, since they will not look good even when painted. Many tire shops have a large collection of suitable rims, so don't be afraid to ask—they are often

Figure 2-2 *The center of the rim is welded in place*

sitting out back rusting away like the ones I salvaged were.

The most useful tool next to the welder is the angle grinder. Together, these two tools allow a garage hacker to yield a power so great that anything can be accomplished. I have built hundreds of bikes and other wild machines over the course of 30 years, and to this day have no desire to own any other tools besides the welder, angle grinder and hand drill. I feel a great sense of accomplishment when veteran customizers who own tools worth more than the house I live in send me a "thumbs up" email on one of my project designs. Working as a minimalist also lets me share my hobby with many who have never hacked a bike up before, and just want an inexpensive hobby where hard work equals results. Learn to handle an angle grinder and a welder like a medieval knight wields a sword, and you will have the power, I promise.

A "zip disc" or cutoff wheel is a grinder thin disc made for cutting, and it can chop out a 3-inch weld like the one in a rim in a few minutes. Figure 2-3 shows one of the rim welds being cut by the 1/16-inch zip disc by running the disc at 45 degrees to the rim center. Try to cut the weld between the rim and center in such a way as to minimize damage to the rim as much as you can, but realize that some damage to the rim will occur. You will be welding up any cuts you make in the rim and then grinding the area clean, which is why it is not important to search for perfect rims, only those that can be sanded smooth and later painted.

When you have cut all the welds holding the rim center in place, you will need to pound out the center by striking the area near each weld, as shown in

Figure 2-3 *Grinding out the welds*

Figure 2-4. A wedge-shaped tool, or a bit of solid steel rod, will do this job, which requires you to bash the living daylights out of the center in order to break it free from the rim. The center was jammed into the rim by a machine, so it's not just going to drop out, even if you have cut clean through all of the welds, so hammer away at each weld, looking for signs that the weld area has begun to separate. Once you see the initial hairline crack on each weld start to form, a few good slams at each weld will eventually drive the center out of the rim, a little bit at a time. You must pound each weld equally, in order to avoid getting the center of the rim stuck at an angle, as it starts to move free from its original position. If a weld is not letting go, recall the grinder for another tour of duty, and eventually you will win the war.

Depending on the width of your rim, the center may have to be pounded some distance before it can be fully removed like the one shown in Figure 2-5. My 7-inch

Figure 2-5 *The rim center has been liberated*

wide rim had a raised edge of approximately 2 inches, so pounding out the center only took a minute or two with my favorite large hammer. As you can see, there is a bit of damage done to the rim as the grinder disc took a bite, but that's nothing a little bead of weld and some careful grinding won't fix. Even if you cut right through the rim, have no fear, as it will all work out in the end. Now do the same thing with the other rim, and then throw the centers into your junk pile for later use in some other evil creation.

Figure 2-6 shows the result of 20 minutes of hard labor with a grinder and hammer—a pair of 7-inch wide rims that actually look like they were designed for spoked wheels. You will probably be amazed at how much lighter the rims are with the centers gone, weighing in at less than half their original weight. Sure, the pair of rims outweighs a typical full bicycle, but dude, we are not

Figure 2-4 *Pounding out the rim center*

Figure 2-6 *A pair of seriously wide bicycle rims*

running the Tour de France now, are we? A chopper pilot gauges performance on how many people gawk at you as you cruise by, not by how many ounces you can shave off your bike by drilling holes in the frame! The funny thing is I think the Gladiator actually rides pretty smooth for a bike with car wheels, and it can glide forever once you get rolling. Now, exercise is finally cool.

Unless you are a magician with the angle grinder, the rims will have some battle damage after cutting away the welds, so you will have to fill and grind the area. Figure 2-7 shows a bead of weld filling the areas where I dug into the rim when cutting out the center welds. Don't be concerned about weld quality here—just fill the holes with a light pass, with the current as low as possible to avoid adding too much weld metal. When grinding away the filler weld, start with a heavy grinder disc and then switch to a smaller disc or flap disc (sanding disc) to finish up the area. Try not to dig into the original rim metal when grinding away the filler weld, or it may show up as a flaw after painting the rim. Take your time, and don't be afraid to go over the area a few times after grinding just to touch up the odd hole. I had to touch up a few missed spots on each weld after grinding away the filler welds shown in Figure 2-7.

With a little effort, the rims will clean up nicely after welding and grinding any damaged areas from the surgery. Figure 2-8 shows my rim pair after filling the holes, grinding them flush, and then spending some quality time listening to 1980s power metal with sandpaper in hand. It took a bit of effort to sand all the rust off the rims, but in the end, they look pretty decent, and will certainly be fine after a fresh coat of paint. If you really have a lot of time and like to build your arm

Figure 2-8 *Repaired and sanded rims*

strength, then keep sanding the rims with finer and finer grits of sandpaper until they look like they are polished. If you get the surface down to a shine, a clear coat could be sprayed on the rims to make them look like chrome or polished aluminum.

Everyone who examines the Gladiator is taken by the fact that it has wheels that look like they are from a car, yet have bicycle spokes. "Where did you get those wheels?" I am often asked. "I made 'em from junk," is my standard response, which usually gets me a look of disbelief, since they look like they were factory made. Those who are interested in bicycles or mechanics will often realize that the wheels are obviously custom jobs and ask about the hubs. They are also amazed when I tell them that the wheels actually have no hubs at all, just a few hand-drilled washers from the hardware store. What you see in Figure 2-9 is all that holds the rim and spokes to the axle, i.e. four standard steel washers!

Yes, indeed, for less than a dollar, you can make your own chopper hubs. You need four steel washers with a diameter of 2 inches, and a center hole with a diameter as

Figure 2-7 *Welding up the rim damage*

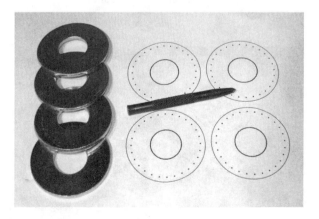

Figure 2-9 *Making the hub flanges*

close to 3/4 inch as you can find. The thickness of the washer should be less than 1/8 of an inch, but no thinner than 3/32, which is pretty much standard for 2-inch diameter washers. You may need to sort through a bunch of these washers at the hardware store to find four that have identical thicknesses, because washers are often made as "random thickness" for some reason which defies all logical thinking. I went to the hardware store and asked if I could pick out four 2-inch diameter steel washers that have a $^3/_4$-inch hole. Tell them you need four with the same thickness and they will probably understand. If you are not good at eyeing up small measurements, just look at the hub flanges on a typical rear bicycle rim and that will give you an idea of what thickness you are after.

Shown in Figure 2-9 is a paper template that can be taped over the washer in order to center punch the 28 spoke holes that need to be drilled in each flange, creating a wheel with 56 spokes. If you would like a copy of this hub template, a copy can be found on our website at www.atomiczombie.com/support/gladiatorhub.jpg

Because there is no way to ensure that the image will print out at the correct size on your printer, I have made it very large, so you can scale it and print it by trial and error until the pattern fits perfectly over your washers. If you want to make your own template using a computer program, or like using a compass, then 28 points on a circle will require a hole every 12.857 degrees. The hole centers should be 1/8 of an inch away from the edge of the flange.

Figure 2-10 shows how the template can be used to center punch the holes that need to be drilled in each washer to create the hub flanges. If you use the provided template, it should be scaled down until it fits perfectly over the washer once carefully cut around

the circumference. Use the center circle as a guide by holding the washer and template up to a bright light, so you can make sure the two are aligned as perfectly as you can get them. Four pieces of clear tape can be used to hold the template to the washer, as you punch out all 28 holes with a center punch.

If you plan to make your own template, or directly mark the washers with a compass, just remember that there are 28 holes, which makes 12.857 degrees between holes (360/28), and that there should be 1/8 of metal between the edge of the washer and the center of the hole. If you hate messing around with computer software, and think using a compass is just too nerdy, then have a buddy who knows how to use a CAD program print you an appropriate template.

I have built several human-powered vehicles using this "hubless" approach now, and it may seem like a lot of work to drill all 112 spoke holes, but like any task, it's no big deal if you do it a little bit at a time. As shown in Figure 2-11, I like to screw the washer down to a bit of plywood and then go to town with the drill. I only have a hand drill, but the job only takes about 30 minutes per washer, with a break every seven holes or so. The holes are drilled using a 3/32 drill bit, which can easily be broken if you press down on it too hard, so take your time. I would recommend purchasing four high quality 3/32 drill bits, because 112 holes is a large number of holes, and you have a good chance of breaking a bit or two along the way. If you have a drill press, then this job will seem a lot easier, but again, this is not a requirement. I am often asked in the Atomic Zombie forum if a slight error on one or two holes is a big problem, and the

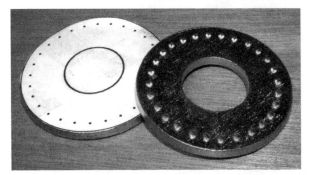

Figure 2-10 *Using a template to punch the spoke holes*

Figure 2-11 *Bolting down the washer for drilling*

answer is no. Once you get the spokes installed, any small drilling error will be taken up by the spokes as they are tightened.

If you take a look at a bicycle hub, even a cheap one, you will notice that each spoke hole has been beveled like the holes on the left washer shown in Figure 2-12. The reason this is done is so that the sharp edge of the spoke hole does not cut or damage the spoke when it is under tension. To bevel each hole (on both sides), simply take a 3/8 drill bit and run it very lightly for a few seconds in each hole, so it just removes the sharp edges of the hole. Try to keep the drill running at the same speed as you do this, so you can get the same bevel each time, and do not press too hard or the bevel will be too deep. Bevel each hole on both sides of all four flanges—this job should only take a few minutes.

Another reason for the beveling of the spoke holes is so the spoke can make it around the bend, as shown in Figure 2-13. If your washers are a bit on the thick side, then you may need to bevel the holes a bit more to get the spoke around the bend without having to force it. The spokes I recommend are 14 gauge with a length of 178mm, but if you plan to use heavier spokes, then the

original 3/32 hole may not be large enough. Before moving ahead, find at least one spoke of the required age, and test several of the holes to make sure no extra drilling will be necessary. Fourteen-gauge spokes are fairly common on most 20-inch rims with 36 or more spoke holes.

The Gladiator has no hubs. The hub flanges will be welded directly to the left and right axles, in order to create a wheel that includes the axle and rim as a single unit. The advantage of this approach is simplicity, strength, and cost. Having a hub machined and drilled could cost a few hundred dollars, and this would be a waste, since it would then have to be held to the axle using a cotter pin or bolt, which would make the bolt the weak link. By creating an integrated hub axle, you get a bulletproof design for mere pennies, and do not have to worry about any ugly lock nut or cotter pin sticking out past the "beauty" side of the wheel. I have used this design on many successful vehicles such as the DeltaWolf, and the Kyoto Cruiser, which has a payload of more than 500 pounds, and I can tell you that it is far superior to any system that retrofits a hub onto an axle. The final design is also very clean looking, as you can see in the glamour shots.

The axles are made from $3/4$-inch diameter mild steel rod, as shown in Figure 2-14, along with the completed hub flanges. Mild steel does not mean that the steel is in some way weak—it just means that it is not hardened, so welding it will not be a problem. Superior ground rod could also be used, but ask your distributor about the hardness of the material, letting them know that you plan to weld a hub to it. Be aware that 3/4 axles will never bend, and can easily support a heavy-set rider, so there is no need to use a larger diameter axle. You could however

Figure 2-12 *Beveling each spoke hole*

Figure 2-13 *Testing the holes with a spoke*

Figure 2-14 *Steel axles and hub flanges*

get away with 5/8 axles, but this would require some planning, especially when it comes to the hub flanges, since washers of 2-inch diameter will be difficult to find with a 5/8 hole. Again, this is not a bike for spandex-clad cyclists, so just go with the 3/4 axles, and you will never have a problem, ever.

You will need at least 36 inches (3 feet) of 3/4 rod for the axles, since there are two axles, each with a length of 18 inches if you have 7-inch wide rims. I would recommend you ask for 6 feet of rod, since you may want to make your chopper a little wider, or use wider rims, and because it is extremely helpful to have a spare 2-foot rod to help center the bearings, as you will see by reading ahead a bit. If your supplier has a pile of rod already cut up, ask them for enough rod to make three 24-inch lengths, and then you can cut them down later. Too much is always better than too little!

Cut your axles into two 24-inch lengths. In the end, my axles were only 18 inches each, but you may want a wider bike to accommodate a wider butt, or wider rims, so it's better to work with more than necessary here.

Since the washer is made to fit over a 3/4 bolt, there is a bit of play around the hole and the axle, which is not a problem, and may even facilitate a better weld. As shown in Figure 2-15, two tack welds are placed on the joint so the gap is approximately equal all the way around the hole. If you can't do this by hand, then wrap the axle in some paper to hold it in place while you make the two tack welds. Try to get the washer tack welded to the axle as accurately as you can, so it appears straight as you roll the axle along your workbench. With only two small tack welds holding the washer, you should be able to tap out any obvious misalignments with a hammer. If the washer is seriously out of alignment, then break the tack weld

and try again. Perfection is not necessary, but obvious flaws should be repaired. I also recommend you tape up the spoke holes when welding to avoid spatter in the holes.

If you are happy with the hub flange alignment, then add two more tack welds and recheck your work. With four solid tack welds, you can then weld the entire joint as shown in Figure 2-16, using a good solid bead all the way around. Don't worry about beauty here—just fill in the area with solid weld metal, and it will be ground flush later.

The inside of the hub flange can be welded once the outer weld is completed. Take your time and weld around the entire joint in three or four passes until it looks like the weld in Figure 2-17. This weld does not have to be extremely heavy, since the outer weld has closed the gap between the washer and the axle.

Figure 2-16 *Outer hub flange welded to the axle*

Figure 2-17 *Inside hub flange welding completed*

Figure 2-15 *Fitting the washer to the axle*

Once both sides of the flange are welded to the axle, grind the outer face flush, as shown in Figure 2-18. You may need to fix a few holes once you grind the face, but with a few pokes of the welding stinger and a little grinding effort, the face will look as though it were machined from the same material that makes up the axle, for a nice, clean, final product. To get the face looking smooth, rough grind with a heavy disc, then take off the last bit of material with a sanding disc. Avoid cutting into the flange, or it will become too thin.

The inside flanges are welded so that they are placed apart at a distance of whatever width your rim happens to be, as measured across the edges with a tape measure. In Figure 2-19, the distance between the flanges is 8 inches because that's how wide the rim was. Yes, a 15" x 7" rim is actually 8 inches wide, but is made for a 7-inch wide tire. This discrepancy is because the tire and rim are measured by the bead of the tire, not

Figure 2-18 *Cleaning up the outer flange face*

Figure 2-19 *Hub axle units completed*

the overall width of the rim as measured with a tape measure.

The inner flange only needs to be welded on the inside, as shown in Figure 2-19. The side that will face the bearing is not welded because it will not be seen, and is really not necessary. If you did weld both sides, this would not hurt anything, though. The same alignment technique is used on all four flanges, making them as straight as you can. Again, do not freak out if there is a slight wobble or hop in the flanges as you spin the axle, as the spokes will take this error up. Obvious flaws that can be seen without having to spin the axle should be repaired.

Each rim has 56 spokes, which in reality form a 28-spoke wheel on each side of the rim. This may seem odd since bicycle rims have the spoke holes in the center of the rim, but due to the fact that the car rim is so wide, it would not make a very strong rim if the spokes were placed in the center of the rim, leaving so much of the rim unsupported. By working with the rim as if it were two rims (one at each side of the rim), the resulting wheel is extremely strong and looks filled in. If you have had a chance to see those fat department store chopper wheels, they look somehow "sparse," as if the designer simply took a standard bicycle rim and made it wider, leaving empty space on each side of the spokes, which is essentially what they did. Personally, I think the beauty of the nice, fat 4-inch rim is lost due to the wimpy-looking hub and spokes, as if the next ride over a curb could fold the wheel like a wet taco.

OK, so how do you mark out 28 evenly spaced holes around the outside of the rim without a math degree? My favorite trick is to drop a 28-hole rim over the car rim and then place a square over each hole to transfer the mark down to the rim. As shown in Figure 2-20, this method is simple and fairly accurate, if you take your time and keep your marker pointing in the same direction as you work around the rim. I like to tape the rims together, then have a helper carefully rotate the rim, as I stand in the same place and make the two sets of points with a sharp marker. The points are made on the little gutter of the rim, just inside the outer edges of the rim. This is the area that is recessed if you look inside the rim, which makes it easy to line up all the holes as you drill them. An alternative method of marking the holes would be by calculating the exact circumference of the outside of the rim, then dividing this distance by 28, and using

Figure 2-20 *Marking the rims' spoke holes*

Figure 2-21 *The drill gets another workout*

Figure 2-22 *Hub axles and rims getting primer*

the resulting number to mark each successive hole. The problem with that system is that each error is simplified by the next hole.

Whatever method you use to mark the rim, just ensure that you take the valve stem hole into account, making sure it is between the holes, and does not end up too close to a hole. Start the first spoke hole near the valve stem hole so you do not forget it is there.

When all of your holes are marked, and you know the valve stem hole is not an issue, center punch each marked spot and then find a nice sharp drill bit and go to town. You can probably get away with a 3/16 bit for the spoke nipple, but if the hole seems a bit snug, move up to a 13/64 drill bit. The spoke nipple should have enough room to tilt slightly so it can point towards the spoke, which is at a slight outwards angle as it heads to the flange. Some spoke nipples are a little larger than others, so try a 3/16 hole first. To drill all the holes, place the rim on the workbench, as shown in Figure 2-21, then use your body to press the drill so you do not get tired as you drill. In this photo, you can clearly see the raised area of the rim that creates the thin well on the inside of the rim—this is the area that is being drilled.

Figure 2-22 shows the completed hub axles and drilled rims with a fresh coat of primer. I normally paint all of the parts of a bike after everything is completed, but since you need the rear wheels to complete the rest of the chopper, they need to be painted ahead of time. The spokes are put into the wheels after they are painted,

since it would look bad if the spokes were the same color as the rim, and taping them all up to paint the rims would be a ridiculous chore. The rear wheels are the focus of this crazy chopper, so take your time and properly prepare and prime the surface so your paint job looks good. A spray can is all you need to do a professionally looking paint job, but you will need to follow the directions to avoid runs and streaks. Primer should also cure for a day before applying any paint.

Wheel building is a skill that can take many years to master, and is often thought of as a dark art by those who have never tried it. Let me assure you that building these wheels is not at all difficult, and will require almost none of the skills a bicycle wheel builder would need or even use. Because these wheels are so different than any normal bicycle wheel, the spoking technique I have developed is extremely simple, and can be done by anyone that can turn a screwdriver. Honestly, follow along, and you will have the wheels done in less than an hour each, without any pain whatsoever.

Once your rims and hubs are painted and left to cure for a few days (Figure 2-23), you can begin installing all 112 spokes: 56 spokes per rim, 28 spokes per side. I recommend you purchase new spokes to build these wheels. New spokes will be made of stainless steel, be all the same length, and of higher quality than some you may find in salvaged kids' bikes. Of course, if your junk pile is well stocked, you may have 112 equal length spokes, but for the cost of new ones, it's almost not worth the effort of unscrewing them all. Go to your favorite bike shop and ask for 112 stainless steel spokes with a length of 178 millimeters and a diameter of 14 gauge. This is not an odd request, and is a fairly common spoke, often used to build 20-inch freestyle rims. If you like getting that bizarre look, then tell the dude at the bike shop that you plan on making a chopper trike with 15-inch car wheels at the rear! Actually, many bike shops are keen on cool customizing, so discussing your wild and crazy projects may get you access to the pile of scrap in the back of the shop.

I am going to make this easy and painless, with no techno-babble or nerdy bicycle science! Start by dropping in two spokes, as shown in Figure 2-24. The spokes are put in so that the head is facing the outer sides of each flange. Both spokes are in holes opposite each other on both the rim and the flange so, essentially, you have hung the hub axle from the rim. Push the nipple on to the spoke thread and give it three turns. Until further notice, you will give all spoke nipples approximately three turns, so that they are all screwed in about the same amount. So far so good, right? I told you this would be easy!

Figure 2-24   *Beginning your wheel-building journey*

Now, insert a spoke in every other hole, so that each side of the rim has 14 spokes installed. Again, all spoke heads are facing the outside of the flanges, as shown in Figure 2-25. You now have a total of 28 spokes in your wheel, and the hub axle is held in place, although it is very loose at this point. If something looks messed up, you probably skipped a hole, so double-check your work to make sure there is a spoke in every other hole on both the rim and the flanges. You are now half done! Say, when does this get hard? Grab the long end of the axle and give it a twist in either direction until all the spokes are tight. As you can see, the spokes center the axle perfectly when forced tight, which is essentially what the next group of spokes is going to do.

As mentioned earlier, the next group of spokes will be installed in such a way that they force the axle to twist, making all of the spokes tighter by increasing

Figure 2-23   *Getting ready to build the wheels*

Figure 2-25   *A spoke in every other hole*

the distance between the nipples and heads of each spoke. If you want to follow along exactly, grab the axle and turn it counterclockwise so the initial spokes tighten right up. You may want a helper to turn the axle for you so you can free up both hands for the rest of the job. Also note that a spoke nipple may be caught outside the hole in the rim, so wiggle the axle or offending spoke(s) until they all drop in their holes and tighten up as you twist the axle.

Now have a look at Figure 2-26 and trace the spoke indicated by the large arrow back to the hub flange. Notice that it is a newly installed spoke with its head facing the other side of the flange. This spoke also points in a direction opposite the other spokes, essentially pulling the hub in the counterclockwise direction, much like what your helper is patiently doing right now as your read this. Have a good look at the photo and try to understand what is happening with this new spoke, as it is the only spoke you have to understand; all others just drop right in from here, I promise.

Assuming that you are holding the axle in a counterclockwise rotation, as shown in Figure 2-26, this new spoke will have inserted six rim holes away from its neighbor directly to the left. The spoke indicated by the large circle is the neighbor to the left, and the rim holes between the two spokes are numbered from 1 to 6. The little arrow on the hub flange is pointing to the new crossing spoke and his neighbor to the left. If this does not make sense, then read it again and study the photo; your helper will just have to be patient!

When you get that Eureka moment and it all makes total sense, insert the remaining spokes based on this "skipped-hole formula," which is an easy task once the first crossing spoke has been determined. The remaining spokes are easy because they simply fit into every other hole as you go around the rim. As long as you installed the first crossing spoke correctly, the others will follow.

Figure 2-27 shows the other crossing spokes installed, based on the "every other hole" technique. It may look technical at this point, but all you did was shove a spoke in every other hole with their heads facing the outside of the flange, then reversed the direction, crossing a set number of spokes as you installed the second group in every other hole facing the inside of the flange. I know that explanations can sometimes seem confusing at first, but if you get that first crossing spoke in the correct place and just keep going, the second time you do this will seem a lot easier. I can build one of these wheels in 10 minutes, but the first time took me over an hour. Be patient, be stubborn!

Figure 2-28 shows what the wheel will look like with all 56 spokes installed. The other side of the rim follows the same rules as the first side, inserting that magic first crossing spoke 6 rim holes away from its leftmost neighbor. Again, you will be turning each spoke nipple three turns, so they are all put in at roughly the same depth. Although the spokes are still somewhat loose, the slack has been greatly reduced, and it may surprise you that if you held the axle and gave the rim a spin, it would run almost true already. A car rim is much too stiff to warp, so the only thing you need to do now is crank all of the spoke nipples the same amount as you work around the rim. Unlike a thin aluminum bicycle rim, small variances in spoke tension will not have much of an

Figure 2-26 *Installing the first crossing spoke*

Figure 2-27 *All crossing spokes are now installed on one side*

Figure 2-28 *The same rules apply on the other side*

effect on the wheel, so go ahead and turn the spoke nipples two full turns each as you work around the rim, starting at the valve hole. Do both sides at the same time, and keep working around the rim until you can no longer make two full turns on each nipple. Basically, try to turn all spoke nipples the same amount by hand. Do not use a screwdriver yet.

The process is, of course, the same for the second rim, and hopefully it will seem a lot easier. In Figure 2-29 I have fully installed all 112 spokes and hand-tightened them all to approximately the same amount by working around the rim, turning each spoke nipple only a little bit at a time. The spokes are not extremely tight, but the rims seem fairly true, even at this point, which is due to the fact that they are extremely stiff, and the spokes are pretty much all at the same tension. To finish up the rims and tighten the spokes, we will mount the axles on a pair of bearings so the rim can be spun and referenced to a stationary object, much like the way a wheel builder would use a truing stand to fine tune the spokes and rim.

Figure 2-29 *Hand-tight spokes*

The pillow block bearings shown in Figure 2-30 will hold the axles to the frame, and we will use two of them temporarily to help finish the wheel-building process by allowing the wheel to spin freely as you tighten each spoke with a screwdriver. These bearings are common items available at many hardware stores or industrial supply outlets, and they are heavy duty enough to last forever in our project. If you walk into a bearing store, just ask for four identical pillow block bearings to fit a $^3/_4$-inch shaft. What you see in Figure 2-30 is what you will get.

Figure 2-31 shows a simple makeshift truing stand made by bolting down a pair of pillow block bearings to your workbench, so you can spin the rim and reference it against the edge of the bench. OK, it's not really a truing stand, but this system works great, and lets you work around the rim, turning each spoke a small amount with a screwdriver as you make sure the rim is coming out true. The rim will most likely be very true as you give it a spin, and as long as you crank down each spoke only a little at a time, the rim should stay that way. A slight hop

Figure 2-30 *Four $^3/_4$-inch pillow block bearings*

Figure 2-31 *A makeshift truing stand*

up and down can be repaired by turning the spokes opposite the "high side" down a little more than those on the "low side." There are so many spokes in each rim that they don't have to be cranked down so they are overly tight. As a reference, feel the tension on a typical 20-inch rim by squeezing a few spokes together—this is your target tension. If you hear that "tink, tink, tink" sound, then you are probably getting a little too much tension in the spoke, so take it easy.

Again, turn each nipple only a little at a time as you work around the rim, and you will most likely not have to even worry about truing the rim, as it should fall right into alignment. If the rim seems to pull to one side, tighten the group of spokes on the other side to move it in that direction. Patience is the key!

Obviously a rim with 56 holes drilled in it will no longer hold air, so you can pop out the original valve stem if you have not done so already. You will need a pair of suitable inner tubes and some duct tape to cover the spoke nipple, so that they do not damage the inner tubes. As shown in Figure 2-32, wrap the area over the spoke nipples with two turns of duct tape, so that there is no chance of the spoke nipple puncturing the inner tube when it is inflated. I also like to put a small amount of air in each inner tube so it can be handled easily, rather than fooling around with it as a deflated floppy mess.

Depending on the size and style of the tires you plan to use, installation can range from easy to brutally difficult, as I have found out in my adventures with car wheels. On my OverKill chopper I manually installed a 15-inch wide drag tire on a rim, and it took me all day, consumed most of my screwdrivers as I snapped them one by one,

and almost tried my patience to the very end of its limits. Of course, I would rather wrap myself in lunch meat and run into a cave full of grizzly bears than give up on a job, so I eventually beat the tire, and then crashed on the couch for two days! Anyhow, you may want to grab a helper, and get a spray bottle full of soapy water, as I have found this makes a huge difference when installing a tire the hard way.

To avoid scratching the paint on your rims, set up the wheel on three buckets, as shown in Figure 2-33. Having the wheel up off the ground makes the task much easier for you and your brave assistant. Now, it may look like I am a long way into the installation in Figure 2-33, but to get to this point is not even a challenge—simply press the tire onto the rim on an angle and it will almost fall on by itself. It's the top bead that is difficult, specifically the last third, as shown in the photo.

When you get to the point shown in Figure 2-33, hold the two screwdrivers on the tire so that the last third of the bead is off the rim, then kneel on the opposite side of the tire so that the bead drops into the rim well (the narrower part of the rim as viewed from the inside). You may want to have your helper force the tire into the well, so your hand can be free to snap screwdrivers, I mean, work the last part of the bead.

The reason you want to force the opposing bead into the rim well is so that it does not have to stretch as far to pop over the last third of the rim. By having your helper kneel on the tire, as you carefully use three screwdrivers to work the bead a little bit at a time over the rim, you

![Figure 2-32](spoke wheel with duct tape)

Figure 2-32 *Cover the spoke nipples with duct tape*

Figure 2-33 *Manual tire installation*

should be able to get to the point shown in Figure 2-34 in a few minutes. This tire was a joke compared to the OverKill tire! I would honestly say that I have had more trouble with some high pressure bicycle tires than I did with this tire, but your experience may vary.

The arrow pointing to the bead in Figure 2-34 is indicating the place where you would have your helper use a third screwdriver to pop the last bit of the tire over the rim. During this process, be careful not to scratch the paint, or puncture the inner tube. Dude, you did install the inner tube first, didn't you? If you are having a difficult time with the last part of the tire, spray some soapy water into the bead and rim edge to make the rubber slippery, as this really helps. I told you that this hobby could be labor intensive! But hey, what would your chop be worth if everyone could build one? You will get rewarded for all your hard work in the end.

With both tires mounted on the rims, your last wheel-building task will involve getting the bead to seat properly. Because the spoke nipples are slightly raised on the inside of the rim, the tire may need to be "coaxed" to pop into the correct place between the spoke nipples and the edge of the rim. First, try inflating the tire to 40 pounds, then let it sit for an hour, as the air pressure may pop the tire into place. If this does not work, use a spray bottle filled with soapy water to wet down the bead and rim edge by letting the liquid run along the tire onto the bead area. With the entire bead lubricated by the soapy water, inflation of the tire should allow the bead to pop into place. On one tire, 20 pounds of air pressure did the trick, but on the other, it took over 40 pounds. Be careful when exceeding a tire's recommended pressure, and do not exceed 45 PSI. Eye and ear protection should be used when working with tires as well. When you have

the tires sat properly, reduce the air pressure to 25 PSI, which is plenty of inflation for such a light vehicle. As shown in Figure 2-35, these rear wheels are seriously cool! Because of the spokes, they no longer look like typical car wheels, and people will often ask you where you found such beastly bicycle wheels.

The Gladiator's frame is based on a simple backbone structure that looks like a pitchfork. As shown in Figure 2-36, this simple design secures the rear axles to the frame's main tube, allowing you a great deal of creative freedom once it is completed. If you are just dying to add your own mods to my chopper design, then fear not—the time is coming close, but I recommend that you create the backbone as it is shown in the next few steps, as it forms a solid foundation on which to add the rest of the tubing and components.

The frame tubing is made from hardware store conduit, which is thin walled tubing, sometimes referred to as EMT. This tubing is lightweight and stiff, and it comes in many diameters with pre-made 90-degree factory elbows available. I used 1.25-inch conduit and two factory elbows to create the frame, as shown in

Figure 2-35 *You want phat? These are phat!*

Figure 2-34 *The battle is almost won!*

Figure 2-36 *The frame backbone laid out*

Figure 2-36, but you could use heavier tubing such as 1.5-inch conduit as well. One-inch conduit would be a tad thin for a large frame like this, so do not use anything less than 1.25-inch conduit. The two factory elbows make a 90-degree corner, and will always be the same size, so you do not have to worry about their dimensions. The main tube that runs from the rear of the frame up to the head tube is a 3-foot (36-inch) length of conduit, which will be later cut down to suit your size and style.

The ends of the two elbows need to be ground out so that they mate with the 3-foot main frame tube, as shown in Figure 2-37. The ends are ground out slightly to create a fish-mouth shape, so there are no large gaps that need to be filled during welding. The angle of each elbow on the rear of the main tube is determined by the distance between the two wheels, so depending on how much "real estate" you take up on a bike, you may want to alter the design slightly. I would recommend you place your two rear wheels together and then sit on a stool or bucket to figure out how much room you think you need between the two tires. I have 19 inches between my tires, which is plenty of room for an average sized rider. If you shop in the big and tall section when you buy your pants, then you may need a little more breathing room.

To ensure that I had 19 inches of room between the rear tires, each elbow should be 9.5 inches from the main frame tube as measured from the center. As shown in Figure 2-38, I am taking this measurement from the furthest point on the elbow, right before it begins to curve back towards the main frame tube. If you read ahead somewhat, you will see that the pillow block bearings mount directly on the elbows where they are furthest away from the main tube, which adds some nice style to the rear of the chopper frame, incorporating the curves

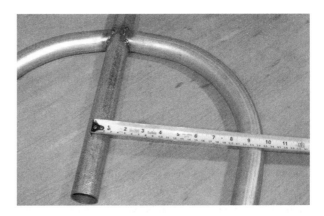

Figure 2-38 *Checking the elbow distance*

into the design. To state this simply, figure out how much room you want between your tires, then make sure the distance between the furthest center points of the elbows is the same as this measurement. The distance across both elbows in my design is 19 inches as measured from their centers.

When you have the elbows where you want them, tack weld them in a few places, as shown in Figure 2-38, with all tubing placed on a flat surface. Recheck your measurements and make any needed adjustments before you add any more tack welds to the joint. If your measurements look good, flip over the frame and tack weld the other sides of the joint as well.

The two elbows are to be fully welded, ground and cleaned up before proceeding any further. It's easy to work on the joint when it is not part of a larger frame, so cleaning up the area, as shown in Figure 2-39, is a fairly easy task. I only ground the top of the joint clean, as the underside will never be seen unless you are riding over someone, so unless you are an ultra perfectionist, there would be little point in spending too much time on the

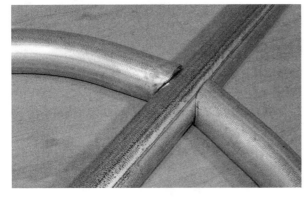

Figure 2-37 *Grinding out tubes for a better weld*

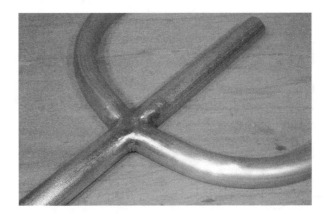

Figure 2-39 *Two elbows welded to the main tube*

underside of the joint. Always be careful not to take off too much weld metal or dig into the original tubing when cleaning up a weld. Start with a rough grinder disc, and then complete the cleanup with a sanding disc.

Earlier in this section, I mentioned that it is a good idea to have 2 feet of extra 3/4 axle material available, and Figure 2-40 shows why. When you place a solid axle through the two outer bearings, they instantly become perfectly aligned, so you can mark the area where they need to mount to the frame. I have built many trikes and go-carts over the years, and I can guarantee you that there is no better way to align your rear wheels than this method. If you try to get this alignment by eye, it may come close, and even look perfect, but any slight misalignment in the rear wheels will cause a toe-in or toe-out condition, which will limit your top speed and create extra rolling resistance. In a car, tire scrub equates to lower gas mileage and excessive tire wear. On an 80-pound, human-powered chopper, tire scrub equates to a lot of rider huffing and puffing!

With your bearing mounted on the axle, drop them onto the frame so that the axle crosses the frame at its widest point on the two elbows, making it 90 degrees to the main frame tube. Pillow block bearings allow the bearing to move around to any angle, similar to a ball joint, so tap the two bearing bodies with a hammer to get their bolt centers onto the elbow center. As you can see in Figure 2-40, the pillow blocks are on a slight angle so they line up with the curved elbows, yet the axle is parallel to the frame, and 90 degrees to the main frame tube. When you have both bearings in the correct place and angles on the elbows, use a marker to draw the bolt holes on the elbows by tracing the entire bolt hole. Don't

forget which side of the frame is the underside—the bearings mount on the underside of the frame.

To mount pillow block bearings to a tube, you would normally drill through the tube and use bolts to hold the bearings onto the tube. I wanted a cleaner look, showing no bolt heads, so I decided to cut off a few segments of threaded rod and weld them directly to the underside of the frame where the bolts would have to go. The nice thing about this design is that you do not have to drill any holes into the frame tubing, there are no bolt heads showing, and the tubing does not have to suffer any crushing from the tightening of the bearing bolts. My pillow block bearings needed half-inch bolts, so I cut eight 1.5-inch segments from the half-inch thick threaded rod, as shown in Figure 2-41. You could also cut the heads off eight appropriate bolts, if you have them handy.

As shown in Figure 2-42, the bearing bolt segments are welded to the underside of the frame in the exact place you marked out the holes when you set up the axle for alignment. Although the bearings will allow a bit of

Figure 2-41 *Bearing mounting bolts*

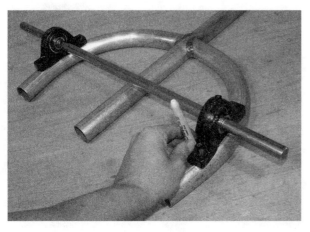

Figure 2-40 *A good bearing alignment trick*

Figure 2-42 *Bolts welded directly to the frame*

adjustment due to their slotted-hole design, try to get the bolts as parallel as you can, and at a 90-degree angle to the frame. The slight movement allowed by the bearings will later be used to pick up any chain slack. To avoid damaging the threads by welding spatter, tape the threaded area when welding.

The beauty of the single-axle alignment system is apparent when you are installing the two inner bearings. Without the axle as a guide, you would have to venture your best guess as to where exactly the two inner bearings needed to be on the frame, in order to ensure perfect rear wheel alignment. Using the axle as a guide, it is impossible to make a mistake. Figure 2-43 shows a test fitting of the steel plate that will be used to carry the two inside bearings. This plate is cut from some scrap 3/32-inch steel plate and only needs to be large enough to carry the two bearings. If, for some reason, the elbows are higher or lower than the main frame tube, due to heat distortion, you may need to add washers under the two outer bearings to compensate. Do not worry about this up-and-down misalignment; it will not induce any extra tire friction, and will not show up in the final design.

The four bolts needed to hold the two pillow block bearings to the plate are first welded, and then the plate is welded to the frame, as shown in Figure 2-44. There is no guesswork involved in aligning this plate to the frame, since the axle is doing it for you. Before you weld the plate to the frame, make sure that both outer bearings are bolted down tightly and in the center of their oblong holes. The same goes for the bearing plate bolts. Try to place the bolt in the center of the bearing holes, so there is a little adjustment available in either direction later, as this may be needed to pick up a slack chain.

Figure 2-44   *Welding the bearing mounting plate to the frame*

Figure 2-45 shows the two rear wheels mounted to the frame. Obviously, the two axles needed a bit of trimming, since they were left long at the beginning just to make sure there would be enough material. The axle hubs should butt up against the face of the outer bearings, and no part of the tire or any spoke should rub on any part of the frame. The distance between the tires should be pretty close to the calculated distance, so check this before you cut down the axles. Roll the assembly across your garage floor, and it should move with almost no resistance, due to the fact that the wheels are perfectly aligned.

From this point on, I turn over the creative torch to you, as you can alter the frame in just about any way you like, now that you have the rear wheels mounted to the backbone. If you like my design, then follow the steps exactly, but don't be afraid to try your own thing. This hobby is all about creativity. If you do decide to make a radical mod, then read ahead to ensure that your changes will allow things like transmission and steering parts to

Figure 2-43   *Aligning the two inner bearings*

Figure 2-45   *Wheels mounted to the frame backbone*

work properly, as with a little planning, anything is possible.

How tall do you want your chopper? On my OverKill chop, I went sky high with the frame rise so my arms would be reaching for the handlebars. On the Gladiator, I planned to install a laidback seat with a backrest, and wanted a more lazy sitting position, with handlebars coming to me, so I kept the frame rise somewhat lower this time. You should really find a seat base or board to plunk down on the frame so you can see for yourself how you may want to sit or reach for the handlebars. The rear end shown in Figure 2-46 could easily be made into 100 totally different looking choppers by simply altering the rest of the frame, so plan the design to fit your style and seating requirements. To decide on a medium frame rise, I placed this 17-inch tall plastic bucket under the main frame tube just ahead of the wheels, as shown in Figure 2-46. I thought it looked good, so all other frame parts will be based on this position. This may seem like a steep frame, but as you will see in the next few steps, I toned it down by using a pair of elbows to make a nice curve up to the head tube. If you like straight lines, then a straight tube up to the head tube will also look cool.

Bending conduit over 1 inch in diameter is just not possible without some serious tube bending equipment, and since I own nothing but a welder, grinder and hand drill, bending the 1.25-inch conduit would be impossible. To keep to the cool, sketched ideas I had for this chopper, I made the S-shaped curve by welding a pair of factory elbows together, as shown in Figure 2-47. You can make just about any type of curve with a few elbows depending on how you cut and join them, but since I

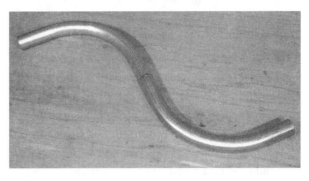

Figure 2-47 *Making curves without a tube bender*

wanted a large smooth S shape, they were simply welded end to end. If you are designing your chopper visually, then only tack weld the parts at this point, until you stand back and think "damn, that's what I want!" For reference, the distance between each end of this S curve is about 24 inches, but don't get too tied up in exact measurements here, as we are not building a helicopter.

After the two elbows were joined together to form the nice flowing curves I wanted, the main frame tube had to be cut down as it was way too long. I hacked the main frame tube to a length of about 27 inches and then welded the curved section in place, as shown in Figure 2-48. By sitting on the exercise bike seat I dropped on the frame, I could reach ahead and imagine how the chopper would feel at this height and length. The head tube would be fairly far ahead of me, but I intended to install some very long handlebars, most likely home built from more conduit and factory elbows. When you lay out a bike visually, you have to imagine how you will be seated when steering and pedaling, so take your time and build it the way you want. For reference, the distance from the very back of the main frame tube to the end of

Figure 2-46 *Choosing a frame rise*

Figure 2-48 *Joining the frame tubing*

the curved section is now 52 inches, although I will most likely be trimming a few inches off the end later.

Once you make a solid weld around a butted tube joint, a little careful grinding can make the tube look like it was made from a single piece. As shown in Figure 2-49, there is no sign of any welding once the weld is ground flush with the surrounding metal. The success of this process involves making a solid initial weld followed by a rough grinding and filling of any leftover holes. The final grinding should be done with a flap disc or sanding disc, so you are not digging into the original metal or weld material between the joint—if you take out too much metal and leave a dip in the joint, it could become seriously weak, so be careful.

The Gladiator is a radical departure from a standard bicycle, but many of the parts are in fact standard bicycle parts, especially those in the steering and transmission. A typical mountain bike head tube is shown in Figure 2-50, ready to be ground clean. A mountain bike head tube is perfect because it is made heavier than a road bike head tube, and will usually have an overall length of at least 6 inches. You could actually get away

with just about any head tube and a matching fork stem, but if you have the junk to pick through, locate a head tube with a length of 6 to 8 inches.

With a little grinder magic, the head tube is free of any original tubing "stubbage" and paint, making it much easier to work with. A sanding disc takes paint off instantly, and will save you from suffering that horrific burning paint smell when you begin welding. Figure 2-51 shows the naked head tube ready for its new life.

A fork stem that matches the head tube will also be needed. As shown in Figure 2-52, simply amputate both fork legs, and then clean up the crown area using the same technique used to clean up the head tube. The little ring (bearing race) above the crown can be tapped off with a hammer before grinding—this way it will not suffer damage. Don't worry too much about the condition of the fork legs; just make sure the stem is not bent. This fork set had been bent in a front end accident, but only the legs were bent, so it was the perfect candidate for hacking, and the stem length matched the head tube.

How long is long enough? Ah yes, the age-old length debate is often a hot topic in the chopper world. I have often heard that you do not need long forks on a chopper,

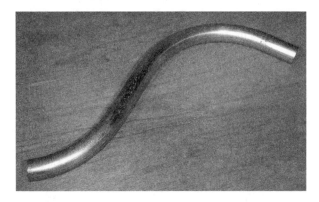

Figure 2-49 *Making a seamless weld*

Figure 2-50 *A head tube cut from an old frame*

Figure 2-51 *Cleaning up the head tube*

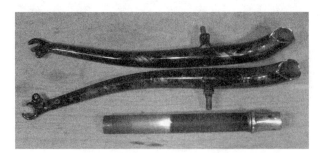

Figure 2-52 *Removing the fork legs*

but personally I think these words are only spoken by those who have to drive a "shorty"! Seriously, though, you can make your forks as long or as short as you want, and I didn't want to use any parts that would make this chopper look stock! I was brought up believing that a proper chopper should have long forks, so I decided to make my fork tubing exactly 4 feet long. Maybe you want to scare more of the roadies off the bike path and think an 8-foot fork would be in order? Certainly, you could do it—this is a trike, so there will not be any balance issue to worry about. Now is the time to choose a fork length, so put your head to the grindstone and hack up some tube. The fork legs are made from the same 1.25-inch thin walled conduit as the rest of the bike. If you want to use some other type of tubing, just make sure it is no smaller than 1.25 inches in diameter if your fork legs are longer than 3 feet.

The fork legs are held in place at the base of the fork stem, as shown in Figure 2-53, using a pair of small 1-inch tubes to fasten them to the heavy crown area. Round or square tubing will be fine for this job, and each piece should be cut to 2.5 inches, then ground to conform to both the fork crown and fork tubing.

Before you weld the lower fork leg support tubes in place, as shown in Figure 2-54, read ahead so you understand the relationship between the angle of the two small tubes and the distance the fork legs need to be placed apart in order to use a standard front bicycle hub. As you will see, having the fork legs cut and ready will greatly help at this stage.

A typical front bicycle hub needs the fork dropouts spaced at 4.25 inches apart. You may have a different

Figure 2-54   *Fork leg support tubes welded in place*

fork design in mind, or plan to use a nonstandard front hub, so consider this when choosing the width of your front forks. You will be able to muscle a half inch either way by squeezing or pulling the legs apart later, but try to get things the way you want now to avoid force fitting parts later. In Figure 2-55, both front fork legs are tack welded to a scrap square tube so their centers are exactly 4.25 inches apart, which is the correct spacing for a stand front bicycle hub, assuming the dropouts will end up in the center of each fork leg. You can also see in Figure 2-55 that the two fork leg support tubes fall right into place, which is how I figured out what angle they needed to be installed on the fork crown, as shown in the previous photo. This scrap tube holding one end of the forks also ensures that they end up parallel when the fork stem is welded at the other end, as will be shown in the next few steps. Again, work on a very flat surface, so all tubing is perfectly parallel.

Before you weld the fork stem in place, as shown in Figure 2-56, you must decide if you are going to have the

Figure 2-53   *Checking the head tube angle*

Figure 2-55   *Setting the fork leg width*

Figure 2-56 *Welding the fork stem in place*

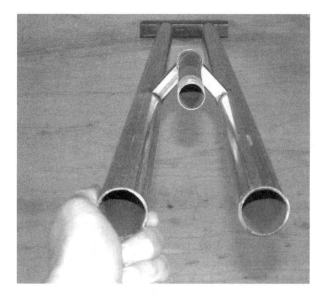

Figure 2-57 *Checking fork alignment*

Figure 2-58 *Checking the fork hardware*

fork legs stick up past the top of the fork threads like my design does, or go with the more conventional triple tree-top plate design. If you look at a typical motorcycle, there are two pairs of clamps holding the fork legs to the fork stem—one at the base of the fork stem similar to Figure 2-56, then a top plate that bolts to the top of the fork stem and to each of the fork legs, essentially capping them all off. This standard motorcycle design can be used here as well, but I make it more stylish to extend the fork legs up past the top of the frame and then mount the handlebars to the side of the fork legs. Not only does this give the chopper a more radical look, it allows more diversity in the handlebar design. I extended my fork legs 6 inches past the top of the fork stem, as shown in Figure 2-56. The fork leg support tubes are first tack welded over the marked areas to make sure all three tubes (fork legs and stem) are perfectly parallel. Two small tack welds on each tube should hold everything in place so you can view it from all angles to check alignment.

Figure 2-57 shows the scrap tube at the end of the fork legs and the tack welded fork stem holding everything in perfect alignment as viewed from the top end. Don't be afraid to break a tack weld and try again if things do not look straight, as this is where it counts, and obvious flaws will effect the steering. The fork legs are exactly 4.25 inches apart as measured from their centers. If you are happy with the alignment, complete the welding around the fork leg support tubing, starting with the top and bottom welds. You may want to add another bit of scrap tubing across the top end of the fork legs, just to ensure nothing moves when welding.

Make sure your fork hardware has no damage. Figure 2-58 shows the two bearing cups, bearings, and

top fork hardware ready for installation. Do not grease any bearing hardware until after the bike is painted, and always remember the bearing orientation rule—balls go into the cups.

As shown in Figure 2-59, the fork hardware should install perfectly if your fork stem matches your head tube. If the threads are too long, trim them down to fit. If there is any excess friction when turning the head tube, you may have installed the bearings backwards, or have the wrong size cups or bearing hardware. There should be very little friction when you twist the head tube. Leave the head tube installed at this point, as it will be needed to take a measurement when setting up the front forks on the frame.

The Gladiator has a unique suspension system which I call the "Inverted Springer." The term "springer fork" refers to a chopper fork that has rods running from the hinged dropouts up to the top of the fork, where a spring

Figure 2-59 *Fork hardware installed*

Figure 2-61 *The pivoting fork legs*

is compressed, creating a spring suspension for the front wheel. I decided to use parts that I already had on hand to build a springer fork, giving it a fresh look and requiring much less work on the part of the builder. As shown in Figure 2-60, the parts are made from a few tabs of flatbar, four mattress or screen door springs, and a few bolts. The springer system is easy to build, works very well, and lets you bounce the front wheel off of the ground while you are parked, just like those hydraulically activated low rider hot rods. Hey, it's all about the show with a chopper!

The tabs shown in Figure 2-60 are made from four 1.75-inch long pieces of 1/8 flatbar with a 3/4 hole drilled at each end. The hole is half an inch from the end of the flatbar, which is then rounded to take off the sharp corners. Read ahead a bit to see how this suspension system is made and you may get some modification ideas of your own. The springs should be fairly hard to pull apart, and you may need two or four, depending on their strength.

The two fork legs shown in Figure 2-61 have been cut from the leftover fork legs taken from the fork stem

used earlier. Each leg is now 6 inches long and has a small tube with a 3/4-inch inside diameter welded at each end. As you will see, these small tubes allow the fork legs to pivot on the bolts and tabs that will hold them in place.

The four tabs are installed onto the fork legs so they hold the pivoting parts, as shown in Figure 2-62. I used an old front hub to hold the fork leg stubs in the correct position as I tack welded the four tabs to the main fork legs. The distance from the tabs to the bottom of the fork legs really depends on the length of your springs, so I will not bother giving you my measurement, as it would mean nothing. The goal is to create a 45-degree angle between the pivoting fork parts and the springs, so as to form a triangle over the fork legs. If this is confusing, look at the next photo.

When the spring is not being pulled, the fork leg stubs and springs form a triangle over the fork legs, as shown in Figure 2-63. This is just a rough guide, and it will really depend on what type of springs you could salvage for the suspension system. My springs came from an old bed frame found at the dump, and I have a bucket full of them. Often, these springs can be found on screen doors

Figure 2-60 *Parts for the springer suspension*

Figure 2-62 *Installing the suspension parts*

Figure 2-63 *Testing the springs*

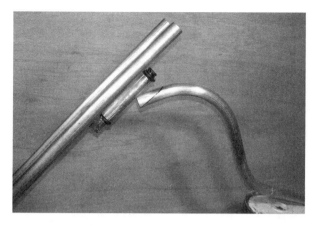

Figure 2-64 *Another design choice*

or at your local hardware store. One last thing to keep in mind when making this suspension system is the clearance between the fork legs and the ground when the springs are pulled to the maximum. If your springs are somewhat weak, the fork legs may strike the ground if you bounce too hard, which is either going to be seen as a bad thing, or a cool way to make sparks at night. You will have to decide! I decided that I will need four springs, as two did not seem to have enough strength to support the front end.

"Fork rake" is another hot topic among bike builders of all kinds. Too much rake, and you always have a chopper; too little rake and your steering feels extremely twitchy, almost too responsive. The good news is that it really doesn't matter much here because you are building a trike, which will always have good control. Seriously, you could drop 20 foot forks on this beast, and it would still be easy to ride! Sure, you would need a six-lane highway to pull a U-turn, but other than that, it could be done. I decided to let fate tell me what angle the forks should be at, since I had already decided on 4-foot long forks and had the rear of the bike propped up on the bucket. Wherever the two parts meet is where they will live, was my thinking. As shown in Figure 2-64, the forks met the frame at about a 45-degree angle, which actually looked great. Of course, anything would have looked great on such a radical ride, so you can't lose.

You will need to have a helper hold the front forks in position with a fully inflated front tire, so you can mark the approximate angle that the front of the main frame tube needs to be cut in order to join the two parts. As shown in Figure 2-64, the black line is the cut I will have

to make in order to seal the deal with my chosen fork and head tube angle. A steeper fork angle would equal a taller frame, and a more relaxed angle would equal a lower frame and longer bike. The choice is yours, but do make sure that you have a full-size front wheel with inflated tire installed on the forks when you mark the frame.

After you rough cut the frame to meet the head tube, it will take some careful grinding in order to get a tight-fitting joint with minimal gap. Take your time and set up the joint properly and the welding will be much easier, resulting in a perfectly aligned front fork. Start with the underside weld first, as shown in Figure 2-65, so you can muscle the head tube into alignment if there are errors. Also, remember that this is one of two tubes that must join the head tube, so leave room for the other tube underneath the top frame tube. With a nice solid weld on the underside of the joint, as shown in Figure 2-65, I recommend that you install the forks then visually inspect alignment. You can leave the forks installed to weld the top side of the joint just to make sure all three wheels are parallel, and there are no obvious alignment flaws.

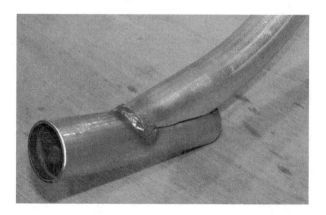

Figure 2-65 *Welding the head tube*

The head tube welding will be one of the beauty welds, so take your time and do it nice, even if you plan to grind the area clean like I did. Figure 2-66 shows multiple weld beads to fill in the low spots and make grinding the area flush a bit easier. It's best to weld only a small bit on each side, changing to the other side often so the distortion is kept to a minimum. It will be very difficult to fix a large error once the head tube is fully welded, so take your time and control that welding heat.

I always clean up large welds that are fully visible as it makes the final product look much more professional. A hidden weld should not be a crappy weld, but it is not usually necessary to grind it flush either. The head tube is very prominent in a chopper, so these welds will be ground smooth to make the tubing flow together as if body filler was used. A lot of grinding, pit filling, and more grinding, produces the smooth joint shown in Figure 2-67, and I think it is well worth the effort. Tiny scratches or dips will be hidden by primer and paint, but a fat ugly weld will not. Sometimes a good weld is an ugly weld, especially if the joint was not well prepared, so the grinder can often become a welder's best friend.

There are two kinds of bottom brackets: the large ones for one-piece cranks, as shown hacked from an old frame in Figure 2-68, and the smaller threaded style for three-piece cranks. You are free to use either type, but as a general rule, one-piece cranks are better for choppers or non-performance bikes because they look much simpler, usually do not include multiple chain rings, and are very easy to find at the dump because they are often used on inexpensive bikes. The smaller three-piece bottom bracket has an axle which the aluminum crank arms bolt onto, and these will always have two or more chain rings. Three-piece crank sets are lighter, made for speed bikes, and just don't seem to look right on a chopper. I guess it's a personal thing, but I just can't picture a chopper with multiple speeds, especially a front derailleur system.

Before you cut any more tubing, read ahead to the rest of the frame build, several steps forward. We are at a point now where you can certainly use whatever materials you have lying around, or make design changes. The only thing you have to ensure is that you can reach the pedals, and that the pedals will not strike the ground—the rest is all up to you. I ran out of conduit elbows, so my frame had to be made with straight sections of tubing. This is OK, and will leave the large top-curved section of tubing as the focal point for the bike. At this point, I decided to use the leftover 1.25-inch conduit and some scrap bicycle frame parts to make the rest of the frame, so there was no plan at all. Yeah, I know this approach spits in the face of all those who like to build things from a massive overly calculated, number-ridden CAD drawing, but that just ain't my style, and would seriously subdue any creative ideas that may come along. Again, you can use the exact tubing and lengths I have used, but do not be afraid to grab some junk and see where it takes you. The bottom bracket was

Figure 2-66 *Completing the head tube welds*

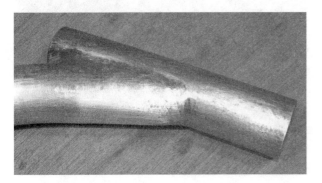

Figure 2-67 *Cleaning up the beauty welds*

Figure 2-68 *Recycled bottom bracket*

welded to a 10-inch length of 1.25-inch conduit, as shown in Figure 2-69. I chose this length because it was lying on my floor, ready to be tripped over. The conduit is ground to conform to the bottom bracket shell and then tack welded at the top and bottom so alignment could be checked.

It's important to ensure the bottom bracket is at 90 degrees to the tube holding it, so that the chain is aligned with the rings at the front and rear of the vehicle. A good way to check alignment before welding the entire joint is by installing the cranks, and making sure the arms are parallel with the tube. Once checked, complete the entire weld as shown in Figure 2-70, starting with the top and bottom of the joint.

Like I said earlier in this section, there was no official plan here, just a pile of scrap tubing on the floor and the knowledge of where that bottom bracket needed to be, so that I could reach the pedals and the pedals had at least 6 inches of clearance from the ground. The tube running from the head tube to the bottom bracket is called the

"down tube." It became 18 inches in length in order to put the bottom bracket where it needed to be, which was 32 inches from where my butt would be on the frame, and 12 inches off the ground. To figure this out, I sat on the temporary seat and had a helper measure the distance from the back of my butt to the bottom of my foot. Subtract the length of the crank arm, and you now know where to put the bottom bracket. See why there are no magic numbers? It's a vehicle built to fit the rider. The other small tube shown in Figure 2-71 can be called the seat tube, and its job is to triangulate the frame for strength. Honestly, it would not be a problem if it was omitted altogether, but I think it looks good on a frame. For reference, this tube is a bit of scrap 1-inch conduit, approximately 9 inches in length. The length of this tube was chosen by what was needed to make the lower tube parallel to the ground.

The frame is almost complete, but at this stage is not ready for the street. The bottom bracket must continue to the rear of the frame, in order to create a fully triangulated frame that will resist bending and become highly rigid. The thin bicycle scrap tubing shown in Figure 2-72 might not look like much, and you could probably bend the thinner tubes by hand, but when added to the current frame, will strengthen it at least 10 times more than it is now. Although the Gladiator is a far cry from a standard bicycle, the frame is still a basic diamond shape. If you take a single tube away from such a frame, it will go from being able to support 400 pounds to collapsing under the weight of a small child. To add stiffness to the current frame, we must join the two rear corners to the open tube at the base to create a basic diamond shape throughout the frame. A dip into my

Figure 2-69 *Installing the bottom bracket*

Figure 2-70 *Bottom bracket welding completed*

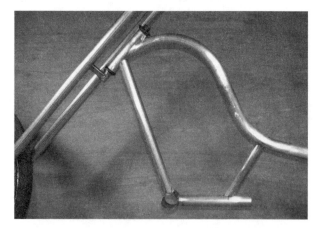

Figure 2-71 *Chaos becomes order*

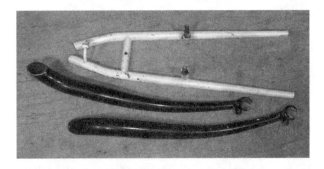

Figure 2-72 *Front control arms*

scrap pile revealed another set of bent fork legs and a seat stay from an old mountain bike. Yep, that would work.

I like working with random bike scrap to make something unique. Often, you can join two bits of tubing and then hide the welds to form something that looks completely alien, hiding any signs of its original life. When I tell people that the Gladiator was built from only the most common bicycle parts available, using only a welder and grinder, they often don't believe me, so I give them a tour by pointing at the parts. See that? Those are washers, not machined hubs, and that's more conduit from the hardware store. Oh, and there is a pair of forks joined to some other frame parts. Personally, I get a massive sense of pride from seeing someone's jaw drop, not only by the look of the final bike, but due to its origins as rusty throwaway scrap. Not much pride in dropping a few thousand bucks at the local machine shop so they can build you a bike, if you ask me!

Anyhow, Figure 2-73 shows some more metal fusion surgery, done by welding and careful grinding to join the fork legs to the frame tubing. I liked the tapered look of the tubing, and the nice curves in the fork legs, which is why I decided to use them. What do you have lying around the shop? Old bed frames, tables, chairs—they all work.

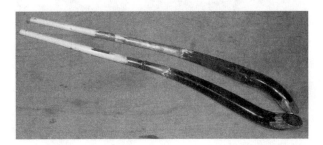

Figure 2-73 *More tubing fusion*

Figure 2-74 shows how the bits of scrap tubing join the rear of the frame to the front section, creating a much stiffer frame. Before the two tubes were added, the frame might have withstood a 150-pound rider, but the weakness would be just behind the curve at the top of the frame, and it would have been a very flexible frame. I'm sure that the front forks would be the weakest link, although nothing besides a head-on collision with a wall or truck would be a problem. Now you can see that spontaneity can really add to your craft, and the results are often better than you would have imagined. I really like the way the frame turned out, even though the lower half was built using only the parts I was too lazy to clean off the garage floor. Having a large scrap pile is better than any expensive tool!

You might have realized that the front forks are not complete yet, because the two fork support tubes would not be strong enough to carry the weight of the loaded bike. All triple-tree forks have a bottom and a top support, but since we extended the forks past the top of the fork threads, the traditional top clamp design is no longer an option. To secure the upper part of the forks to the top of the fork stem, this simple clamp system does a great job, and is made by cutting the clamp from a standard bicycle gooseneck and replacing it with these two small pieces of flat bar. Figure 2-75 shows the gooseneck after removing the clamp, as well as the two bits of flat bar that will connect to the front fork tubes using a pair of screws.

Figure 2-76 shows the completed fork mounting system, which is built by dropping the gooseneck stem into the top of the fork stem to figure out how long and at what angle these two bits of flat bar need to be cut. As you will see in the next photo, the holes drilled at the

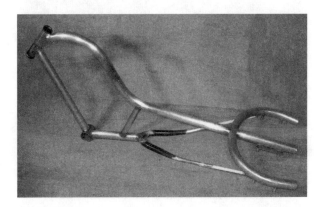

Figure 2-74 *The frame is now complete*

Figure 2-75   *The top of the fork mounting system*

Figure 2-76   *Fork mounting part completed*

ends of the flat bar allow a woodscrew to hold the unit to the fork legs.

With the top clamp holding the fork legs in position, as shown in Figure 2-77, your forks are now ready to take the full weight of the loaded chopper. You can now safely handle the vehicle or sit on the frame without any worry of bending any part of the frame or forks, since they are completely rigid now. Of course, there is a bit more work

to do before you can cruise down the center of the road like you own it.

When building a chopper using car parts, you are presented with several challenges in the transmission and brakes department, since bolt-on bicycle parts will not work. A trike also adds another dimension to the problem because you have to find a way to drive one of the rear wheels from a standard bicycle crank set. I decided to solve both problems with one simple part which I call the "remote coaster hub." A coaster hub is a single speed hub that allows the pilot to pedal backwards to activate the brakes. This design was widely used before the derailleur on most bikes, and is still commonly used on most children's bikes and some beach cruiser-style adult bikes. I liked the coaster brake idea because it would eliminate all ugly looking cables and levers from the bike, something that would look out of place on a chopper. The problem was how to transfer the motion of the rear-drive axle directly to the coaster hub, so it would act as a brake and still allow the pilot to stop pedaling and coast.

The answer was actually pretty simple—weld a sprocket to the coaster shell, then run a chain to another sprocket on the rear-drive axle. This way the coaster hub would appear to be directly connected to the rear wheel, so it would act as if it were doing its usual job. To create this transmission system, you will need two steel sprockets of approximately equal size (number of teeth) and a steel coaster hub. All parts shown in Figure 2-78 came from discarded children's bikes—the ones with really tiny wheels. Often, these bikes have perfectly good coaster hubs installed in tine wheels with 16 spokes, so they are perfect for salvage and have probably had almost no wear on them. The two 28-tooth sprockets are also from old kids' bikes, taken from the crank set by

Figure 2-77   *Front forks completed*

Figure 2-78   *Mid-drive transmission parts*

unscrewing the nut that holds them onto the small one-piece crank arm. You could probably use any steel sprocket and matching chain you like, but try to keep the number of teeth in the two large sprockets as close as possible.

If you read ahead, you will see how this mid-drive/brake system is going to work, and you might be wondering if you can alter the design and use speed bike parts. Sure, if gears is your thing, then just about anything is possible, but then you will have to run a cable to a front brake, or come up with some other method of adding a brake to the rear wheel. This is not a racing bike, so a single brake would be no problem, but do some planning ahead. Figure 2-79 shows how the steel sprocket from the kid's crank set is welded to the non-drive side of the coaster hub. This sprocket had "spokes," so cutting it to fit was easy. Some kids' bike chain rings will be solid, so I will show you how to deal with them in the next step.

The sprocket shown in Figure 2-80 does not have spokes that can be easily cut, so a larger hole has to be made in order to get it to fit over the coaster hub shell. By drilling a bunch of small holes around a circle drawn with a marker, you can save yourself a lot of hand filing. Drill as many holes as you can, then cut the leftover metal out with wire cutters. Now you can hand file the rough area to get the hole perfectly round. You will also notice that the sprocket in Figure 2-80 is offset about 1 inch, a design that may actually help you line up the chains if the coaster hub is too narrow. When you are building this transmission, remember that all chains must be parallel to their sprockets, or you will have difficulty with derailments or friction. A little planning ahead and

Figure 2-80 *Making a hole in the sprocket*

test fitting before doing any permanent welding is always a good idea.

Figure 2-81 shows the newly installed sprocket on the non-drive side of the coaster brake hub. You will notice that the sprocket is welded to the flange and then the flange is welded to the hub. The flange needs to be welded to the hub because it is only press fit onto the shell and not designed to take any real stress. Huh? How does a wheel stay together then? you might ask. It's actually the tension of the spokes that keep the hub flanges together on a press fit hub, nothing else. If you don't put a few beads of weld on the hub and flange, it will eventually fall right off. I was actually able to pull the flange right off with my bare hands. A few small welds is all that is needed, but try not to overheat the hub too much or you will have to regrease it later. There is a good chance the welded sprocket will not be perfectly true due to heat distortion. Fix this with a hammer. A ding here, a ding there, and it will be running true in no time.

The remote coaster hub must act like it is directly connected to one of the rear wheels, so the newly

Figure 2-79 *Adding the sprocket to the coaster shell*

Figure 2-81 *Sprocket added to the coaster shell*

installed coaster hub sprocket will connect to a fixed rear axle sprocket of the same size. Now, whatever the rear-drive wheel does, that is what the coaster hub will also do, making it seem like the rear wheel and coaster hub are one piece. To get the sprocket connected to the rear-drive wheel (the left wheel), we need to fix it to the left axle. This is done by welding the sprocket to a small piece of round tube that just happens to fit snugly over the ³/₄-inch axle, as shown in Figure 2-82. Where do you find such a tube? Try the tubing from a mountain bicycle handlebar, as it is the perfect fit. A 2-inch segment of the handlebar tubing should be plenty to mount the sprocket and the bolt that will hold it to the axle.

The completed rear-drive axle sprocket is shown in Figure 2-83, ready to mount to the left axle. When you install your crank set and mid-drive transmission, you will see that the left axle is the only logical place where you can mount this sprocket in order to get the two chains to line up. The two bolts will secure the sprocket to the axle, but do not drill out the axle until you know your chains are all going to line up. Some adjustments

Figure 2-82 *The rear axle sprocket*

Figure 2-83 *Rear-drive axle sprocket completed*

may be needed later. Also notice that most of the axle mounting tube is on one side of the sprocket. This is done because the sprocket will need to be as close to the bearing as possible when you get those chains lined up. This will become obvious as you progress.

As mentioned earlier, the rear-drive sprocket is placed on the left axle so that it is as close to the inner bearing as possible without any sprocket or chain rubbing. The coaster hub is just wide enough to deliver the chain from the crank set on the right side of the bike to the left axle at the rear of the bike, which is why this is done. The coaster drive sprocket and crank ring always have to be on the right side of the bike, or the brake will not function. Figure 2-84 shows the drive sprocket bolted to the axle with minimal clearance between its face and the inner bearing. I would advise waiting until all of the chains are installed before you drill the axle holes, just in case you need to make some adjustments or alterations.

When you mess around with bicycle parts, there is one tool that is well worth the 10 bucks you will shell out for it—the chain tool. In the old days, I would bash chains apart by using a punch, or a finishing nail to bang out the link pins, but after turning so many fingernails blue, I decided to invest in the chain tool. Seriously dude, putting chains together is so easy with the chain tool that it is baffling how I got by without it. Figure 2-85 shows the chain tool and a ball of chain about to be made into the dual chains needed to propel the Gladiator.

Figure 2-86 shows how the remote coaster hub transfers the drive across the hub shell to the rear left axle. It makes no difference at all where you place the mid drive, or how you attach it to the frame, as long as the chains line up and they do not rub on any part of the frame. I cut a pair of rear dropouts from an old speed

Figure 2-84 *Rear-drive sprocket installed*

Figure 2-85 *Get one of these tools!*

Figure 2-86 *Installing the mid drive*

bike, bolted them onto the coaster hub, then cut and installed two chains with the least amount of slack possible. From here, I could mess around with areas to weld the two dropouts, in order to permanently mount the mid-drive hub. I found that placing it in the middle of the frame, as shown in Figure 2-86, was optimal, which is why I ended up calling it the mid-drive hub. You could hide it directly under the rear sprocket at the far rear of the bike, or move it to the front of the bike if you wanted, as long as the chains are happy. When I did find a place that kept the chain slack to a minimum and had no chain interference, I tack welded the two dropouts directly in place while holding on to the hub. This system works well, but be careful not to hit the chain or sprocket with the welding stinger. In Figure 2-86, the two dropouts have just been tack welded to the frame. It took a bit of fiddling around to get the two chains the right size, but that's why you have the chain tool. Also, that arm sticking out of the non-drive side of the coaster hub must

be secured to the frame for the brakes to work. If you do not secure the arm, the entire hub will twist around the axle, loosen all the bearings, derail the chain, and then seize up solid on you. Don't forget about this before your first test run, or you will look like a dork as you push your chopper home.

Do you want to lie back in a relaxed, recumbent position, drinking a coffee as you cruise down the street, or hang on for dear life, crunched over in a most uncomfortable position? Yes, both of these seating positions are totally cool with the chopper crowd, so it's up to you which one you choose, or some combination of the two. Because a trike is a stable vehicle that will allow you to stop without putting a foot down, I went for the relaxed seating position, which would also include a nice, tall backrest. Because I am a high ranking member of "Club Laziness," this laidback position was my choice. I planned to sit back, and then let the handlebars come to me, but if you already have a firm plan for the handlebars, you may want to leave the seat until the end. From here on, you can finish the chopper in any order you like; the only thing you have to be mindful of is how far you need to be away from those pedals for a comfortable ride.

Figure 2-87 shows my randomly planned high-back seat, made by slicing and dicing some 3/4-inch thin square tubing to conform to my back as I sat on the temporary seating. The only advice I have is to not make a headrest, or at least a headrest that is always on the back of your head. This may seem like a good idea at first, but you will find it highly annoying as the bumps from the rear wheels are amplified and sent to the back of your head. A neck rest is OK, but not a headrest, unless it is simply there for looks.

Figure 2-87 *Seating is a personal touch*

After making an entire chopper from inexpensive raw materials, I wasn't about to spew up a hundred bucks on some fancy chrome ape hangers that would probably be too flimsy or short for the bike anyhow, so I decided to make my own. It was one of my design goals to not have any chrome on my chopper since this is a worn-out tradition, so making the handlebars would be fine, as they would be painted anyhow. That 3/4-inch EMT conduit is exactly the same diameter as bicycle handlebar tubing, and it was a bit heavier, so welding parts together would be a snap. Figure 2-88 shows two more factory elbows and some lengths of 3/4-inch conduit to make my ultra stretch handlebars.

More welding and more grinding is all you need to create seamless joints between two pieces of metal, as shown in Figure 2-89. The two handlebar halves are nothing more than 12 inches of conduit welded to a factory elbow, but they are perfectly suited for this bike, so they are better than anything that can be purchased, in my opinion.

I think it's cool to have the handlebars come directly from the front forks, like they are on Figure 2-90. This system is strong, simple, and screams out customization.

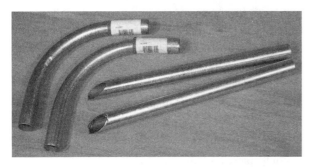

Figure 2-88 *Ape hangers the poor man's way*

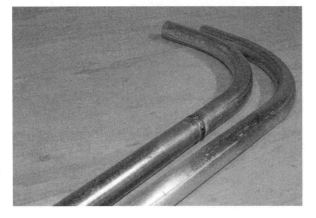

Figure 2-89 *Weld and grind to hide seams*

Figure 2-90 *A unique handlebar mounting method*

As you can see, the handlebars are quite long due to my relaxed seating plan, so it would have been difficult to find a bicycle handlebar to reach. How about hacking some furniture tubing into rounded handlebars? Yes, look in the most unlikely places to find your needed materials.

This chopper is so close to getting painted that I can already smell the fumes! It's almost painful to spend time capping the ends of the tubes and cleaning up any messy welds, but this is where true craftsmanship shines. I took all of the large tube ends (forks and frame) down at a 45-degree angle, then capped them off with some oblong sheet metal discs, as shown in Figure 2-91. Sure, it took a while to weld and grind, then reweld and grind again until all five caps were perfect, but damn did they look good afterwards. If you plan to add the bells and whistles, do it now before painting.

The capping of the five open tubes took longer than the building of the rear wheels, but the results were worth the effort. You can't purchase tubing like this, so you really have a true custom. The handlebar welds and most other welds that were visible were all cleaned up and

Figure 2-91 *Don't forget the trimmings!*

ground for appearance as well since the next step was primer. Take a few sheets of emery cloth and go to town on the frame, cleaning up the metal and burnt paint as much as you can so the primer coat goes on smooth. Old paint does not have to be totally removed, just sanded smooth so it blends into the bare metal at the edges. I use only spray paint from a can, but the results can be very good if you take your time and apply the paint according to the directions, which is usually in short strokes held at 8 to 10 inches from the metal. I hang my frames from a clothesline in the backyard when I paint, and usually apply primer in two coats to avoid any runs or streaks. If you are new to painting, ask a fellow builder who knows how to wield a spray bomb, or search the Internet for a spray can painting tutorial, as there are many good sources of information.

The rear rims are metallic blue, so what would be the point of throwing on a front chrome rim? I had nothing better to do while the primer was drying overnight, so I pulled out all of the spokes in this old steel mountain bike front rim so I could paint it to match the rear rims. Yes, you could try to tape up all the spokes to paint the rim, but that would be a tedious job that would probably leave you with a rim full of paint runs, and that would not add to your skills. As shown in Figure 2-93, I removed the 36 spokes by loosening them two turns each around the entire rim, so they would then come out by hand turning the nipples the rest of the way. Now, I could properly sand the rim as well as the hub to give them the same metallic blue paint job that I gave the rear rims.

Building a bicycle rim is really not as hard as some would have you think, and since you already have all the spokes and can see how they were put in, you only have

Figure 2-93 *Matching the front rim to the rear*

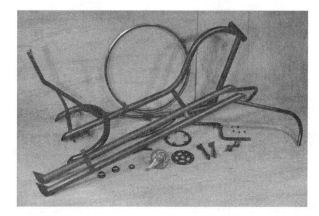

Figure 2-94 *Painted and curing*

to copy what was already done. These big old steel rims are easy to paint, easy to spoke, and can be trued up fairly easily, so don't be afraid to go all the way.

The parts shown in Figure 2-94 are freshly painted and need to cure for a few days to avoid easily scratching the paint. I know this is the most difficult thing in the world to do, but resist the urge to assemble the chopper while the paint is still tacky. Hey, you still need to build a seat, so let's move on to that for the rest of the day. Step away from those parts! Oh, and if you want to see what color the Gladiator is, then you will have to visit AtomicZombie.com to see it in full living color.

I mentioned earlier that I found that two springs seemed a little spongy for my liking, so I added four springs to the suspension system (Figure 2-95). Here is a close-up shot of how I joined the spring to the pivoting fork leg, in case you want to use this design as well. The four springs are much stiffer, and let me hop the front wheel up and down off the ground while I am parked at a red light, just like those crazy low riders. This suspension system is actually quite smooth and functional.

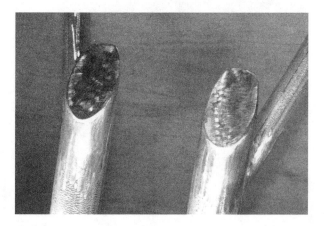

Figure 2-92 *Capped tubing*

Figure 2-95 *Dual spring detail*

If you do not have a ready made seat for your chop, then grab that jigsaw and some scrap wood. It is not difficult to fashion your own crude seat. I took two bits of half-inch plywood, one 10 x 10 inch piece, and a 10 x 14 inch piece, then rounded the ends off using a dinner plate. The three bits of flat bar shown on the seat boards in Figure 2-96 are drilled for a few woodscrews, then bent to whatever angle you want to hold the seat boards together. You may also want to hack up an office chair, or find a fishing boat seat to modify. The possibilities for making a chopper seat are as endless as the number of wild customs there are.

I had some random bits of foam padding left over from a few other bike builds, so I used up the scraps shown in Figure 2-97 to pad the seat. The seat base is made from a 1.5-inch thick bit of high density foam and the back is made from rigid 1-inch foam. Stiff white foam used to pack appliances also works great for seats, but do avoid the really spongy stuff you might find in furniture cushions, as it's way too soft. The foam is rough cut and then glued to the seat board using spray adhesive.

Figure 2-96 *Making a seat*

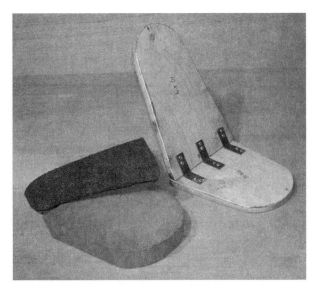

Figure 2-97 *Choosing seat padding*

Once the rough cut foam is glued to the seat boards, I like to take the grinder and run it along the board perimeter to make a perfect foam cut, as shown in Figure 2-98. Be careful when using a grinder on foam, as it may snag and throw the seat away from you if you try to press too hard when grinding away the foam. You could also trim with a long sharp knife, but you know I love my grinder, and would probably use it to shave if my electric razor ever burned out.

Figure 2-99 shows the seat getting a nice covering of black spandex material. You can cover your seats in anything you like, but try to find something that will not

Figure 2-98 *Cutting the foam edges*

Figure 2-99 *Recycling spandex pants from the 1980s*

Figure 2-101 *Making a rear seat cover*

wear out or soak up water easily. Originally, I planned to use jeans material to cover the seat, but decided there was enough blue in the rims and accessories already. Cut a piece of material larger than you need and then start going to town with your staple gun.

You should try to stretch the material a bit at a time to avoid wrinkles as you staple the perimeter on the rear of the seat, as shown in Figure 2-100. Use plenty of staples, and pull out any that are not tight enough. This is not the best way to upholster a seat, but it requires no sewing, and that's good enough for me.

What do you do about those ugly staples and plywood on the back of your seat? How about cutting a chunk of plastic or some thin sheet metal like I did in Figure 2-101? Now you can paint the back of the seat to match the bike, and it won't look like a chunk of wood from the neighbor's doghouse. A few woodscrews or thumbtacks will hold the seat cover in place and protect the wood from the elements. I also added a backrest from an old car seat into the design, placing it on the end of the

long tube that holds the seat back. It actually became one of the most comfortable seats I have ever added to a human-powered vehicle.

Yes, indeed, "Whoa!" is about all you can say when you look at the final product in Figure 2-102! Even though it was built from rusty car parts, a few dead bicycle bits, and some hardware store electrical conduit, this chop is sick. See what can be done with junk and a few basic tools? The only ingredient you need to add is your imagination and some sweat, and you can have the ultimate custom chopper trike. Words cannot describe how good it feels to roll down the street on this machine, having people run after you to ask you questions. "Where d'ya get that awesome chop?" or "Can I buy it?" or "Will you make me one?" Dude, it never ends.

Figure 2-103 shows the Gladiator Chopper ready to take on the world, sending mass produced choppers back to their assembly lines whimpering in fear. So, how does a chopper with two rear car wheels ride? Well, great, actually! Once you get rolling, it seems like you have some evil force propelling you and you can glide forever.

Figure 2-100 *Seat material stapled in place*

Figure 2-102 *"Whoa!"*

Figure 2-103 *"Yes, I do own the road"*

Figure 2-104 *Words cannot describe...*

Sure, it takes a little muscle to get up to speed, but if you were afraid of a little exercise, you wouldn't be building a human-powered vehicle now, would you? Of course, with the sheer amount of attention this ride commands, it's tough to get more than half a block before you are hijacked by someone needing a closer look. What a showpiece!

On my first long test run, I took the Gladiator for a 10-mile ride. Everything worked perfectly, and the suspension was remarkably effective over small potholes and bumps. The seat was very comfortable, the steering was good, and the brakes had enough power for all but the really steep hills, which I avoided anyhow. This was one of the first times I have rolled out of the garage and not come home for several hours without needing some type of small tweak or minor modification. The Gladiator really rode as good as it looked right from the second the paint was dry. Sha-weet! Again, I am posing it up on the chopper in Figure 2-104.

Although I try not to abuse something I spend so much time building, it is hard to resist pounding the front wheel up and down, so it bounces right off the ground when I am waiting for a light to change or at a stop sign. The springer fork can really get bouncing pretty high, and it almost looks like there are hydraulics in the forks, an idea that I may actively pursue when it comes time for modifications. Yes, no custom is ever complete! Look at that sick profile in Figure 2-105.

It's easy to forget about all the healthy exercise you are getting when you are rolling around on such a cool machine (Figure 2-106). After three hours, your legs feel like they have done some real work, but that's not so bad, as it just means you will be alive longer to enjoy your newfound hobby! I thought of adding a small electric

Figure 2-105 *Rides like a dream*

Figure 2-106 *Relaxing after a long cruise*

motor to the Gladiator, but it's just so crazy to have such a full-sized chopper under only human power, as if throwing the bird to all those who make a huge pile of noise with their petrol burning engines and spit oils all over the place! Yes, we take rebellion one step further by rebelling against the mainstream rebels. After all, riding a chopper is all about being different!

Well, I sure hope you enjoyed this project, as I sure did! It's almost painful to put the Gladiator away at the end of a ride, but my legs can only take me so far, and I know the chopper will never give up. This project is proof that you do not need to spend a lot of money, acquire high-tech tools, or learn any elite skills in order to make your mark in the custom community. I look forward to seeing your version of the Gladiator Chopper, and hope you will drop by our website to share your photos, ideas, and stories of pain and glory with the rest of the avid garage hackers. Well, I am off to hit the wide open road on the ultimate chopper trike shown in Figure 2-107, so as the gladiators would say: live free, fight hard! The next project isn't as labor intensive as the

Gladiator, but is another cool chopper with an "old skool" design.

Figure 2-107 *Until next time*

Project 3: Old Skool Attitude

The Old Skool Attitude chopper brings back the golden age of chopper building, where most bicycles had only a single speed, one brake, balloon tires, and nice flowing lines in the frame. Your daddy would remember the days when he and his buddies would take that old bicycle, extend the forks by hammering another pair end to end, replace the front wheel with a smaller one, and then strip all unnecessary parts from the frame. Those builders who had access to a welder may even go as

far as modifying the frame or forks to adjust the rake, which is what I plan to do here. The fun thing about this project is that all of the parts are true to the era, right down to the 1970s exercise bike that I used for parts. Well then, crank up the Black Sabbath, grab a cold drink from the "Harvest Gold" colored refrigerator, and let's travel back in time for some Old Skool chopping!

Figure 3-1 shows the vintage 1970s single-speed frame that I found at the dump one day while scavenging for parts. I was just a young "whippersnapper" when these bikes were on the road, but I do remember seeing them chopped and modified, ridden by garage hackers of the day. I thought it would be cool to salvage this old frame from the mud and give it the life it always wanted, keeping true to the style of the era from whence it came. I had an old 1970s exercise bicycle sitting in the back of my garage, as well as a banana seat, an old single-speed wheel, and a vintage bicycle light, so this project was certainly possible.

I planned to add longer forks to the chopper, but did not want it to lean back too much, so a simple frame extension would be necessary in order to increase the rake and length of the frame, which

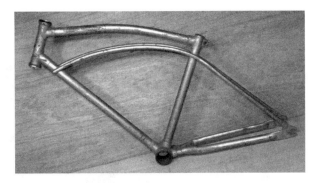

Figure 3-1   *A vintage 1970s single-speed frame*

would keep the bottom bracket roughly at the same height to the ground. As shown in Figure 3-2, the original head tube was removed and two pieces of conduit, equal in diameter to the original frame tuber, were fit in place. The original head tube could not be reused due to the lugged construction, which leaves huge holes in the head tube, and contaminates the joint with brazing filler. Another head tube from some scrap frame was used to replace the original head tube.

Figure 3-3 shows the top tube extension being tack welded to the frame. This 8-inch long piece of conduit has the same diameter as the original frame tubing, so it

would be easy to seamlessly integrate it by grinding away the weld material. Head tube angle was also increased to allow the new forks to have quite a bit of rake, so that the frame would not be lifted at the front, creating a skyscraper style chopper. When doing these ad-hoc frame modifications, it is a good idea to place the parts and wheels on the ground to make sure your final product will have the angles and dimensions you are after. Bottom bracket height is another matter of concern, as you don't want your pedals to hit the ground around corners.

The first extension tube was completely welded at the top tube joint, as well as the head tube joint, as shown in Figure 3-4. The head tube was checked for vertical alignment with the seat tube just after the first few tack welds were made.

To make the final product look like it was born to be a chopper, the frame welds are ground completely flush and then carefully cleaned with a flat file. The resulting tube extension, as shown in Figure 3-5, is completely seamless. When making a seamless joint, rough grind the

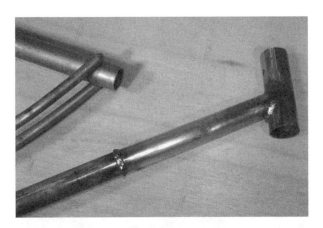

Figure 3-4   *First tube completely welded*

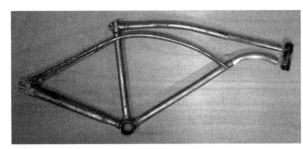

Figure 3-2   *A simple frame modification*

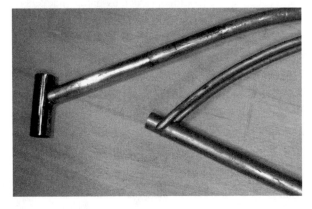

Figure 3-3   *Installing the top tube extension*

Figure 3-5   *Cleaning up the welded area*

weld material with a grinder disc, being careful not to create a low spot, and then finish up with a hand file, so you do not accidentally take away any excess material. The seamless joint will be just as strong as the original weld, as long as your penetration and welding technique are adequate. Chances are that you will have to fill in a few tiny holes and the grind a little more, but it is all for the love of the craft.

The missing section of down tube will be filled in with another piece of electrical conduit: this time a section of factory elbow. As shown in Figure 3-6, the elbow is cut so that the curve joins the original down tube as a butt joint and then curves up to meet the head tube. If you have a tube bender, you could also make your own curved extension tube.

Figure 3-7 shows the completed frame modifications after cleaning up the down tube welds at the seam. Once painted, there will be no detectable seam where the frame was cut and welded, making the final product look very professional. I'll bet your dad never had a chopper frame of this quality!

To keep in with my silly idea of only adding vintage parts to the Old Skool chopper, I dug out this vintage 1970s exercise bike from the huge pile of scrap at the back of my garage. I could almost hear ABBA songs playing in the background, as I gave the cranks one last spin before heating up the zip disc for total annihilation of the unit. I decided to use the handlebars of the exercise bike as chopper forks, integrating the actual handlebars and fork legs as one smooth flowing line. Those massive cranks would also fit the new chopper perfectly, as they were also the correct style for this project.

The two front tubes that form the handlebars and front stand on the old exercise bike were hacked from the frame, as shown in Figure 3-9. I also cut the legs off some really old kids' bike forks so I could weld them to the end of the new forks, creating that nice tapered look that all forks had back in the day. Also shown in Figure 3-9 is a fork stem that will fit into the chopper frame's head tube, allowing me to create the new long forks.

To create the new flowing front forks, the two amputated fork legs were bolted onto the front wheel axle and then tack welded to the exercise bike tubing. Figure 3-10 gives you a clear image of how I planned to integrate the curved tubing into one continually flowing

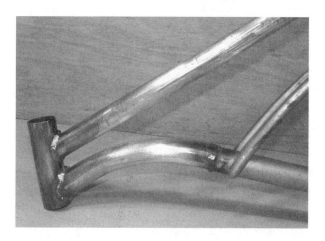

Figure 3-6 *Adding the down tube extension*

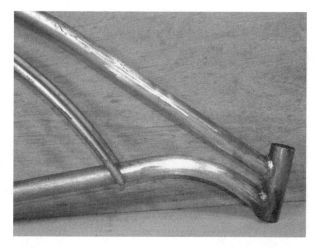

Figure 3-7 *Frame modifications completed*

Figure 3-8 *Official vintage exercise gear*

Figure 3-9 *These parts will become the new forks*

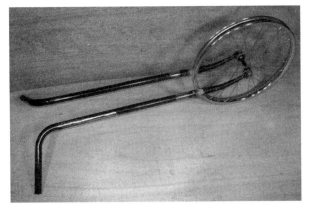

Figure 3-10 *Using the front wheel as a guide*

handlebar and fork unit. The front wheel is used as a guide to make sure each fork leg and handlebar half has the same shape and angle.

To join the new long forks to the chopper, a fork stem would need to be connected to both tubes in order to create a useable pair of forks. The fork stem and two small tubes shown in Figure 3-11 will be used to complete the assembly, securing the two fork legs to the head tube. The two smaller tubes will be welded to the fork stem at 90 degrees to each other, and are at whatever

Figure 3-11 *Adding the fork stem*

length is necessary to place the fork legs parallel to each other, as will soon be shown.

The two small tubes are welded to the base of the fork stem, as shown in Figure 3-12, so there is a 90-degree angle between them. Because the front wheel is being used as a guide to keep the two fork legs parallel to each other, it is easy to figure out how long each of the two small tubes needs to be. The ends of both tubes are ground out to conform to the fork legs, making the joint easy to weld.

Figure 3-13 shows how the new fork stem joins the two fork legs together, so that they run parallel to each other once held in place by the front wheel axle. The position of the fork stem will determine the length of the forks and the height of the handlebars in relationship to the frame. I decided to keep the forks as long as possible, placing the handlebars very low for a "low and lean" look. The fork stem should also run perfectly parallel to each fork leg,

Figure 3-12 *Fork tubing welded*

Figure 3-13 *Installing the fork stem*

in order to keep the forks aligned with the rest of the frame.

The new forks can now be installed, as shown in Figure 3-14, although it is not a good idea to put any weight on them just yet. The two small tubes that connect the fork legs to the fork stem are not strong enough to take any real weight, so some extra support will have to be added between the top of the fork stem and the fork legs. This dual support system is often referred to as a triple-tree fork. Also, notice how low my handlebars are in relation to the rest of the frame. The rider will have to hunch over to reach the handlebars, but whoever said a chopper should be comfortable? It's about attitude, not comfort!

The top of the fork stem needs to have some kind of support, in order to keep the two lower tubes from bending once weight is on the frame. I found two reflector mounting brackets, as shown in Figure 3-15, which, when bent the appropriate way, would fit perfectly between the top of the fork stem and the fork legs, assigning the needed support. You could also make

a triangular plate with a hole for the fork stem to do this job as well.

The two reflector brackets are shown installed on the fork stem in Figure 3-16, adding the needed support for the fork legs. A small hole will be drilled in each fork leg, so a machine screw can secure the brackets to the fork legs, giving the entire fork assembly much greater strength. Now the fork legs can take the weight of a rider safely.

Well, that's pretty much it. To keep true to the era, I added a banana seat, single-speed coaster brake rear wheel, vintage bicycle light, and the crank set from the old exercise bike. Oh, and most choppers from back then had a front wheel that was much smaller than the rear, so I added a 20-inch front wheel. Figure 3-17 shows the Old Skool chopper before painting.

Although you can't see the colors in Figure 3-18, the frame was painted a nice faded green, and the forks were sprayed using what is referred to as Gypsy chrome, a silver paint. Notice how the frame shows no signs of

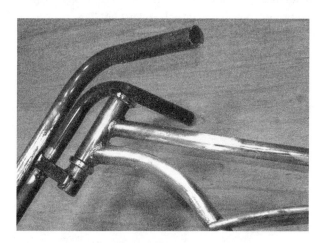

Figure 3-14   *New forks installed*

Figure 3-15   *More recycling of parts*

Figure 3-16   *Top support brackets installed*

Figure 3-17   *Staying true to the era*

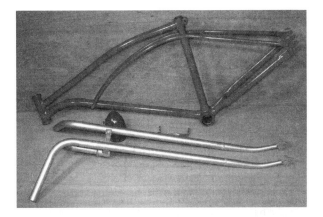

Figure 3-18 *Faded green and Gypsy chrome*

Figure 3-20 *Bad attitude—Old Skool style*

Figure 3-19 *Just like your old man used to ride*

modification due to the careful cleaning of all the joints.

The finished Old Skool chopper is shown in Figure 3-19, complete with a rear balloon tire, and retro brown handle grips. Since most bicycles of the 1970s had fenders and a chain guard, chopper builders of the day

would remove them to be rebellious, which is why this chopper has neither.

The Old Skool chop actually rode very well, considering the ultra low handlebars and extended forks. With that huge balloon tire and coaster brake rear wheel, you could lock up the wheel, and fishtail like crazy around the corner, leaving a thick, black trail of rubber on the road. The wheelie ability of this chopper was intense, thanks to the banana seat, allowing the rider to shift the center of gravity over the rear wheel. Just look at how cool that Old Skool chopper looked—Figure 3-20. Next time you find a bunch of retro junk at the dump, bring some attitude back from the past. Hang onto your welding hats, kids. The next project will appeal to the young chopper enthusiasts.

Here is a chopper for your young whippersnappers. It's only fair that the little ones have cool wheels, especially if you have built yourself a chopper. This simple chop can be made in an afternoon using an old children's bike and a few scrap bits of tubing, twisting a normally mundane set of wheels into a cool mini chopper that will be the envy of all young rebels on the block.

These tiny kids' bikes are so common at the city dump scrap pile and at yard sales that you could probably go into business making kids' choppers! Figure 4-1 shows one of the dozen or so identical kids' bikes with 10-inch wheels that I had in my scrap pile. It should be an easy job to make a chopper out of this bike, requiring only a longer set of forks and a few basic frame modifications.

The entire bike was disassembled so all the parts could be cleaned and checked over for damage. The bike came apart in minutes, as shown in Figure 4-2, revealing parts that looked almost brand new once degreased and cleaned a bit. Since kids' bikes often have little mileage, rust will be your only enemy.

The plan was to replace the original forks with a pair from a full-sized mountain bike, giving that "extendo fork" chopper look. Of course, simply replacing the front forks would also make the front end a lot higher, creating a very difficult-to-ride bike that leans back so much that it would wheelie instantly. To accommodate a much longer set of forks, the frame will need some extension and head tube rake adjustment. To make the head tube adjustment and stylize the frame, the entire middle section is cut and removed, as shown in Figure 4-3. The head tube is then ground clean so it can later be welded.

The head tube will be placed at a more relaxed angle and a little further from the seat, so that the young chopper pilot does not have the handlebars too close to his or her knees. This modification will also keep the bottom bracket at the original height, avoiding the wheelie effect. Figure 4-4 shows the general frame

Figure 4-1 *A typical kids' bike*

Figure 4-2 *Taking inventory of the parts*

Figure 4-3 *Hacking up the frame*

Figure 4-4 *Using some bent conduit*

modification plan, using a bit of scrap bent conduit to replace the missing center frame tubing. The curved tube made the frame take on a whole new stylish look. You can use just about any scrap tubing to make these mods since the frame will not be under much stress.

As shown in Figure 4-5, the original head tube will be reused, so take your grinder and remove the excess frame tube left over from the cutting. If you plan on replacing the front forks, make sure that the new head tube matches the forks you intend to use. Some kids' bikes are so cheaply made that the head tube does not even include

any bearing hardware, so you may have to replace the original head tube.

Before you go crazy with the welder, install the wheels and lay out the frame, as shown in Figure 4-6. You can ensure that the rake and bottom bracket height are what you want. The best ride will come from a chopper that is modified in such a way that the original bottom bracket height and handlebar position remain the same, regardless of how long the forks are. As you can see in Figure 4-6, the rear of the bike is almost at the same angle that it was originally.

To make the best weld, the tubing should be ground in such a way that there are few or no gaps in the entire joint. As shown in Figure 4-7, the new frame tube conforms to the shape of the bottom bracket, making the joint much easier to weld.

Figure 4-8 shows the joint between the original frame tubing and the new length of conduit that will extend the frame. If you are joining two different sized tubes, it may help to squeeze the ends of the wider tube flat so there is no gap.

Figure 4-6 *Figuring out frame angles*

Figure 4-5 *The original head tube*

Figure 4-7 *Grinding the tubing for a better weld*

Figure 4-8 *The upper frame joint*

When you have figured out the optimal position and angle for the head tube, the tubing must be ground to make this joint possible. Figure 4-9 shows how the end of my new frame tubing is cut to hold the head tube at the angle I calculated while laying out the frame and wheels. Make a few tack welds on the head tube and then reinstall the forks and wheels, just to ensure that your angles are good. The head tube should also be parallel to the seat tube so that the forks are properly aligned.

The main part of the frame extension is shown tack welded in Figure 4-10. It is usually best to tack weld all the frame tubing in place, and then install the forks and wheels just to make sure alignment is proper before completing the welds. At this point, the frame would certainly be strong enough for a kid to ride, but I thought the single tube looked like it was missing something, so I decided to add another frame tube.

Figure 4-9 *The head tube joint*

Figure 4-10 *Tack welded frame*

An old fork leg was added to the frame, as shown in Figure 4-11, filling in what seemed to be an empty space. The new frame now seemed to do a Figure 8, giving it a really nice flowing style. At this point, the alignment between frame and forks seemed to be good, so the rest of the welding could now be completed.

To avoid any side-to-side distortion due to welding heat, start with the top and bottom of all joints, and then fill in the sides. There will always be some distortion after welding, but an experienced welder learns how to work around this, planning each weld ahead of time. Figure 4-12 shows the completed frame, completely welded around every joint.

The original handlebars had a crash pad, giving them that rugged motocross look, but that is not really a good style for a chopper so the crash pad bar was removed, as shown in Figure 4-13. The leftover material is then ground smooth, removing any sharp edges from the handlebars.

The chopper parts are ready for painting—a job that should be done by the young rebel to get them into the hobby. Figure 4-14 shows the frame and accessories freshly painted and curing overnight for reassembly in the morning.

Figure 4-11 *Adding one more frame tube for style*

Figure 4-12   *Completing the frame welds*

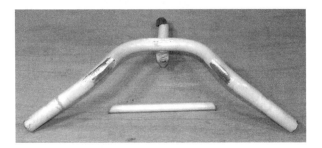

Figure 4-13   *Chopperizing the handlebars*

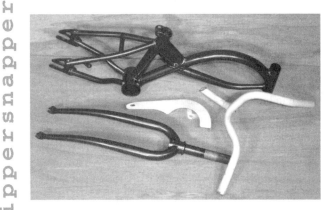

Figure 4-14   *Wet paint—do not touch*

The completed chopper shown in Figure 4-15 will be just as easy to ride as the original bike it was made from, due to the bottom bracket and handlebar position remaining almost the same. Of course, kids have a natural ability to learn

Figure 4-15   *The completed Whippersnapper*

Figure 4-16   *"Dude, I like your bike"*

things in a hurry, especially when they are about to look so much cooler than the other kids on the block!

Brayden pilots the Whippersnapper chopper with ease, making all the other kids jealous of such a cool bike. You'd better plan for spending a lot of time in your garage, because there is a good chance that every kid on the block will bring his or her bike to you for choppification!

Now that you've created some chops with attitude, let's move on to some custom trike projects designed for speed and comfort.

Speed and Comfort

Project 5: StreetFox Tadpole Trike

The StreetFox Recumbent Trike is a fast and comfortable version of the classic "tadpole" style trike (two wheels in the front), which includes rear suspension, three-wheel braking, and a 26-inch rear wheel for optimal gear range. The StreetFox looks and handles as well as many commercially produced trikes, and can be completely built using commonly available and inexpensive bicycle components, and some square tubing. Not one single machined part is needed. The StreetFox has an adjustable bottom bracket, so it can fit riders of most heights from 5 feet to well over 6 feet tall. Due to the rugged construction and use of a sturdy 14-mm axle and 48 spoke front wheels, the StreetFox can accommodate riders over 250 pounds. There is plenty of room for customization due to the simple mono boom frame, so you can alter the design to suit your own style or needs. Let's get building!

The good news is that the entire rear of the trike is an unmodified stock rear triangle from a typical suspension

Figure 5-1 *A 26-inch suspension frame and rear wheel*

mountain bicycle. As shown in Figure 5-1, you will need the complete rear triangle, the suspension spring, and at least the part of the main frame that contains the tube that connects the rear triangle to the frame, allowing it to pivot. The front part of the frame is not important, so it can be in any shape as long as the rear of the frame is perfectly functional. Since you can purchase an entire bicycle for about $100, I decided to work with new parts, as this would ensure the suspension, wheel, and brake parts were all in perfect working order. I ended up using the front half of the mountain bike for an electric project called "The LongRanger Scooter" (Section 5), so every part of the inexpensive, department store bike went to good use. If you decide to consult your scrap pile or the local dump for parts, ensure that the rear brakes are working, and that the rim is not damaged in any way.

One of the main concerns when designing a tadpole-type trike is how to connect the two front wheels to the steering components. The obvious answer would be to simply find a pair of typical 20-inch front wheels, then move the threaded hub axle to one side and bolt it onto something strong enough to support it. This answer is partially correct. The problem with this approach is that a typical bicycle hub will not have an axle strong enough for connection on a single side, and your axles would bend around the first corner, or before you made it to the end of your driveway if you are on the larger side. Sure, you could probably build a kids' trike using the largest mountain bike hub axles you can find, but this would still be a risky design, as the first bounce up a curb would certainly bend the axles. Another factor in choosing trike wheels is their lateral strength. Since a trike cannot lean into the corners, the two front wheels will be under massive stress around any corner at higher speeds. For this reason, almost all tadpole-style trikes use 20-inch or 16-inch front wheels with at least 36 spokes due to their excellent strength.

You may be thinking that this project calls for a pair of expensive rims mounted to a set of custom made hubs, but the good news is that you can purchase a pair of sturdy 20-inch trike wheels from any bicycle shop in the form of 14-mm axle, 48-spoke BMX wheels, like the pair shown in Figure 5-2. Unless you have an unlimited

Figure 5-2   *A pair of 20-inch BMX wheels*

budget, or really know the dynamics of bicycle wheels and plan to make your own hubs, the BMX wheel route is the best way to go. You can expect to pay between $100 and $200 for a pair of suitable wheels complete with smooth tread tires from a bicycle shop.

As you can see in Figure 5-3, the difference in axle size between a common bicycle hub and the 14-mm BMX hub is amazing. The 14-mm axle on the BMX hub is made for serious abuse, and if you have ever seen a BMX freestyle competition, then you know what I mean when I say these wheels are indestructible. The smaller axle would be barely enough to hold up a kids' stroller if held by only one side, but the BMX axle would easily survive trike use with a rider at roughly 250 pounds. I have made several trikes using these hubs, and I can assure you that the crazy stunt riding I did went far beyond what most "sane" riders would ever consider doing on a trike! So, as a general rule, do not use any threaded axle less than 14 mm in diameter, and do not

use a wheel with less than 36 heavy gauge spokes. Follow this rule, and you will never bend an axle or "taco" a wheel.

When you are sourcing steering parts for the front of a tadpole trike, you will basically need two of everything when it comes to the front wheels. As you can see in the completed StreetFox photos, the steering system is made of bicycle components, so you do not need any expensive machine shop work done. The resulting steering system is much smoother than the traditional kingpin bolt-style steering system often found on go-carts. Your goal is to find two identical head tubes and the fork hardware to match. When I say identical, I am not referring to color or brand, but the overall length and width of the head tube, so that both front wheels will be at the same height when the parts are installed. As shown in Figure 5-4, I have two identical head tubes and two matching fork stems that fit into the head tubes.

All of my parts are from at least three different bicycles, but they are all the same size, and that is all that really matters. If you don't have access to a large bicycle scrap pile, then you could trim a head tube down an inch or two to match the other one, but keep in mind that you will also need to cut the fork threads, and could end up with not enough thread to install the fork hardware by cutting away too much of the threaded tubing. Also, do not worry about the condition or size of the fork legs, as they are not being used. Just make sure that the fork stem is not bent.

It doesn't matter if the head tube is from a 1960 beach cruiser and the forks are from a competition BMX bike, as long as the hardware fits together properly and the forks spin around without friction. I ended up using a 20-inch and a 26-inch fork, but again, this doesn't matter,

Figure 5-3   *Hub axles—yes, size does matter*

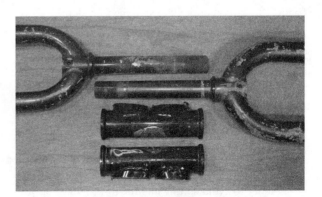

Figure 5-4   *Finding two identical head tubes*

Figure 5-5   *Checking the fork hardware*

Figure 5-6   *Remove the bearing race before cutting the fork legs*

Figure 5-7   *Cutting the fork legs from the stem*

Figure 5-8   *Cutting down the fork crowns*

as the fork legs are soon to be cut from the fork stem. As shown in Figure 5-5, fork hardware consists of two bearing cups, a pair of bearings, and at least two locking nuts. Bearings are inserted into the cups' balls first, the fork stem should spin without friction from the bearings, and the cups should be the correct size. There are at least three different cup sizes, so the wrong bearing will cause friction, making it difficult to spin the fork stem.

Before you start amputating the fork legs from the fork stem, first tap off the small bearing race (Figure 5-6), which sits just above the fork crown, by tapping it with a small hammer. Tap all the way around the ring until it falls off, so you can work on the fork legs without risk of damaging the bearing race with your hacksaw or angle grinder.

How you cut the fork legs from the fork stem is up to you, but try to avoid cutting into the fork stem when you are doing this job on both forks. I like to cut around the

fork leg using a cut-off disc in my grinder, leaving about $1/4$ inch, as shown in Figure 5-7. I then finish up the surface grinding with a rough grinder disc and then clean the area with a flap (sanding) disc. Keep the fork legs around for some other project; they make good headrest supports or seat frames.

At this point, you have two perfectly clean fork stems, as shown in Figure 5-8. The length of the crown area may not match on both fork stems, but this does not matter since we will be cutting them down to 1 inch anyway. A bit of tape around the crown area will make a good cutting guide, or you can draw a line around the crown using a marker. The cut line should be 1 inch from the top of the crown, where the underside of the bearing race would have been sitting.

Figure 5-9 shows the two fork stems after cutting each crown down to 1 inch. The bearing races have been reinstalled by tapping them down the same way they were originally removed. Now both fork stems are exactly the same length and have the same amount of crown material. If you look closely at the bottom of the

Figure 5-9 *Fork crowns cut to 1 inch*

Figure 5-10 *Cleaning up the head tubes*

Figure 5-11 *Putting all the fork hardware together*

original fork crown piece that you have just cut, then you can see that it has a small bead of weld around its edge (shown in Figure 5-9). This is because the crown is actually made of two pieces of tubing that fit together, which might be apparent if you look closely at the area that you have just cut. You should run a small bead of weld around this area, as it was done on the part you just removed in order to secure the two pieces of tubing back together. If you look ahead to Figure 5-11, you will see that I have done this, as shown on the underside of the fork crown.

The head tubes can now be cleaned up by grinding away the excess tubing left over from the original bicycle frame. As shown in Figure 5-10, a little work with a grinder disc and flap disc will bring the metal tubing back to its original, ready-to-weld state. Removing the paint will spare you from the horrific smell that will fill your garage or shop when welding painted bicycle tubing, so it is worth the extra effort. Again, avoid cutting into the head tube metal when cleaning up the area, and

take this opportunity to weld up any holes that have been drilled into the head tube where the other tubing was welded. A few passes around the outside of the hole will fill it with weld metal so you can grind the hole flush.

Now you can clean all of the fork hardware and reassemble the parts, as shown in Figure 5-11. Notice the small bead of weld on the underside of the fork crown where the cut has been made. Now you have two identical parts to make your lightweight and friction-free steering system. This steering kingpin system is far superior to the bolt and bushing system, which often has unacceptable play and friction. The bicycle head tube was designed to steer a bicycle wheel in a friction-free manner, so it only makes sense to use it to do this job on your trike. The bolt and bushing system is easier and would be less expensive to manufacture, but you don't have to cut corners when it comes to your own hard work.

The 14-mm axles will be held to the head tubes by a pair of axle mounting tabs (one for each wheel). Originally, I had made a pair of machined tubes to do this same job, but that increased the cost of building the trike, and my goal was to keep the cost down and not use machined parts, so I came up with this system, which I think is easier to work with. Your goal is to make two 1.75" x 1.25" tabs with a thickness of half an inch so you can secure the axle to the fork crown. The next photo, Figure 5-13, shows the exact dimensions and drilling for these two tabs. Since half-inch thick plate is hard to come by as well as cut, I decided to make my axle mounting tabs by welding two $1/4$-inch parts together to form the required $1/2$-inch total thickness. You can use any scrap metal, plate or flat bar to make these tabs, as long as the required dimensions and thickness are met. Figure 5-12 shows the four pieces I cut from the scrap $1/4$-inch plate to make the two $1/2$-inch axle mounting tabs.

Figure 5-12 *Cutting the axle mounting tabs*

Figure 5-14 *Welding the two half tabs together*

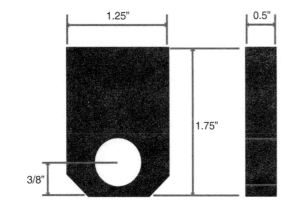

Figure 5-13 *The axle mounting tabs*

Figure 5-15 *Finishing the $^9/_{16}$-inch axle holes*

Figure 5-13 shows the size and thickness of the axle mounting tabs, as well as the drilling for the 14-mm axle hole, which is placed $^3/_8$ inches from the end of the tab. Obviously, a 14-mm drill is about as common as a comfortable upright bicycle seat, so rather than try to find one, just use a $^9/_{16}$-inch drill bit for the 14-mm hole. 14 mm equals .551 inches, and $^9/_{16}$ is .563 inches, so the resulting hole will be perfect for the 14-mm axle. If by chance you have a bit of scrap $^7/_{16}$-inch thick plate, you can make the axle mounting tabs from that, but do not make them any less thick than $^7/_{16}$ inch. Half an inch thickness is your best bet as the weight saving is almost nothing.

Since I made my axle mounting tabs from $^1/_4$-inch plate, I had to weld two pieces together in order to achieve the required $^1/_2$-inch thick result. As shown in Figure 5-14, the axle mounting tab on the left has been welded and ground clean, and has a pilot hole drilled for the axle hole. The axle mounting tab in the right of the photo is fresh from the vice and has yet to be ground, but you can see that it is nothing more than two bits of $^1/_4$-inch scrap plate welded together to form a $^1/_2$-inch part.

To drill the $^9/_{16}$-inch axle mounting holes, start by drilling a small pilot hole using a $^1/_8$-inch drill bit, and then gradually work your way up to the $^9/_{16}$-inch bit. I work my way up the drill bit set, skipping every third bit, so the final hole is exactly where it should be. If you try to go straight for the big drill bit, your holes will probably not be in the same place on each axle mounting tab, or you will jam the bit and end up with the part flying out of the vice. I only own a hand drill, so a drill press is not required, although it would certainly make the job easier.

With a $^9/_{16}$-inch hole drilled in the axle mounting tab, it should fit snugly over the 14-mm axle with just enough play to not have to force it on. The tab and the axle should form a 90-degree angle, and there should be plenty of axle length to get the axle nut installed, as shown in Figure 5-16. Another advantage of the axle mounting tab system over the hollow tube mounting method I previously used is that the axle does not need to be taken out and moved to one side of the hub. Not all axles are threaded all the way through, so this became a bit of an issue.

The other side of the axle can be cut flush with the locking nut, as shown in Figure 5-17. There is no point in

Figure 5-16 *Testing the tab on the axle*

Figure 5-17 *Trim the other side of the axle*

Figure 5-18 *Mounting the axle tabs to the fork stem*

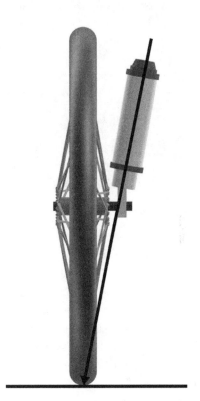

Figure 5-19 *Center point steering geometry*

having the axle sticking out past the outer side of each front wheel, as this would be a hazard to pedestrians in a close encounter, and also very annoying while trying to move your trike through a narrow doorway. Just use a hacksaw or zip disc and cut the axle flush with the locknut. Now would be a good time to make sure that all of the axle nuts are tightly secured as well. You may also want to paint the bare metal where the axle has been cut, or add a bit of lock-tight just to be safe.

There are a few angles that we need to get right concerning tadpole trike steering, so take your time and fully understand the next few steps before doing any welding. Nothing here is rocket science, so if you take your time, and fix any obvious errors, the trike will handle like a dream. Shown in Figure 5-18 is one of the axle mounting tabs tack welded to the base of the fork crown at some weird angle. This angle is no mistake, as it creates what is called "center point steering," so that the wheels turn directly on their axis when touching the road. This concept is shown in Figure 5-19, with the line that travels down the fork stem directly to the center of the

tire as it hits the road. Without this center point geometry, the vehicle would swing from side to side as you turn the front wheels, causing a dangerous inertia at high speed that could launch you from your seat. Only extremely slow moving vehicles can operate without center point steering, but since it is generally good practice, even lawn tractors and many kids' toys seem to have it. Read ahead a bit, so you can understand how this axle mounting tab angle is derived directly from your wheel.

With center point steering, the tire will pivot directly on its center, so there is no tire scrub or side-to-side inertia when making turns at high speed. Figure 5-19 shows an imaginary line drawn through the center of the head tube directly to the road under the center of the tire, which is how the angle of the axle mounting tabs is to be calculated.

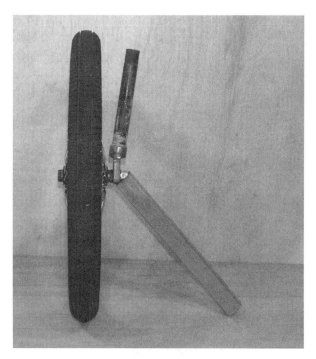

Figure 5-20   *Getting the correct axle tab angle*

Figure 5-21   *The second part is a copy of the first*

Figure 5-20 shows how I derived my axle mounting tab angle. By using a wheel with a fully inflated tire, I initially held the fork stem in place and made a mental image of the required angle. The axle mounting tab was then removed from the front wheel and I tack welded it to the fork stem using my "best guess" from the image I had in my mind. Chances are you will get pretty close on your first attempt, but this needs to be checked by bolting the wheel back onto the axle mounting tab, and then inspecting the angle by drawing that imaginary line through the fork stem to the point of contact under the tire. You can also hold the wheel at 90 degrees to the ground and try turning the fork stem to see if the wheel actually pivots on its center axis. With a single tack weld holding the axle mounting tab to the fork crown, you can easily force the part into another position if necessary, but don't be afraid to break the tack weld and try again if your center point angle is obviously wrong. I did the tack weld three times to get to the angle shown in Figure 5-20. You only have to do this guesswork once, as the other side can be made by copying the angle you get on the first side. As you can see, specifying an angle here would be useless, as it totally depends on your tire size, hub width, and even the type of hardware on your axles. When you have the axle mounting tab tack welded to the fork crown at the correct angle, secure it by welding the tab at both ends. Do not weld the face of the tab (hole

side) until you have rechecked your alignment using the wheel once more.

Figure 5-21 shows the two axle mounting tabs securely welded to the fork crows by welding the sides of the joint. I copied the angle from one part to the next so I would not have to go though all that guesswork again. While laying side-by-side, the fork stems and axle mounting tabs should look identical. At this point, you can weld the face of the joint, making sure you do both parts in the same sequence. If you first weld the hub side of part number one, then start part number two in the same manner, this way any slight distortion due to welding heat will affect both parts in the same way.

Figure 5-22 shows the completed welding all the way around the axle mounting tabs on both parts. Since I used the same amperage setting and sequence to weld both parts, any slight distortions in the axle mounting tab angles have affected each part the same amount. A final test shows that my center point steering angles are almost perfect. Two small holes have also been drilled in the axle mounting plates to lighten them up a bit. Of course, losing a few ounces of weight to drill shavings is not really important, but the parts do look more professional like this.

Figure 5-22   *Axle mounting tabs fully welded*

Figure 5-23 *Steering parts ready to go*

Now that you have two identical wheels mounted to identical steering tubes, as shown in Figure 5-23, much of the hard work is now done. Eventually, you will need to mount each head tube to the cross boom at the correct angle, but other than that, the rest of the frame is pretty basic, and we will be working on the easy stuff for a while now.

The only part of the bicycle frame that you need besides the rear swing arm is the small tube that carries the plastic plugs allowing the swing arm to pivot up and down. Not all suspension frames are the same, but you can easily figure out which parts you will need, even if they do not look like the one shown in Figure 5-24. Basically, you will be taking any parts needed to refasten the swing arm to the trike frame, so cut them from the original mountain bike frame.

The entire StreetFox frame is made from 1.5-inch mild steel square tubing with a 1/16-inch wall thickness. This tubing can be ordered from most steel suppliers and is very easy to cut and weld. Steel suppliers like to rate tubing wall thickness using a gauge number, so 1/16-inch wall tubing will be called 16-gauge tubing, and although

there is a slight difference between the two, your best bet is to tell them you want a length of 1.5-inch square tubing with a wall thickness as close to .0625 inch (1/16) as you can get. If you are a heavy rider (over 300 pounds), then you should ask for the next size after that, which will likely be 14-gauge tubing, or .078-inch wall tubing. Round tubing should be avoided for this project, as it will yield a very flexible frame, which will most likely be out of alignment after all the welding is completed.

The main boom runs from the pivot point of the rear swing arm all the way to the front of the trike. The main boom also carries the seat back, the two cross boom tubes, and allows the sliding bottom bracket to adjust for practically any leg length. The overall length of the main boom is determined by the size of the tallest rider that will use the StreetFox, so it can be cut down to a maximum length based on rider height, rather than having it stick out way past the bottom bracket. To start, cut a 6-foot length of the square tubing to become your main boom. Yes, 6 feet is more than you will ever need, and you will be cutting it down once the bottom bracket has been set up for your tallest rider. It's better to start with an extra long main boom than end up an inch too short because you forgot to add the extra few inches added by the seat padding or something similar. If you plan on adding a full fairing to the StreetFox at some later time, the extra long boom may be helpful in supporting the front of the fairing, so consider this as well. Figure 5-25 shows a section of the 6-foot long, 1.5-inch square tube boom that I have just cut.

The suspension swing arm pivot tube that was cut from the mountain bike frame needs to be connected to the end of your main boom, so you will have to cut the appropriate "fish-mouth" shape at one end of the tube, as shown in Figure 5-26. A good way to do this is to use the pivot tube as a guide to mark the area to be ground; this way you don't take out too much material.

Figure 5-24 *Remove the swing arm bearing tube*

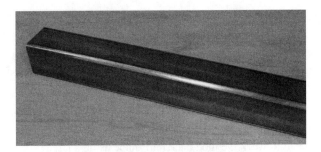

Figure 5-25 *Cutting a length of tubing for the main boom*

Figure 5-26 *Installing the swing arm pivot tube*

Figure 5-28 *Pivot tube fully welded*

Figure 5-27 *Pivot tube tack welded in place*

Figure 5-29 *Rear triangle installed onto the main boom*

The goal here is to make a good joint that will have no gaps once the part is welded in place. Again, your swing arm parts may differ, but the idea is the same. Before welding, tap out the swing arm bearings, as they are probably just plastic plugs and will melt during the welding process.

Figure 5-27 shows the swing arm pivot tube tack welded to the main boom on all four corners. Once you have the part tack welded and it has been checked for alignment, you can continue welding the joint, starting with the top and bottom and then the sides. To ensure that the tube is welded in line with the boom, place a straight edge across one end of the pivot tube and check it against the length of the boom.

The entire joint should be welded, as shown in Figure 5-28. Once the parts have cooled, you can tap the plastic bearings back into the pivot tube and then bolt the rear suspension triangle onto the main boom. The pivot tube may have been warped slightly due to heat distortion, so it may take a bit of effort to get the bearings to fit back into the tube. If distortion is extreme, you may need to work the tube back into shape using a large hammer—you know the routine!

The rear triangle is bolted onto the pivot tube, completing a large chunk of the building process. As shown in Figure 5-29, the installed rear triangle can pivot up and down freely, and it should be in line with the rest of the boom when viewed along the length of the main boom. Currently, there is nothing to connect the suspension spring to, but that is going to change in the next few steps.

The seat tube is a 14-inch long piece of 1.5-inch square tubing that will serve the function of connecting the suspension spring, as well as becoming a support for the rear of your seat. This tube is cut to a total length of 14 inches, and then one end is cut off at 45 degrees, as shown in Figure 5-30. Forty-five degrees is a comfortable seat angle, but if you have other plans and know how this angle will affect the rest of the frame, then feel free to alter the seat angle to suit your needs. A more upright position will place your head up higher, but could become uncomfortable if increased too much, making it difficult to pedal or even reach the handlebars. At the other extreme, a very laidback seat will make the trike

Figure 5-30   *The seat tube has been cut*

Figure 5-32   *Capped tubing looks so much better*

feel more like a low racer, making it difficult to see properly in traffic or even over your own feet.

The top of the seat tube is cut off at 90 degrees and needs to be capped off to keep out moisture and anything else that may want to crawl inside. It is a rule of mine to never leave an open-ended tube as this looks totally unprofessional, allows moisture into the frame, and could be a hazard due to the sharp edges of the tube. A simple cap can be made by cutting out a 1.5-inch square section of some scrap tubing, as shown in Figure 5-31. The cap is simply welded along the seams, using only enough heat to fill the joint with weld metal.

After welding the cap to the open end of the tube, use a flap disc or sanding disc to clean up the joint, so the tube will look like it never had an open end. Figure 5-32 shows the capped seat tube after a few minutes of joint cleanup. These little details add up to take your final product out of the home built zone into the professional looking zone. It feels much better to be asked "how much did that cool trike cost?" rather than "how long did it take to build that?"

The seat tube is placed on the main boom in a position that will allow the suspension to travel properly when the main boom is placed at the required height from the

ground. I could just tell you to measure 9.5 inches from the rear end of the main boom and mark a line, but depending on the overall shape of your rear triangle and the length of your suspension spring, the measurement that I used may not work properly for you. Of course, like all of my plans, I like to derive parts-dependent measurements by using the "set it up and look" method, which ensures that your final product works out, regardless of small differences in your donor parts. There are two conditions that need to be satisfied here, and to ensure that you get the correct placement of the seat tube, you will need to install the proper rear wheel with a fully inflated tire, and have the suspension spring you plan to use installed on the rear triangle.

The first requirement is the clearance from the underside of the rear boom to the ground, which should be 9 inches. For this, I simply found a bucket that was 9 inches tall and placed it under the main boom, as shown in Figure 5-33. With the boom supported and rear triangle populated with the wheel and suspension spring, it is easy to see where the suspension spring needs to connect to allow at least 2 inches of spring travel (requirement number two). Try to choose a seat tube position that allows the rear wheel to be lifted off the

Figure 5-31   *Never leave an open-ended tube*

Figure 5-33   *Installing the seat tube onto the main boom*

Figure 5-34 *Looking for the optimal seat tube placement*

Figure 5-35 *Tack welded seat tube*

ground two inches, but keeps the seat tube as far back as possible to keep the wheel base short. The next few steps will make this clearer.

If you place the seat tube way ahead of the rear triangle, and try to connect the suspension spring to the very top of the seat tube, you will have added an extra 6 to 8 inches of unnecessary length to the frame. Sure, this would indeed work, and might be a good way to earn some battery or luggage space, but unless that is your goal, the seat tube should be placed as far back as possible, putting the suspension spring mount lower on the seat tube. In Figure 5-34, I have the seat tube taped to the main boom as I try to find the optimal mounting position, which was about 9.5 inches from the center of the pivot tube to the rear edge of the seat tube. If I lifted the rear wheel off the ground two inches, the top of the triangle had about half an inch of clearance before it would strike the top of the seat tube, so this was just fine for my suspension spring, which had only 1.75 inches of total travel. When you find the magic place, just park the main boom so you can weld the seat tube in place.

Figure 5-35 shows the seat tube tack welded to the optimal place on the main boom. A small tack weld is first placed in the center of the joint, so you can ensure that the seat tube is in line with the rest of the main boom. View the two parts from all angles, and manipulate the seat tube if an alignment adjustment seems necessary.

When welding square tubing, always start with the two joints that will add the least amount of heat distortion into your work and then complete the other two joints. Here, the alignment is more critical from side to side, as it is on the top of the boom where the seat tube makes the 45-degree angle with the main boom, so the top joints on

the main boom were welded first, followed by the two joints on each side of the boom. The step angle on the opposite side of the 45-degree angle is a little tricky, so take your time and get it right. A little more amperage into that joint might keep your arc welder from sticking, if you are just starting out your welding career. When the seat tube is completely welded to the main boom, there should be very little side-to-side deflection if viewed from the top of the frame or from either end. The suspension spring mount will be installed after the front wheels have been attached, which is the next step.

The two front wheels are held to the main boom by two tubes I call the "front arm tubes." On some trikes, these two tubes are a single length, which would be called a "cross boom," but to ensure that shorter riders can enjoy the StreetFox without having their ankles rubbing on the frame, I decided to go with two tubes placed on an angle to give shorter riders a lot more foot room. The standard track (measurement across tire centers) is 30 inches, but

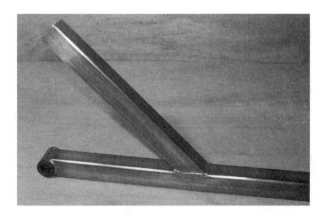

Figure 5-36 *Seat tube completely welded*

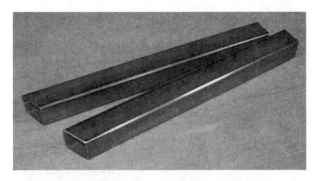

Figure 5-37 *Front arm tubes allow for more footroom*

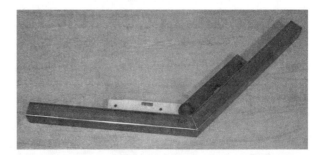

Figure 5-39 *Joining the two front arm tubes*

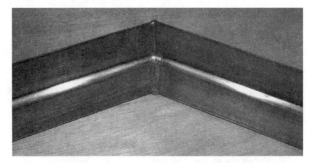

Figure 5-40 *The side that will mate with the main boom*

this is another easily modifiable aspect of this project. If you are on the heavier side, you may be more comfortable with a 32-inch track, or may have radical plans to make the StreetFox into a low-racing speed machine, so a 27-inch track might be your goal. For everyday riding and good stability, a track between 30 and 32 inches seems optimal, so I decided on a 30-inch track. Again, measurements may change slightly depending on the width of your front hubs, length of your head tubes, and type of hub hardware, so we will make all cuts a little longer than needed and work backwards, using your parts as a guide. Start by cutting the two front arm tubes to a length of 16 inches, as per Figure 5-38.

As shown in Figure 5-38, the two front arm tubes are 16 inches in length, with a 22.5-degree angle cut off one end of each tube. You will note that 22.5 degrees is half of 45 degrees, so if you fold a piece of paper diagonally twice, you will have a handy guide to check the 22.5-degree angle.

The two front arm tubes are welded together at the angle cut ends, so that they form a total angle of 135 degrees when welded together. An angle checker verifies that I am very close to the desired 135 degrees (Figure 5-39). Welding is done on the front and rear of the joint and then only on the top of the joint, leaving the bottom of the joint untouched so it can sit against the main boom tube perfectly flat. If you were to weld all the

way around the two arm tubes, alignment of the main boom tube would be difficult, since the weld would create a high spot. As you will soon see, the side of the joint that sits on the main boom tube does not need welding, as it will be secured by the welding between the rest of the front arm and main boom joints.

Figure 5-40 shows the side of the newly joined front arm pair that was left unwelded so it can sit flat against the main boom tube. If your welding has created a high spot on this side of the joint, grind off any of the excess welding bead until you can place a flat object directly over the joint so it lays perfectly flat.

The top side of the front arm tubes is fully welded, as shown in Figure 5-41. Since the front arms are now joined to create one single piece, I will now refer to them simply as the "front arm tube."

Whenever a non-critical weld is clearly visible, I try to grind the weld metal flush with the original tubing for aesthetic purposes. You might think that the weld holding the two arm tubes together is one of the most strength-critical welds on the entire frame, but as you will soon see, the joint between the main boom and the unwelded side of the front arm boom is what really holds the front arm boom to the main boom, so you can easily afford to clean up the top weld. Figure 5-42 shows the results of some careful grinding and touchups using a

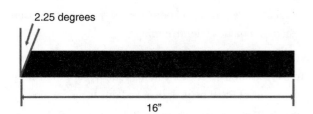

2.25 degrees

16"

Figure 5-38 *Front arm tubes*

Figure 5-41   *Front arm tubes joined*

Figure 5-42   *Front arm tube joint cleaned up*

Figure 5-43   *Measuring the head tube to tire center distance*

Figure 5-44   *Calculating track based on the arm boom*

flap disc. While cleaning up welds, be careful never to dig into the weld material, creating a valley in the joint. A little hill over the joint is much better than a little valley, which could seriously compromise the strength of the joint.

Now that you have the two front arm booms joined together to form the front boom, you must trim them down to achieve the desired track, since the initial measurement given was longer than you need. Track is measured across the fully inflated tire centers, so you will need to have your front wheels installed on the head tubes with fully inflated tires and all the head tube hardware installed. As shown in Figure 5-43, you need to know the distance from the center of the head tube to the center of the tire (or rim), in order to determine how much material needs to be trimmed off the front arm boom in order to end up with the desired track. Again, 30 inches is a good track width to shoot for unless you have other requirements. It may be difficult to measure the distance from the center of the head tube to the rim center with military accuracy, so just try to get a decent estimate within a quarter inch or so. My measurement was 2 inches, but yours may differ slightly.

Once you have determined how far it is from a head tube center to the rim center, you can use that measurement to figure out how much length needs to be cut from the arm boom to get the desired track width. We took the measurement from the center of the head tube, since this is approximately where it will end up once you cut the fish mouth out of the arm boom to weld it in place. Now, run a measuring tape across the arm boom assembly from center to center, as shown in Figure 5-44, so you know what the track would be right now if the arm boom was left uncut. The distance across my arm boom assembly from center to center is currently 28.5 inches, so if I add the extra width given by the two head tubes ($2 \times 2 = 4$ inches), I would end up with a 32.5-inch track. Obviously, I will need to chop off 1.25 inches from each arm on the arm boom assembly to end up with the desired 30-inch track. Therefore, the distance across my arm boom assembly needs to measure exactly 26 inches after trimming.

Figure 5-45 shows a closeup of how the arm boom is to be measured across its width. A small mark indicating the center of the tube is drawn, and this is where the measurement is taken at each end of the arm boom assembly.

Figure 5-45 *Measuring the overall arm boom width*

Figure 5-47 *Cutting the fish mouth for the head tubes*

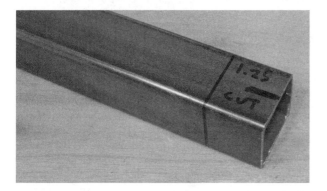

Figure 5-46 *Trimming the front arm boom assembly*

Figure 5-48 *Joining the arm boom to the main boom*

After a little calculating magic, I determined that the overall width from point-to-point on my front boom assembly needed to be reduced by 2.5 inches, so I marked a 1.25-inch cut line at the end of each arm, as shown in Figure 5-46. The resulting center-to-center distance of my arm boom will soon be 26 inches, creating a trike with a 30-inch track by adding the distance from each head tube to the wheel center (4 inches in total).

The head tubes are welded to the ends of the front arm boom at such an angle that the tires end up at 90 degrees to the ground. Since we are using center point steering, the head tube angle is not the same as the axle or the wheel, so this needs to be taken into account when cutting the fish-mouth joint on each end of the front arm boom. For now, simply trace a line using a head tube (or something of equal diameter) on the top of the front arm boom assembly, as shown in Figure 5-47. The top of the arm boom is the side with the weld, opposite to the non-welded side.

The front arm boom will mate with the top of the main boom at a distance of 12 inches, as measured from the point of contact to the 45-degree corner formed between the seat boom and main boom. This distance allows

plenty of room to mount your seat base, and places the handlebars in an easy-to-reach, comfortable position. Figure 5-48 shows the first tack weld made at a distance of 12 inches from the seat corner as described. A single tack weld is made so that the angle can be checked and easily adjusted if needed, in order to get the two arms at the same angle in relationship to the main boom tube. An adjustable square or angle finder is used to compare each side, making slight adjustments until they are both equal. At that point, you can place another tack weld on the front of the joint to lock down the angle.

Figure 5-49 shows the front arm boom tack welded in two and placed along the top to lock it in position after the angles on each side have been checked to be equal. Notice that there is plenty of room (12 inches) for your seat base, so you can use very thick seat padding if you like. A mesh seat with a frame could also be installed as there is plenty of room. Also notice that the arm boom is welded on the top side of the main boom, which will help keep your chain line simple and clean.

Figure 5-49 *Front arm boom in position*

Figure 5-50 *Welding the front arm boom to the main boom*

With the angles between each arm and the main boom equal, complete the joint by welding the two top sides first. You will want to use a clamp or vice to hold the boom securely to the frame to avoid any side-to-side distortion caused by welding heat. Figure 5-50 shows the completed welds on the front and back side of the arm boom and main boom.

Once the top of the arm boom and main boom joint is welded, you can finish up on the underside of the joint, as shown in Figure 5-51. Use the smallest bead of weld you can in order to keep heat distortion to a minimum, and make sure you weld both sides using equal amperage and travel speed so any deflection is equal on both sides.

It's time to play "good news, bad news." The bad news is that the next step is by far the most complicated part of the entire build, and will most likely require more than one try to get right. The good news is that, after this, you are home free, as the rest of the build will feel like smooth sailing. Head tube angle is your battle now, and it is by far the most difficult part of a tadpole trike build, which is why many first-time builders go for the delta-style trike instead. You have to get the head tube

Figure 5-51 *Welding the underside of the arm boom*

angles correct in two ways: first, the center point angle, and then the caster angle. Oh, and you have to do this twice, making sure everything is the same on both sides! OK, if you are still reading, then that means you have decided to build a tadpole trike, no matter what, so let's get going and begin the fun! I recommend that you use another spare head tube or some equal diameter tube as a guide, so you can easily check your progress by placing the tube in the joint, as shown in Figure 5-52.

As shown in Figure 5-53, you need to get the head tubes on such an angle that the wheels are standing at a 90-degree angle to the ground, and the head tubes are angled slightly forward, creating the desired 75-degree caster angle. The 75-degree angle allows the wheels to self center, creating a more stable ride at high speed with a good degree of response. If you relax this angle too much, your wheels will tend to flop from side to side, and you will increase your turning circle by reducing the amount of steer. If you make the head tube angle too steep, then your steering will be instantly responsive and

Figure 5-52 *Figuring out the head tube angles*

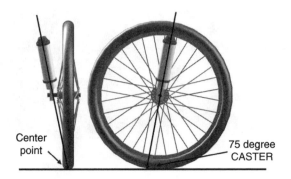

Figure 5-53 *Head tube angles of concern*

somewhat "twitchy," especially at higher speeds, where you will tend to wobble back and forth as you pedal hard. Seventy-five degrees seems to be a good angle for a trike of this wheelbase, but feel free to alter this angle if you feel that you know what you are doing and how it will change the handling characteristics.

The alignment of each head tube will be determined by the shape of the fish mouth cut at each end of the front boom arm, which is why the top cutout was made earlier as a starting point to work from. You should take a small amount of metal out a bit at a time and check your angles often, in order to avoid taking out too much material all at once. The odd looking fish-mouth cut shown in Figure 5-54 is what I ended up with after spending an hour or two with a small grinding disc and a round hand file. There is no magic formula or easy method I know of to get these angles in a hurry, so be prepared to spend some quality time with old-fashioned hand tools. What about a frame jig? Don't professional frame builders use those? They sure do, but how do you make sure the jig has the correct angles? Or how do you ensure the jig is going to be rigid enough to withstand the forces

of welding? Wood is certainly out. In the end, it would be twice as much work to create a proper jig, and a professional custom jig for this project would cost 10 times the price of a manufactured trike imported from across the globe. Dude, you will need to have to open a can of patience on this one!

Using my favorite bucket trick, I raised the front boom up to the correct height to check my work each time I took a little metal out of either of the fish-mouth cuts. The front arm boom should meet each head tube as low as possible, leaving about half an inch of head tube material sticking out past the underside of the boom arm, so you can weld the joint without melting the edge of the head tube. This height is not critical, since it is really determined by the height of the head tubes, but on my trike, this height was approximately 11 inches, slightly ahead of the boom joint where my bucket is placed in Figure 5-55.

The process of grinding a little, then checking angles takes a while, but if this was easy, then everyone would have a cool custom trike, and the results of your hard labor would then be worth a lot less! If you want to keep the head tubes locked on the fork stem, you can add some tape around the bearing cups, or simply remove the top bearings, which will keep the head tube from rolling all over the place when you are using the wheels as a guide, as I am in Figure 5-56. No welding is done until both wheels pass visual alignment inspection from every possible angle.

After your grinding, filing, and inspecting ordeal is over, it is time to weld the head tubes to the arm booms so you can get on to the easy stuff, like riding your trike. Again, I cannot offer any magical solutions that will ensure an instantly aligned head tube pair with minimal effort, so be prepared to fiddle, break tack welds, and try your patience. I prefer to place the frame up high enough

Figure 5-54 *Adding the proper fish mouth to each end of the arm boom*

Figure 5-55 *Checking my head tube angles*

Figure 5-56 *More visual angle inspection*

Figure 5-58 *Two tack welds secure the head tube*

Figure 5-57 *Welding the head tube in place*

so I can stand back and compare the wheels to the ground, and the wheels to each other. Figure 5-57 shows the main boom held securely and parallel to the ground using a workbench vice. At this point, I also placed one small tack weld on each head tube, hoping that my fish-mouth cuts were close enough that I could force each head tube into alignment and continue welding. I was lucky with the left side, but the right side was visually out of alignment enough that I decided to do a little more filing to correct the error. This is one stage in the build where correcting your errors will make a huge difference to the look and feel of the finished product.

I normally do not weld a frame with my good wheels or bearings installed, but this is one time where it is impossible to do otherwise, as you really need those wheels up on the bench for close visual inspection. An old towel or cloth placed over the tire and rim will keep welding spatter off the wheel, and always place

your ground clamp on the frame, not the head tube hardware, or current will surge through the bearings and destroy them or the cups. With a single tack weld on the head tube, you should be able to manipulate the head tube enough to get the wheel aligned the way it needs to be, with a focus on the wheel being at 90 degrees to the ground. When the wheel is where it should be, another tack weld on the other side of the joint should hold that head tube securely enough to later remove the fork hardware once both sides are looking good. Remember, this alignment counts, so break a tack weld and file away if you need to.

With both head tubes securely tack welded in place, perform one last vigorous alignment quality check because the next step is the point of no return. I placed the frame and wheels back on the ground and then put both wheels in the straight ahead position, so I could check their 90-degree alignment directly from the front, as shown in Figure 5-59. If both wheels seem to be standing at a perfect 90 degrees to the ground, then there is a good chance that both your head tube caster angles are also close to the desired result and equal as well. Look through the spokes at both head tubes to see if they are both at the same angle, and a very slight difference is not nearly as bad as a wheel not being at 90 degrees to the ground. If one wheel seems to be "gimpy," then get busy and fix your work, as this will surely be seen by others and make your work look poor.

Welding the head tubes in place is done like any other welding, starting along the top of the joint, then the

Figure 5-59 *One last quality check before welding*

Figure 5-60 *The battle is now over*

bottom, and finally the sides. Weld both parts using the same sequence, so that any distortion is copied to both parts equally, and ensure that there are no holes or gaps in the weld. It's a good idea to tap in an old head tube cup on the lower side of the head tube when you are welding, as this will help keep the head tube from becoming oval from heat distortion. Figure 5-60 shows one of the head tubes fully welded around the joint.

Figure 5-61 shows the nearly completed StreetFox frame after an epic adventure with multiple and duplicate angle cutting and welding. If you enjoy bicycle building as a hobby, then every minute of the building process so far was pure fun, but you can sure learn to appreciate the high price tag on a tadpole trike when you consider how much effort it takes to get here. If this were a two-wheel recumbent bicycle, I would have been on my third

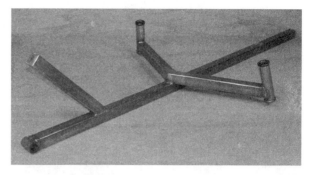

Figure 5-61 *The hard part is now completed*

hundred mile ride by now, but the final product will be worth it nonetheless.

Once your suspension spring is mounted to the frame, you will be able to place all three wheels on the trike and roll it around your garage or shop—a great milestone in any bike build. The suspension spring is held to the rear of the seat tube using a pair of $1/8$-inch bits of flat bar, or you could simply slice them out of your donor mountain bike frame like I did. As shown in Figure 5-62, the suspension mounting tabs can be cut from the donor frame, bypassing the need to drill the appropriate holes for the mounting bolt.

The suspension mounting tabs are shown in Figure 5-63, salvaged from what was left of the original mountain bicycle frame. If you decide to use flat bar to make the tabs, then just make sure you make them both the same size and drill the appropriate hole in each for the suspension spring bolt.

Since you have already figured out where your suspension spring needs to be mounted, it's just a matter of installing the tabs on the suspension spring, then using it as a guide to make the first few tack welds. The spring can then be removed and the tabs completely welded to

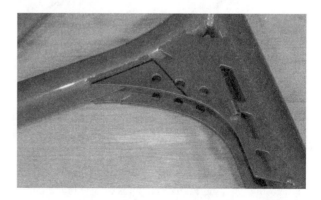

Figure 5-62 *Salvaging the suspension mounting tabs*

Figure 5-63 *Suspension mounting tabs ready*

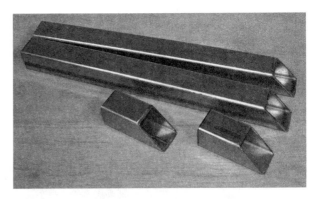

Figure 5-65 *Cutting the brake support arms*

Figure 5-64 *Suspension spring mounted*

the rear of the seat tube, as shown in Figure 5-64. I would recommend that you install the front wheels and fork hardware, and then get out that 9-inch tall bucket to prop up the rear of the frame, just so you can ensure that your ground clearance is exactly what it should be. When your suspension spring is secured to the seat tube, you should be able to jump up and down on the frame and have the suspension take it all in stride. No part of the suspension spring body should rub on the frame as the spring compresses and expands.

Braking on a tadpole trike is another puzzle that is a little more complex than a delta trike. Personally, I would be just fine with trusting one good rear brake, and often I will leave it at that and allow the builder to decide if he or she wants to install standard caliper brakes on the front of a bike. Of course, standard front brakes are simply not an option on this trike, since there is no place to mount them, so this will be addressed. Some tadpole trikes have hard-to-find custom drum brake hubs or disc brake hubs with custom axles, but the goal here is to avoid all

expensive or non-standard parts, so I will avoid those two options. Another puzzle is what to do with three brakes and two levers, but there are always answers. To get a brake on each front wheel, you will need to make a brake support arm that allows you to place a standard bicycle brake in the appropriate place around the wheel. To do this, a pair of L-shaped brake support arms will be made from some 3/4-inch square tubing with a 1/16-inch wall thickness (same thickness as the boom tubing). These tubes are shown cut in Figure 5-65.

The brake support arms need to wrap around the inside of each front wheel to allow the installation of a standard caliper-type bicycle brake. Because of the tight half-inch clearance between the brake support arms and the tire, you should read ahead and consider making the length of the tubes slightly longer so you can work backwards, especially if your front tires are quite wide or have aggressive tread. The measurements given in Figure 5-66 are based on a 2-inch wide, 20-inch tire with minimal tread, which is a common choice of tire for this project. Much wider and taller tires are certainly available.

The two brake support arms are shown in Figure 5-67 after tack welding them together for a test fitting. The brake arms will connect to the fork crown (shown ahead in Photo 64), and should come around the tire so that there is half an inch between the widest part of the tire sidewall, as well as between the top of the tire and the

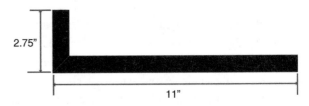

Figure 5-66 *Brake support arms*

Figure 5-67 *Tack welded brake arms*

Figure 5-69 *Installing the brake support arms*

small 90-degree portion of the brake support arm. Basically, you are making a "half fork" to emulate the way a caliper brake would hug the tire on a typical front fork.

If you are satisfied with the position of the brake support arm in relationship to your front tire, then completely weld around all of the joints and then cap off the open ends of the tubing as it was done for the seat tube. Avoid weakening the joint by grinding away too much weld metal if you decide to clean the welds flush with the base metal like I have. Figure 5-68 shows the brake support arms being capped and cleaned right before permanent installation onto the fork stems.

The brake support arms are connected directly to the fork crown at a 90-degree angle, as shown in Figure 5-69. You need to ensure that the clearances between the arm tubing and the tire are not less than half an inch, so a few scraps of half-inch plywood in between the tubing and the inflated tire can be used to help while you are tack welding the parts together. You will notice

that my brake support arm is slightly larger than the base of the fork where it is welded, but this was accidental, and you should not have this problem if you trimmed the fork crown as directed. The brake support arms are only tack welded in one spot at this point, so they can be aligned properly with the wheel.

Figure 5-70 shows how the brake support arm should look when viewed from the top of the wheel. Notice that the arm comes from the fork stem at a bit of an angle, as it heads towards the outer edge of the tire. The result is half an inch of clearance between the top and side wall of the tire and the brake arm tubing. If you think of the brake support arm as a one-sided chain stay, then the angle and tire clearance makes sense. The brake will be bolted over the edge of the tire, just as it would be done on a typical bicycle.

Figure 5-71 shows the brake support arm as viewed from the end. Notice the small tube is at 90 degrees to the

Figure 5-68 *Capping the open ends of the tube*

Figure 5-70 *Checking brake arm clearances*

Figure 5-71 *Brake arm view from the end*

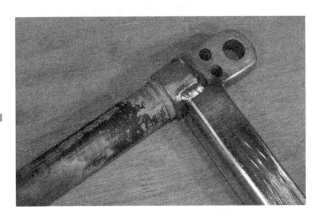

Figure 5-72 *Welding the brake support arms*

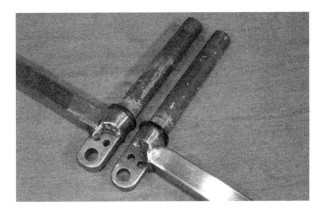

Figure 5-73 *Brake arms fully installed*

Figure 5-74 *Steering almost completed*

tire rather than the fork stem to which it connects. This is necessary as the brake also needs to be mounted at 90 degrees to the wheel so the brake pads hit the edge of the rim squarely.

The brake support arms can be completely welded to the fork stem crown, as shown in Figure 5-72, once you have checked the alignment with the tire. Welding should be solid and gap free, as the brake arms will be put under considerable stress during a fast stop from high speeds. As mentioned earlier, I cut the fork stem a bit too short, so I had to weld part of the brake arm tubing to the axle mounting hardware as well, but this worked out fine.

The same process must be done to both brake arm support tubes, making sure alignment with the wheel and rim is correct for proper braking contact. Figure 5-73 shows the two completed parts ready for installation.

At this point, you can roll your trike forward and the two front wheels will fall into line, and automatically track together due to the caster angle of the head tubes (Figure 5-74). Of course, you can't expect this to be safe on the road, and if you try to move the trike in reverse, the two wheels will point together and instantly lock up the vehicle. Like any vehicle with dual front steered wheels, the two wheels must be made to turn together, taking into account that the wheel on the inside of a corner must actually turn sharper than the wheel on the outside. This might sound like a complex problem best solved by some computer controlled hardware system, but is actually a very simple problem to solve. The name for this type of steering is "Ackermann Steering," and you can learn a lot about how it works by hitting your favorite search engine for more information. The good news is that I have figured out all the proper steering geometry by hours of trial and error, so you will not have to, although knowing how it actually works is always a good thing.

The two front wheels will turn together because they are joined by a pair of steering control arms, a pair of aspherical rod ends (ball joints), and a control rod. The position of the control arms in relationship to the pivot point of the fork stem actually determines how much one wheel increases its angle on the inside of a corner, so we can say that Ackermann Steering is just a factor of control arm placement. If you have too much Ackermann effect, then the inside wheel will start to scrub or completely lock up as it flops all the way against the frame. If you have too little Ackermann, or none at all, then one of the wheels will have to scrub around every corner, causing you to slow down and wear out your front tires in a hurry. Figure 5-75 shows the two control arms, which are nothing more than 2-inch long pieces of 1-inch wide flat bar with a thickness of $^1/_8$ to $^3/_{16}$ inches. The holes are drilled approximately half an inch from the ends, and are the size needed to connect the ball joints that you decide to use.

Mark a point on the underside of the brake support arm 5 inches from the welded joint, as shown in Figure 5-76. This location is where you will install the control arms for proper Ackermann steering compensation. The X marked on the far side of this line indicates that the control arm is on that side of the line, not on the center of the line. To determine this control arm location, I spent

Figure 5-75 *Steering control arms and bolts*

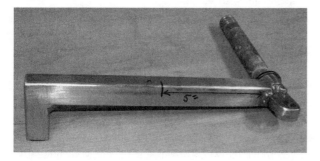

Figure 5-76 *Control arm position*

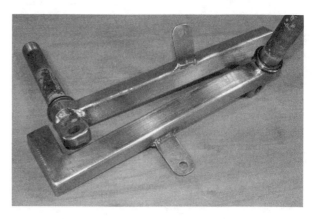

Figure 5-77 *Control arms installed*

the day trying random places along the brake support tube, using many different length control arms until I could make every possible turning circle with little or no tire scrub. This took at least 20 attempts. There are formulas and charts available on the Internet that can help determine this for you, but I found that "trial and error" is the best method. Old-fashioned hard work is always the way to win any battle!

Figure 5-77 shows the two control arms welded to the brake support tubes, so the distance between the edge of the control arm and the fork stem is 5 inches. The $^3/_8$-inch holes drilled in the control arms match the ball joints I found at the local hardware store.

Ball joints are actually called "spherical rod ends," so remember that if you are asking for them at a bearing store, as they like to use the proper terminology. Most of us just call them "ball joints," so I will continue to do so from now on. The two ball joints and the control arm are shown in Figure 5-78. My ball joints have a bolt on the body, and require a bolt to hold them down, while some may be the opposite, including the bolt on the ball part instead. It really is not important which type you choose, as long as the bolt or bolt hole is no smaller than $^3/_8$, and the ball or bearing is not made of plastic. The control rod

Figure 5-78 *Ball joints and connecting rod*

Figure 5-79 *The bolt fits inside the control rod*

Figure 5-80 *Completing the steering control rod*

Figure 5-81 *Installing the control rod and ball joints*

is just a half-inch hollow tube from the hardware store with a $^1/_{16}$-inch wall thickness. The $^3/_8$ bolt just fits into the tube, so this allows a nut to be welded at each end of the tube to make the steering alignment adjustable.

Figure 5-79 shows how the $^3/_8$-inch bolt fits right into the hollow control rod, using a nut to allow the ball joint to be adjusted about half an inch up or down. This nut will be welded directly to the end of the control rod, and then another nut will be used to lock down the ball joint once the two front wheels have been perfectly aligned. The length of the control rod is determined by installing all the front wheel hardware and the rear wheel, then walking the trike forward until both front wheels have tracked together for at least a 4-foot distance. At this point, try not to disturb the front wheels as you measure the exact distance from the center of each hole on the control arms that you just welded in place. You can then subtract whatever length is needed from this measurement, so that the ball joint holes will be the same distance apart as the holes in the control arms. The ball joints should be set up so that they can be screwed in or out the same amount, allowing optimal tire alignment fine tuning in each direction.

Figure 5-80 shows what the completed control rod should look like once the nuts have been welded at each end of the tube, and the ball joints screwed in and locked down. At this point, the center-to-center distance between ball joint holes is exactly the same as the center-to-center distance between the control arm holes after rolling the trike forward. I did not even have to do any fine tuning of the steering alignment after this point, as the steering was tracking perfectly and friction-free, after assembling the trike for the very first time.

Figure 5-81 shows the control rod installed by putting a bolt through all of the joints and the two control arms. I recommend that you use good quality bolts with plastic lock nuts for all steering hardware, so that the nuts don't work their way loose over time due to vibration. If you lose the control rod at high speed, one wheel may flop to the side and cause the vehicle to roll or steer wildly out of control. Don't take chances with low quality parts on steering or braking components.

The ball joints are placed on top of the control arm, as shown in Figure 5-82, so they are easy to inspect every once in a while. You could certainly place them on the other side of the control arm, but then they are always hidden from your view.

Once the steering system is ready for testing (Figure 5-83), the trike should roll in a straight line without any resistance, and steer around most sharp corners without any tire scrub. Obviously, steering too far in one direction will cause the inner wheel to travel past the steering center and literally flop all the way over, but in normal steering conditions, this will never happen. If you move the wheels from side to side, you can really see how the Ackermann steering affects the inside wheel,

Figure 5-82 *Ball joint and control arm closeup*

Figure 5-83 *Testing the steering system*

causing it to steer at a tighter angle the farther you move the outer wheel. If your trike seems to roll with any friction or obvious tire scrubbing, then you will have to unbolt one of the ball joints and adjust it properly. Remember, rolling the trike forward and allowing the two front wheels to self center is a good way to find the perfect alignment point.

It would be very difficult to guess the exact placement of the bottom bracket for your inseam without making a small error. A fixed bottom bracket also means that you could never alter the seat, or adjust the StreetFox for anyone but yourself, so an adjustable bottom bracket is the only option. If you cut the bottom bracket from an old bicycle frame, and leave a 12-inch or longer section of the seat tube connected, you can create a very functional, adjustable bottom bracket with nothing more than a pair of steel plates. The bottom bracket and included seat tube will be mounted to the main boom

Figure 5-84 *Creating an adjustable bottom bracket*

with a clamp arrangement that allows the assembly to be placed practically anywhere on the boom, allowing riders of varying heights to ride the bike. The two plates shown in Figure 5-84, along with the bottom bracket and seat tube stub, will form the sides of the clamp that will hold the bottom bracket against the side walls of the main boom, as they are forced together by two bolts.

The bottom bracket and front derailleur tube move together along the main boom, so you do not need to worry about the position of the derailleur over the chain ring. The chain is cut to the perfect length once the bottom bracket has been adjusted for the main rider, but will still allow a few inches of adjustment in either direction without a chain alteration due to the spring-loaded rear derailleur. Of course, the difference between a 6-foot, 5-inch person versus a 4-foot tall person will likely require some chain adjustment, but this can be done in less than a minute with an inexpensive chain link tool. Cut two $3/16$-inch thick pieces of plate or sheet to make the shapes, as shown in Figure 5-85. The round area is cut to conform to whatever size bottom bracket you decided to go with,

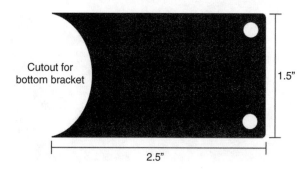

Figure 5-85 *Bottom bracket mounting plates*

and the two bolt holes will be drilled for ¼-inch bolts. The bolt holes are drilled at the edge of the plates, so that there is ⅛ inch of material between the hole and edge of the plate. These two plates will form a "vice" that will clamp the bottom bracket to whatever position on the main boom you want.

Trace around the bottom bracket shell to draw a line on the plates, as shown earlier in Figure 5-84. The marker line should meet the corners of the plate to ensure that you don't cut out too much off the plate when grinding out the rounded area. If you take too much metal away from the plate, there will not be enough room between the bottom bracket and the clamping bolts for the main boom. When grinding out the traced circular area for the bottom bracket, as shown in Figure 5-86, place both plates in the vice so they can be done at the same time; this will ensure that both plates are exactly the same size. Try to make the gap as small as possible around the bottom bracket shell, so that there is less distortion as the plates are welded. If the bottom bracket becomes overly distorted, it will be a serious challenge trying to screw the hardware back into the bottom brackets without stripping the threads.

In order to get the two plates aligned with each other and at the correct spacing, a bit of frame tubing can be clamped between the two plates, as shown in Figure 5-87. This will allow you to set the bottom bracket into the correct position for tack welding, as well as to ensure that both plates are aligned with each other so the bolt passes easily through both holes. I also placed a sheet of paper (folded in half) between each plate and the side wall of the tubing, in order to allow a little

Figure 5-87 *Setting up the plates for welding*

56 Degrees

Figure 5-88 *The proper derailleur tube angle*

clearance between the clamp plates so that the unit is not so difficult to slide when making adjustments. Without the small gap, it is easy to scratch the paint if you plan to move the bottom bracket around a few times.

As shown in Figure 5-88, I used an adjustable square set to the required 56 degrees, in order to set up the bottom bracket and derailleur tube to place two small tack welds at the top of the joint. It may be a bit tricky to get the first tack weld in place while holding the bottom bracket in position, but once this is done, the rest is easy. With the piece of frame tubing as a guide, you will automatically have the correct distance between the plates. Now, you will need to ensure that there is approximately equal distance between each plate and the edges of the bottom bracket. In other words, the bottom bracket will be in the center of the main boom as viewed from above.

Figure 5-89 shows the first tack welds done at the top edge of the plates in order to secure the parts together. Ensure that the derailleur tube angle is close to 56 degrees and that the bottom bracket is at 90 degrees to the square

Figure 5-86 *Shaping the bottom bracket plates*

Figure 5-89 *Tack welding the plates to the bottom bracket*

Figure 5-91 *Capping the end of the derailleur tube*

tubing. You can also move the clamps to hold the plates in a position that will allow you to weld around the entire joint, but do not begin welding without them as there may be distortion. Also, welding is only done on the outside of the joint, not between the plates. If the inside of the joint is welded, the weld bead will not allow the bottom bracket to sit directly on the frame tubing and could cause alignment problems. Because of the way the clamp system works, there would be no added benefit from welding the inside of the joint anyway.

The two plates will form a vice-like grip on the frame tubing when the ³/₈-inch bolts are tightened. In order to drill the holes into the plate, draw a line that represents the bottom corner of the frame tubing, as shown in Figure 5-90. Now you can punch and drill the holes using the line as a guide that will allow you to ensure that the holes stay below the line. If the holes are not below the line, you will not be able to pass the bolt through both plates because it will hit the frame tubing instead.

With the derailleur tubing cut to length, it is a good idea to cap off the open end of the tubing so that it looks

better and keeps any moisture out. As shown in Figure 5-91, a washer is welded over the end of the tube and then completely filled in around the joint and the hole. To fill in the washer hole, several passes are made around the edge until completely covered in weld metal.

After welding the washer in place and grinding the area clean, you can't even tell that the tube shown in Figure 5-92 once had an open end. Going the extra mile can make you final product look very professional, possibly even better than some mass produced machines, since they might have simply capped the hole with a plastic plug to save time.

Figure 5-93 shows the sliding bottom bracket secured to the frame. Notice that the bolts are fairly close to the underside of the frame tubing, thanks to the guide lines used to drill the holes. The derailleur tube is very close to the 56 degree angle as well. Tighten the bolts enough so that you can't force the bottom bracket to move by pulling on the derailleur tube, but not too tight, or you may actually crush the frame tubing. Don't bother

Figure 5-90 *Marking a line to drill the bolt holes*

Figure 5-92 *Completed adjustable bottom bracket*

Figure 5-93 *Testing the adjustable bottom bracket*

greasing up the bottom bracket bearings just yet, as you will want to do that after the priming and painting is completed.

You are almost ready to cut the proper length of chain and connect your transmission system. The obvious problem right now is that the chain will rub on the underside of the right front arm tube, so an idler pulley needs to be installed in order to guide the chain under the tubing. Almost all tadpole trikes have at least one drive side idler pulley. This is unavoidable due to the low seat height and position of the front chain ring. Fewer chain idlers and less of a bend in the chain is always better, since idlers and sharp chain bends cause extra friction in your drive line, robbing you of your top speed. The StreetFox has only a single idler and it only bends the chain slightly, so the transmission is very efficient. Figure 5-94 shows the V-belt idler pulley that is used to guide the chain—it is a commonly available 3-inch diameter $1/2$ V-belt idler pulley with a $3/8$ ball bearing mounted in the center.

You can find these pulleys at any lawn and garden supply center and many hardware stores. A hard nylon V-belt pulley will also work, but make sure that the bearing in the center is a good ball bearing, not some cheap brass bushing, as that will add massive friction. The cut-up bolt, also shown in Figure 5-94, is a lawn-mower wheel-mounting bolt, which just happens to fit the $3/8$ bearing hole in my idler pulley. Any decent quality bolt will work for this, as the goal is to mount the pulley about one inch off the side of the main boom. Some trike builders have used skate blade or skate board wheels with the center ground out to take the chain, but I would avoid that system as it causes extra friction due to the softness of the wheel material, and the diameter of the pulley should not be less than 3 inches when installed on the drive side of the chain. The skate wheels system also allows the chain to jump out of the track easily, forcing you to add some type of chain-stopping plate, so it really is not worth the hassle just to avoid spending $5 on a good quality V-belt idler pulley.

The chain idler pulley is mounted to the right side of the main boom just behind the right front arm tube, as shown in Figure 5-95. Hold the pulley against the frame so you can determine the best place to install the idler pulley bolt. When you are done, the pulley should be as close to the arm tube as possible without rubbing, and the pulley should stand off from the frame about 1 inch, which is why the lawn-mower wheel bolt worked perfectly for this job. A bolt and a few nuts could also be used if you can't find a decent shoulder bolt like I did.

Figure 5-96 shows the chain idler pulley guiding the drive side of the chain around the right front arm tube at a very slight angle. At this point, you can install your

Figure 5-94 *A V-belt idler will guide the chain around the frame*

Figure 5-95 *Installing the idler pulley bolt*

Figure 5-96 *Chain idler pulley in action*

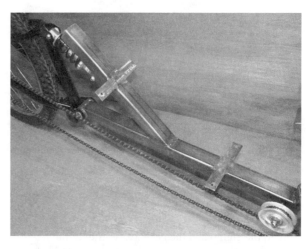

Figure 5-98 *Installing the seat mounting tabs*

front cranks, rear derailleur, and fiddle around with your chain tool until you get the chain cut to the correct length. A chain link tool can be purchased from any bicycle shop for about $20, and will save you countless bangs on your thumbs if you try to break and connect a chain using a punch and hammer. You will probably need two full bicycle chains, and possibly a small part of a third chain to make the trike chain, so try to ensure that all chain segments are the same size and type. Bicycle chains are not expensive, so it might be best to work with new materials here.

Seating is a personal choice, and since the StreetFox has such a simple frame, you can install just about any type of seat you would like, from a simple board seat to a high-tech adjustable mesh and frame style seat. I prefer the simple board and pad seat over mesh seating, so I will show you how to fashion a comfortable and simple seat using some plywood, foam, and fabric. To use a board-style seat, you will need to bolt it directly onto the frame, but since drilling holes into the main boom is a bad idea, a few seat mounting tabs like the ones shown in Figure 5-97 need to be made. Four lengths of 2-inch long, $1/16$-inch thick flat bar have been cut, and will be

welded to the main boom to hold down the two plywood seat boards.

The two-piece plywood seat has a base and a back, which are bolted to the seat mounting tabs. The tabs are welded to the main boom and seat tube, as shown in Figure 5-98. The tabs are flush with the top of the square tubing, and the welds are ground flush as well. You can now mount any type of plywood seat, or even a fishing boat seat to the frame. Later, I will show you how I made my minimal, yet comfortable, wood and foam seat.

The steering system is now complete, but you have no place for your hands. The handlebars need to be created so that you can get out and test ride your new trike. Handlebars are another piece of hardware that can be modified to suit your own style, so feel free to experiment with angle, size and shape using the basic design shown here. No matter how you decide to fashion your handlebars, you will need two gooseneck stems that can fit into the fork stems installed on your trike. Figure 5-99 shows the two standard steel bicycle goosenecks I have decided to cut up to make my handlebars. Do not worry

Figure 5-97 *Seat mounting plates*

Figure 5-99 *Two standard bicycle goosenecks*

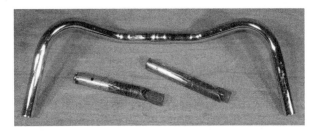

Figure 5-100 *Trimmed goosenecks and a bicycle handlebar*

Figure 5-102 *Testing the handlebar position*

about the condition of the gooseneck clamp, as my simple handlebar design only requires the stem.

Trim the clamp and the neck away from both goosenecks, leaving nothing but two equal length gooseneck stems and the mounting hardware, as shown in Figure 5-100. You can now find a set of bicycle handlebars to cut in half and weld directly to the gooseneck stems in order to create your handlebars. It may take some experimenting to find the most comfortable angle and length, but a good place to start is with a pair of these common "Granny bike" handlebars, as shown in Figure 5-100.

The handlebars need to be cut into two equal pieces to form a left and right handlebar for your trike. As shown in Figure 5-101, the handlebars are cut in half, then ground out to mate with the gooseneck stems at a position that feels comfortable and allows a full range of motion and easy reach of the brake levers. Most likely, this optimal position will take some experimentation to get right, so it's a good idea to have your seat installed, or at least a few boards, so you can sit on the trike and get a feel for where your hands feel most comfortable.

With your seat, or a few boards installed on the frame, test your handlebars by only tack welding them to the gooseneck stems at first, so that you can make sure you

like the position and angle of the bars. I found that having my hands just slightly inside the edge of the front wheels at the position shown in Figure 5-102 seemed just right. Close your eyes and let your hands find the most natural position, as this is probably the position that best suits you.

When you do find that perfect handlebar position and angle, weld one of the handlebar halves with a few more solid tack welds and then install it back on the trike for one final test. If the handlebar seems good, fully weld the joint and then do the same to the other half, trying to make it exactly the same as the first. Remember, there are right and left sides, so you can simply copy the fish-mouth cut from one bar to the other. Figure 5-103 shows the two handlebar halves after completely welding around all the joints and checking the weld for errors.

If you choose to make a simple wood and foam seat like I did, you may also want to make an adjustable backrest as well. This is optional, and you could choose to make your seat back a lot taller instead. I would

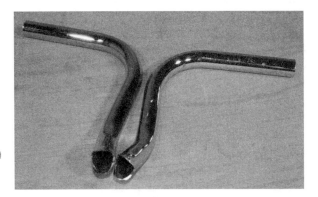

Figure 5-101 *Cutting up the handlebars into two pieces*

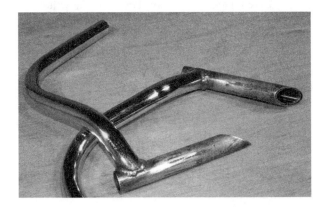

Figure 5-103 *Handlebar halves completed*

Figure 5-104 *Making an adjustable backrest*

Figure 5-105 *Backrest tube and adjustment clamp*

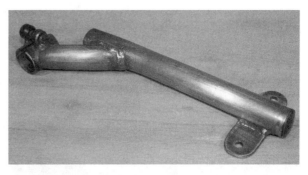

Figure 5-106 *Backrest mounting tabs*

Figure 5-107 *Installing the adjustable backrest*

complete the seat before committing to a backrest, in case you find that it really is not needed. Another thing to avoid is a headrest, as any road vibration will be fed directly to the back of your head, making travel on bumpy pavement really uncomfortable. Using some leftover mountain bike handlebar tubing and a pair of those knuckle protectors (Figure 5-104), I came up with a simple yet functional adjustable backrest system.

I knew where the backrest needed to be placed by simply sitting on my seat, then I cut the old mountain bicycle handlebar tubing to the correct length, and then welded it to a section of one of those knuckle protector things I had in my scrap pile. Figure 5-105 shows the resulting assembly, which looks like some type of long, bizarre gooseneck.

The backrest will be made just like the rest of the seat, so a pair of seat mounting tabs are added to the backrest support tube, just like they were added to the main frame earlier (Figure 5-106).

To allow the backrest to adjust for comfort, a small section of the handlebar tubing was welded to the top side of the seat boom tube, as shown in Figure 5-107. Now the clamp can be tightened to hold the backrest tube at whatever angle feels best. Plastic handlebar plugs have been used to cap off the ends of the tubing, since they fit perfectly.

Yes, you are almost ready for a test ride! You may have noticed that the return chain (non-drive side) is hanging pretty low to the ground, and can easily flop around, bouncing off the rear wheel. Sure, you can leave it like that, but there is nothing uglier than a flopping return chain to ruin the nice lines of your trike. A hanging return chain makes your vehicle look like a pregnant fish from the side, and looks so unprofessional. To keep the chain under control, all you need is a small idler wheel made from just about any lightweight pulley, skateboard wheel, or even an old rear derailleur, like the one shown in Figure 5-108. There is absolutely no stress or tension

Figure 5-108 *Making a chain slack pickup idler*

Figure 5-109 *Removing most of the derailleur parts*

Figure 5-111 *Installing the front brakes*

on the return chain, so you do not have to use a ball bearing pulley, or worry about the diameter. Even a bit of garden hose can do the trick.

Figure 5-109 shows what is left after removing most of the derailleur parts: a perfect idler system, ready to be welded directly to the main boom. The only goal here is to pick up the slack chain so it does not flop around or hit the ground when you are hitting the trails on your trike.

By welding the derailleur part to the rear of the main boom, as shown in Figure 5-110, the problem of an ugly, flopping chain has been solved. A bit of garden hose, skateboard wheel with a slot cut, or even another V-belt pulley, would also do this job, so dig through your junk pile for ideas if you do not have an old derailleur to butcher.

By drilling a hole in the brake support arms you can simply drop in a pair of stand bicycle caliper brakes for your front wheels, as shown in Figure 5-111. The hole should be directly in line with the center of the tire, so that both brake pads hit the rim at the same position. The

only downside to this system is that one cable is on the inside of the trike and the other will be on the outside. This will not cause any problems, but being a symmetry freak, I wish I could avoid this.

Now your trike has three brakes, which will require three cables. The problem is that you only have two hands, and therefore will only have two brake levers. Obviously, one lever will have to pull both front brakes, and there are three solutions that I know of to solve this puzzle. The cable doubler shown in Figure 5-112 is the worst option, but it is also the cheapest. This little adapter lets you pull two cables with one cable by clamping them into the small metal block shown in the right of the photo. The device does work, but it is ugly and adds a lot of friction to your braking system, so you need to apply a lot of force to get the front wheel rakes to engage. Most bicycle shops can sell you an adapter like this, as they are used on older freestyle BMX bikes to help with something called a "gyro."

A better option is a readymade cable that looks like one cable into a small housing with two cables coming out. Again, this is a common BMX part that most bicycle shops will know where to get.

Figure 5-110 *Installing the slack chain idler*

Figure 5-112 *A basic cable splitter*

The best option, which is the one most used on tadpole trikes, is a brake lever that is already set up to take two cables. You might have to search the Internet for "dual brake lever," but there are several companies that make them. With a dual brake lever, you can go about your cabling as if you were cabling a standard bicycle.

I like a minimal seat on most of my vehicles, so the plywood, foam, and fabric system tends to be my first choice. You should really consider your seating wisely, as everyone has a different opinion of what is comfortable, and my system may not be the best for you. My seat is 10 inches wide, with a base length of 10 inches, and a back height of 16 inches. The two parts are shown being cut from some $3/8$-inch thick plywood in Figure 5-113. The rounded front and top were drawn by tracing a dinner plate and a bucket lid.

To join the seat base to the seat back, I cut out a pair of 3-inch long pieces of $1/16$-inch thick angle iron, and then hammered them out to the angle of the seat. Figure 5-114 shows the two angle iron pieces with woodscrew holes drilled into them.

Seat foam material should be comfortable to sit on, but not so soft that it does nothing. Brown sponge foam is

Figure 5-115 *Adding foam to the seat*

useless, and hard Styrofoam is no better than bare wood. For my seat, I used some quality 1-inch thick seat foam for the back of the seat, and some scrap 2-inch thick packing material foam for the base. Both pieces of foam were glued to the wood using adhesive spray (Figure 5-115), and then trimmed along the wood using the angle grinder.

As shown in Figure 5-116, the angle grinder makes nice work of trimming the edge of the foam around the edge of the wood. Be careful when grinding foam with your grinder though, as the foam tends to grab the disc if you try to take off too much at once or don't hold the seat securely when doing the work.

Seat padding is cut to fit, then stretched over the foam to be stapled on the back of the seat. A real seat is sewn together as a top cover with sides, but this "poor man's" upholstery system works equally as well and takes very little time. I am not big on cutting corners, but if you ride

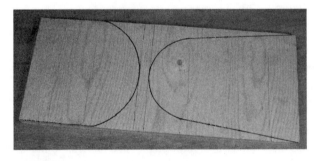

Figure 5-113 *Making a wooden seat*

Figure 5-114 *Seat board clamps*

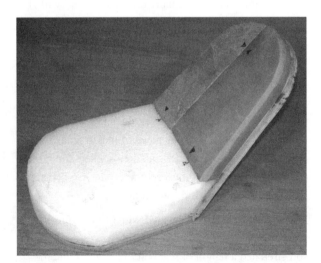

Figure 5-116 *Trimmed foam ready for material*

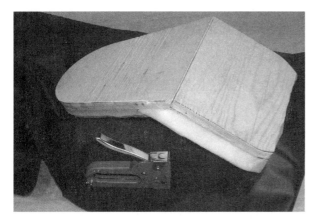

Figure 5-117 *Adding the seat material*

Figure 5-118 *Stretch and staple*

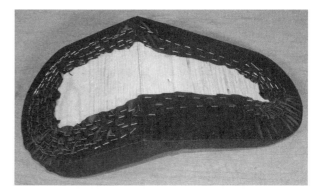

Figure 5-119 *Completed seat upholstery*

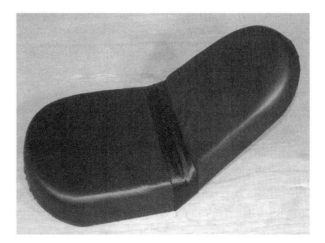

Figure 5-120 *Securing the seat material*

a lot, the seat material will wear out after a year anyhow, so why get too fancy? Figure 5-117 shows the beginning of the staple intensive journey.

If you stretch the material as you add each staple, you can probably get most of the wrinkles out of the material. A thicker material will be more difficult to work with, but the results can be decent if you take your time. Figure 5-118 shows how many staples keep the material stretched around the corners to reduce the wrinkles.

After you fight with the material and empty your staple gun, your material will likely be wrapped around the seat to look like the seat shown in Figure 5-119. To cover up the back of the seat, another piece of material can be cut to the shape of the seat back and then stapled around its edges. Plastic or very thin sheet metal can also be used to hide the ugly staples and bare wood.

Because the seat was covered with one continuous piece of material, the seam at the center was not perfect, so I stretched a length of seat belt material around the seam and stapled it to the wood. Figure 5-120 shows the

completed seat, which is not perfect, but not all that bad either.

The backrest is made exactly like the rest of the seat, and then covered with a piece of sheet metal (Figure 5-121) to hide the wood and staples. I decided to use metal rather than more staples and material to finish up the backrest, because it would be much more visible on the trike, so it should look professional.

Figure 5-121 *Making the backrest*

The last thing that needs to be done is trimming off the excess main boom length. With your seat and cranks installed, have the tallest rider sit on the trike (wearing shoes) and set up the adjustable bottom bracket to the correct place on the main boom. Now mark the boom, leaving a few extra inches, and cut away any extra length that will not be necessary. If you plan to add a fairing or body later on, you may consider leaving the boom extra long. Of course, you have to cap off the open end of the boom tube just like you did with the seat tube, so it looks clean and solid, as shown in Figure 5-122.

Waiting for paint and primer to dry is a painful experience, but don't lose patience and try to get your trike on the road before the paint or primer has cured. Primer often needs a full day to dry, and paint will not be ready for handling for at least two days. If you read deep into the application directions, you might even see that proper paint curing time is a week or more, so the paint will be easily scratched if you are not careful. I always use department store spray cans, and usually the results are very good if I can avoid the urge to assemble a wet paint frame. Figure 5-123 shows the frame freshly primed with the smaller parts painted and curing.

After two days of paint curing, I eagerly assembled the entire StreetFox in about two hours. All bearings were cleaned and greased, and new handlegrips, tires, cables, and chain were installed. The completed bike shown in Figure 5-124 really looked like a factory job, and it handled as well as it looked!

A cycle with three brakes, two levers, and two shifters has a lot of cable routing and planning that needs to be done. I really don't enjoy running cables, but this is a

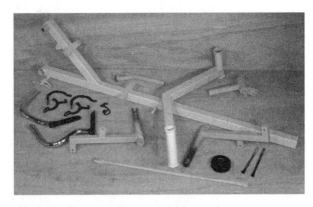

Figure 5-123 *Primed and awaiting a paint job*

Figure 5-124 *Finally, we are ready to launch*

necessary evil that has to be done, so take your time and plan ahead before cutting any cable to length or adding any zip ties. Avoid sharp bends in any cable, make sure the wheels can turn from side to side without putting a cable under great tension, and always replace any rusty, frayed or stiff cables with new ones. Figure 5-125 shows the plethora of cables running from the handlebars, as well as the low budget one, into the two-brake cable adapter.

Figure 5-122 *Cutting down the excess main boom length*

Figure 5-125 *The art of bicycle cable routing*

Both derailleurs will need to be set up properly by adjusting the high and low limit screws, so the chain does not travel past the largest or smallest chain ring when shifting. Adjust the limit screws before the cables are installed by simply moving the derailleurs by hand to set the minimum and maximum travel distances. Figure 5-126 shows the front derailleur being set up before the cables were installed.

The brakes need to be fine tuned so that they don't rub when not in use, but at the same time, grip the rim with optimal power when the lever is fully engaged. Each brake lever has an adjustment screw, and it should be screwed all the way in when you initially set up the brakes. If the brakes don't grab hard enough, slowly turn the lever adjustment screw counterclockwise to stiffen up the brakes. A tiny bit of brake rubbing is OK, as long as you can spin the wheel freely. Figure 5-127 shows the

Figure 5-128 *Front brake detail*

rear of the StreetFox, with the rear brake cable fully installed.

Setting up the front brakes is done in the same way as it was done for the rear brake. There is an adjustment crew on the actual brake body, just like the one on the brake lever, so both should be screwed fully clockwise when first setting up the cables. Figure 5-128 shows a closeup of the front brake and the adjustment screw.

Go over all of your cables one last time, making sure that you can turn the handlebars at least 45 degrees in each direction, without having a cable under tension or the front wheels rub on any cable. Also check the chain line, ensuring that no cable is rubbing on the chain, or any moving part. You can jack up the rear and test the shifting, or simply hit the road to see if everything is working the way it should. It's also a good idea to always check the front axle bolts before any ride, in case they have worked loose from excessive vibration. Figure 5-129 shows the completed StreetFox trike, calling out for its pilot to jump into the driver's seat.

Figure 5-126 *Setting up the derailleurs*

Figure 5-127 *Setting up the rear brake*

Figure 5-129 *Ready for launch*

Figure 5-130   *The StreetFox is a beautiful machine*

Figure 5-132   *The journey is always better on a recumbent*

Figure 5-130 shows what can be done with a little hard work and some commonly available bicycle parts. The StreetFox looks and rides as good as any commercially available trike, yet costs less than the shipping charge on a factory unit. Speed and braking are very good, and steering is comfortable and stable at any speed. Hill climbing is not a problem due to the full range of gearing, and the suspension makes for a very smooth ride. Most people who encounter the StreetFox think it was store bought, and give a strange look when told it was made using hand tools from typical bicycle parts.

Kathy (Koolkat) hits the trails on her new StreetFox, taking on pavement, steep hills, gravel roads, and bush trails without any problems at all. Riding a comfortable recumbent trike is so much easier on your body than an upright bicycle, so the long journeys are truly enjoyable. The wonderful scenery shown in Figure 5-131 is Kakabeka Falls, near our home base in Northwestern Ontario, Canada.

Figure 5-131   *Kat's new wheels*

Figure 5-133   *The StreetFox Tadpole trike*

Because you can ride in the most comfortable seated position with never a need to put your feet down, a recumbent trike is like your favorite living room chair and the entire world is your television set. Often, we ride our recumbent trikes to the beach or somewhere to have a picnic, and we never have to bring chairs along! Figure 5-132 is another shot of Kathy enjoying the scenery at the Kakabeka Falls campground, with the Kam River in the background.

We hope you enjoy building and riding your recumbent trike as much as we have! The results of your hard work will certainly be worth it, and nothing can compare to the feeling of riding your own quality-built recumbent trike. Figure 5-133 shows the StreetFox taking another great photo in front of the tree line. We hope to see your work or progress in the Atomic Zombie forum, and would certainly be happy to post a photo of your completed trike in our gallery for other builders to see. The next project is another Atomic Zombie Extreme Machines favorite—the DeltaWolf Racing trike—Brad's main ride these days.

Riding a trike is completely different from riding a normal two-wheeled bicycle. Obviously, balance is effortless since the trike cannot tip over during "normal riding," so you can concentrate on the scenery in a more relaxed state. And while stopped, you do not have to put your foot down or shift your comfortable position. Load carrying and trailer pulling are easier with a trike, and you can climb hills at practically any speed without ever having to get off and push your vehicle up the hill. Some disadvantages of a trike include their weight, they tend to look somewhat "unsporting," and they can become quite unstable around corners at any decent speed. Now, this is not true for all trikes, but many of the "delta" style (two wheels in the rear) trikes do take on an appearance of large and clunky "granny shopper" vehicles. However, the Atomic Zombie DeltaWolf aims to change all of the shortcomings usually plaguing delta-style trikes.

Why did I choose to build a performance trike using the delta layout (two rear wheels) rather than the generally accepted tadpole layout (two front wheels)? There were many factors that made this project come to life. First, I really think that the two large rear wheels and extremely low pilot's seat look very sporty and unique. Another reason was to have the ability to design a simple hub/axle setup that would allow the use of 26-inch rear wheels rather than the typical 20-inch size used on almost all delta trikes. With 20-inch drive wheels, you really limit your top speed due to the reduced rim radius and final gear ratio, so what you end up with is a load pulling machine. Another reason for this project was

because it could be done using easy-to-find parts, most of which come from standard inexpensive scrap bicycles, unlike a tadpole trike, which requires many expensive or machined parts to build the complex front end and steering system.

The final product is a very slick, one-of-a-kind, delta trike low racer that not only performs as well as it looks, but is a true joy to ride on the long hauls. There is very little that can go wrong with the simple design of the DeltaWolf, and the parts can be easily found at any bicycle shop or from scrap bicycle parts. Like all my designs, I like to keep things as simple and effective as possible, requiring only common parts and basic tools, and there is plenty of room for experimentation and alternative design based on your style or needs. It's best to read the entire project over first before you start cutting, so that you can understand where the next step will be going and review all the alternative methods and sources for parts.

Most of the DeltaWolf frame is made from 1.5-inch mild steel square tubing with a $1/16$-inch wall thickness. This tubing can be ordered from any steel supplier and is very easy to cut and weld. Steel suppliers like to rate tubing wall thickness using a gauge number, so $1/16$ wall tubing will be called 16-gauge tubing, and although there is a slight difference between the two, your best bet is to tell them you want a length of 1.5-inch square tubing with a wall thickness as close to .0625 inch ($1/16$) as you can get. If you are a heavy rider (over 300 pounds), then you should ask for the next size after that, which will most likely be 14-gauge tubing, or .078-inch wall tubing. Round tubing should be avoided for this project, as it will yield a very flexible frame which will most likely be out of alignment after all the welding is completed.

The drawing in Figure 6-1 shows what I will refer to as the rear of the DeltaWolf. It is made from the same 1.5-inch square, 16-gauge (or $1/16$-inch wall thickness) tubing that will make up the entire main frame. This part of the frame will determine the overall width of the final trike, and it can be widened for larger riders if necessary. As you can see, the overall width of the rear frame is 24 inches, and this is based on a typical adult shoulder

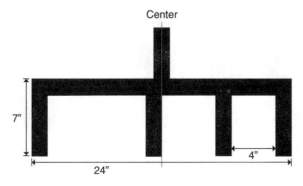

Figure 6-1 *The rear part of the DeltaWolf frame*

Figure 6-3 *The five parts that make up the rear frame ready to weld*

width of between 16 and 19 inches. If you are a large rider with a shoulder width larger than 19 inches across, then add 1 inch to the overall width of the frame for every inch wider than 19 inches your shoulders are. So, if your shoulder width is 21 inches, you should make the rear of your frame 26 inches wide to give yourself a little extra elbow room.

Normally, elbow room is no concern on a trike, but since you will be sitting directly between the two rear wheels, you should build your trike wide enough for the largest rider who intends to pilot the DeltaWolf. For the record, my shoulder width is 18 inches, and I chose to make my trike 23 inches wide, as I wanted to have the smallest track (width between wheels) as possible. Read ahead a bit before you start to cut tubing, and the odd placement of the four short pieces will make a bit more sense.

The four small lengths shown in Figure 6-2 are called the "axle mount tubes," since they will hold the four pillow block bearings and the two rear axles in place on the frame. To avoid having to plug up as many open tubing ends as possible, the two end pieces and the main

rear tube are cut on a 45-degree angle. The two center tubes are cut straight across, taking into account the 1.5 inches that must be subtracted from them to achieve the required 7-inch length (shown next). It is always best to make 45-degree cuts when creating a 90-degree joint, as this closes the tube completely to make a much stronger part.

The five rear frame tubes are cut and laid out in their basic shape, as shown in Figure 6-3. Notice that the main tube and two end tubes are cut into a 45-degree joint, and that the two center tubes are cut straight across and to a total length of 5.5 inches, in order to achieve the desired 7-inch total length. If you remember, try to make cuts to the tubing so that the seam will end up hidden from view in the final design, as it makes sanding and painting a little easier.

Start by welding the two end tubes to the main rear tube, as shown in Figure 6-4. It is important to maintain proper alignment, so place or clamp your work on a very flat surface, make small tack welds at first, and then check the inside 90-degree joint with a square. If the two end tubes are not exactly parallel, and close to 90 degrees with the main rear frame tube, then your wheels could end up misaligned, which will cause excessive tire wear and loss of power. This alignment does not have to be done to precision, but try your best to get the two end

Figure 6-2 *The four axle mounts: two for the end and two for the center*

Figure 6-4 *Welding the two end tubes in place*

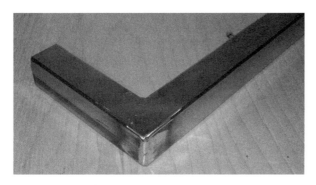

Figure 6-5 *The end tubes are completely welded and ground*

Figure 6-6 *Installing the drive-side axle mount tube*

tubes as parallel as possible so they look perfect from a visual inspection. If your 45-degree cuts were a little off, just tack weld the ends of the joint, and tap the two parts to 90 degrees with a small hammer; you can later fill in the gap with welding rod.

Weld the top joint at both end tubes before you flip the part over, so that all three pieces will remain perfectly flat on whatever surface they are placed. Once square, the far outer end joints can be welded followed by the opposite (bottom) side of the joint. Save the inside corner welds for last, and make them with a little heat and welding rod as necessary in order to reduce distortion.

Once you have the two axle tubes welded to the ends of the main rear frame tube, clean and grind the welded area as much as necessary, as shown in Figure 6-5. Avoid taking off too much weld material, or else you will weaken the joint. There should be a small smooth hump over the welded area, rather than a low spot after you are done grinding with whatever finishing disc you prefer. The reason we are cleaning up the welded area at this stage in the build is because the part is easy handled on the workbench, and this makes the job much easier than if it were a complete frame.

The DeltaWolf is driven by a freehub installed on the left-side axle, and braked using a disc brake installed on the right-side axle (or both axles if you so choose). For this reason, the drive-side axle mount tube shown tack welded in place in Figure 6-6 needs to be placed on the left side of the center line (refer to the drawing in Figure 6-1). As you can see in Figure 6-6, I made a marker line on the main rear frame tube at the exact center, then placed the tube directly to the left of this line, making sure it was at 90 degrees to the main rear frame tube before tack welding. Since the multi-speed freehub will be connected directly to the end of the left-side axle,

this places the freehub in a perfect place for alignment with the front-drive chain ring on the crank set; in other words, it creates a transmission much like that on any normal bicycle.

The reason the left wheel is chosen as the drive wheel is because all freehubs have the ratchet set up this way, and it would require some very complex or expensive custom built parts in order to drive the right wheel. Driving both wheels is an even more complex task, requiring a special differential which would greatly increase both the weight and cost of this project, yet add no real benefit. Most delta trikes only drive the left rear wheel, and this is completely unnoticeable by the rider while riding under normal conditions. On taller "granny style" delta trikes, there is a tendency to have the drive wheels slip a little bit when accelerating around a lefthand turn, since the high center of gravity will make the left wheel try to lift away from the road, but since these trikes are not built for any real speed, this is not a real problem. On the DeltaWolf, the rider is placed so low that the center of gravity is such that a wheel lift is almost impossible under all but the most extreme circumstances, and drive wheel slipping is not an issue at all.

The drive-side axle mount tube is completely welded and ground clean using the same careful methods as before to ensure alignment at 90 degrees. The last tube to be installed in the rear frame is the brake-side axle mount tube, and it is placed 4 inches away from the tube on the right side of the frame so that there is a 4 inch distance between them, as shown earlier in the drawing in Figure 6-1. This 4-inch space will be used to place the disc brake hardware, and will leave plenty of room for whatever make and model brake you choose to install. Tack weld, then check alignment carefully before completely welding any of the joints, and don't be afraid to cut and reweld if you find a problem—this is one of

Figure 6-7   *Welding the brake-side axle mount tube in place*

Figure 6-8   *Completed rear frame section*

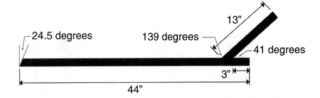

Figure 6-9   *The main boom and seat tube drawing*

Figure 6-10   *Cutting the main boom and seat tube*

the most important parts of the entire frame, so it should be done to the best of your abilities.

Once completely welded, ground, and cleaned, your rear frame section will look like the one shown in Figure 6-8. All of the four axle mount tubes will be perfectly parallel, and there should be a very close 90-degree angle between all the joints. If you place the part on a flat surface so that the open ends of the axle mount tubes face the ceiling, you may notice a very slight bow shape to the frame. This is due to the distortion caused by heat on the welded area of the inside joints. Do not worry too much about this, as I have worked this out in advance, and the way the axle is installed later on will take care of this problem, producing a perfectly aligned set of rear wheels.

Now that you have made the rear part of the frame, the rest of the main frame will come together quickly, as there are only three more tubes needed: the main boom, the seat tube, and the front tube. Shown in Figure 6-9 is the main boom (the longer tube) and the seat tube (the tube your seat will lean against). The length of the main boom will allow riders from 5 feet in height to 6'-6", and if you are taller than that, just add a few inches to the recommended 44-inch length. The 3-inch bit of the main boom sticking out past the seat tube joint is there to allow mounting of the drive chain pulley. Again, read ahead a bit to understand how the main frame will come together before cutting any tubing, just in case you want to make your own modifications or additions.

Cut the main boom and seat tube from the same 1.5-inch square tubing as the rest of the frame, as shown in Figure 6-10. You can either mark and cut the angles directly from your length of tubing, or simply cut both tubes off at their respective lengths, 44 inches and 13 inches, dealing with the angles for each piece later to make working with the tubing a little easier. If your cuts are a bit off the mark, simply file or grind the ends of the tubing until your angles are within a degree or two of the required dimensions. Again, you do not need precision—just try to get as close as you can to the correct lengths and angles—a half inch too much or too little tubing will not throw this part of the frame out.

The seat tube is welded directly to the main boom so that there is a 3-inch length of tubing extending past the top rear joint, as shown in Figure 6-11. A marker line around the tube makes this easy to see when you are setting up both tubes to be welded together. I like to tack weld the top of the joint (where your butt will be) first so that I can doublecheck the angle and alignment before going further. If the angle is good, the two parts can then be held in a vice to ensure alignment, as you weld across the top and rear of the joints. The sides of the joint are welded last, by welding a small portion at a time and then switching sides to minimize distortion from heat.

Figure 6-13   *Joining the rear of the frame to the main frame*

Figure 6-11   *Getting ready to join the main boom and seat tube*

Figure 6-12   *Main boom and seat tube become one*

Figure 6-14   *Cleaning the welded area is easy at this stage*

In Figure 6-12, I have completed the welding of the main boom and the seat tube. Grinding this joint will not be necessary since it will be completely hidden from view by the seat. I only grind welds that are highly visible; this way you do not inadvertently weaken the joint. By the way, this is the most stressed joint in the entire trike, and to further strengthen it, a gusset will be added later in the build.

The rear part of the frame will now join the main part of the frame by welding it directly to the top of the seat tube, as shown in Figure 6-13. The two parts will form a perfect T-shape, with the seat tube meeting the rear frame tube directly in the center. Start by tack welding only a single spot on the top and bottom of the joint, as shown in Figure 6-13—this will make it easy to get that perfect 90-degree angle between the rear frame tube and seat tube.

As shown in Figure 6-14, the 90-degree angle between the seat tube and rear frame tube is very important, in order to keep the rear wheels aligned with the front wheel. Check both sides using a square and tap the pieces into perfect alignment with a small hammer if necessary. Weld a little more of the top and bottom joints, checking alignment as you go, but do not weld the side of the joint yet, or your frame will end up out of alignment. The side of the joint will be done after the two truss tubes are in place, as shown in the next few steps. Your goal is to set up the main frame and rear frame to a perfect 90 degrees, using a few solid tack welds so that you can install the truss tubes. Once the truss tubes are in place, welding the sides of the joint can be done without any alignment problems.

The truss tubes form a pair of triangles between the seat tube on the main frame and the rear frame tubing, which makes the frame extremely strong and resistant to bending. The frame would certainly be strong enough to support your weight without these tubes, but if you ever

Figure 6-15 *Two truss tubes cut from 1-inch square tubing*

Figure 6-16 *Installing one of the truss tubes*

Figure 6-17 *Both truss tubes tack welded in place*

hit one of the rear wheels into a curb or wall with enough force, the rear of the frame may bend at the seat tube joint. These two small tubes (Figure 6-15) will not only make the frame extremely durable, but they will also keep the tubes straight while you complete the welding process. These tubes are not made out of the same 1.5-inch diameter tubing as the rest of the frame, since they are only needed for their tensile strength and will not carry any weight. Thinner tubing from 1 inch down to 3/4 inch will be fine, and this tubing can take any shape you like.

The angles needed at the ends of each tube are found by measuring the distance from the tube wall just below the seat tube and main boom joint to the corner of the rear frame tube. The exact length and angle measurement is not given because it depends on the width of your frame, and is much easier to figure out by simple trial and error rather than trying to get it perfect on the first try. I like to take a rough measurement, cut an inch or two longer than necessary, and then grind the end down until it fits perfectly. To make the second tube, transfer the length via a marker and cut again.

If your truss tubes are square, then try to get them cut so that the flat side faces the same direction as the rest of the frame tubing, as this just looks proper. It's best to cut both truss tubes, then tack weld them in place so that they can be visually inspected before completely welding them, just in case one tube is too long or too short. Figure 6-16 shows the first truss tube being tack welded in place. I am using thin 3/4 tubing purchased from a local hardware store, but many sources of tubing can be used, including tubing from furniture or even 3/4 thin walled electrical conduit.

Figure 6-17 shows the two truss tubes tack welded in place and ready for the final welding. Don't forget to do a thorough visual inspection before welding, or you will be spending a lot of time with that cut-off disc later! Actually, if both truss tubes are the same length, it's almost impossible to end up with a misaligned frame at this point.

The truss tubes can now be welded in place, as shown in Figure 6-18. You may not be able to get the welding rod into the joints that form the small angles top and bottom, but this is not a problem as long as at least three sides of the tube at each end get welded. Once you have the truss tubes securely welded in place, you can finish the welding between the inside joint of the rear frame and seat tube, without any risk of misalignment due to welding heat distortion. The rear of the frame is almost complete, but those ugly open-ended tubes must be

Figure 6-18 *Completing the rest of the rear frame welding*

Figure 6-20 *Welding the end caps in place*

capped off or your completed DeltaWolf will have "home built" written all over it. Open tubing is not only very unprofessional looking, but it leaves very sharp edges that can cut a person handling the trike, and leaves your frame open for attack from moisture or any critter that feels like taking up residence inside the tubing. Let's cap those tubes!

It may seem like a lot of work to cap off those open tubes, but in reality it is as simple as running a cut-off disc along a small sheet of 16-gauge sheet steel a few times, then spending a few minutes with your welder and grinder. There are plastic plugs you can get to cap off the tubing, but those will not enhance the strength of the tubing, and this is important since the pillow block bearings will be held very close to the open ends of the tubing. Without the steel caps welded in place, the force of the nut and bolt at the end of the tube may crush the walls. Figure 6-19 shows the bit of scrap steel I used to cut my five 1.5-inch end caps from. The sheet metal should be as close to the thickness of the tubing wall as you can find, so 16-gauge sheet metal will fit the bill perfectly. Cut the squares any way you like. My favorite method is to run a cut-off disc along the marker lines, then bend the parts back and forth until they snap apart.

There's nothing special about this job; it just takes a few minutes of light welding to seal the tubing

Figure 6-19 *Five 1.5-inch squares to be cut*

Figure 6-21 *Capped tubing looks so much more professional*

(Figure 6-20). Before you cap off the last tube on the rear frame, give it a shake to release any slag or junk that may have fallen into the tube while you were working.

Figure 6-21 shows the welded and ground tubing ends, which will now keep your tubing sealed from the outside world and remove all sharp edges from your frame. I usually rough grind the area with a heavy grinder disk, then finish the cleanup with a sanding disk to avoid taking off too much weld material. The results are really worth the extra effort.

Because the rider will be sitting directly over the joint that connects the main boom to the seat tube, some reinforcement may be necessary depending on the weight of the rider. If you weigh less than 150 pounds, you could probably just skip the gusset altogether, but considering it is simply a bit of 1.5-inch square tubing that adds almost no extra weight to the trike, it's a good idea to include it anyway. This gusset will transfer some of the

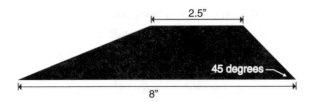

Figure 6-22 *Cutting the under seat gusset*

2.5"

45 degrees

8"

loading further up the boom so that it can absorb minor flex in the length of the boom, instead of at the welded joint. If you were to hit a large speed bump while moving at a considerable speed, you will be glad that you have the under seat gusset. It will stop the frame from bending at the joint, and take some of the "after bounce" away from the impact. The gusset is made from an 8-inch length of the same tubing that was used in the rest of the frame. Figure 6-22 shows the basic shape of the under seat gusset—it is not critical.

Mark out the cutting lines for the under seat gusset on a bit of leftover frame tubing, so you can cut it with a hacksaw or cut-off disc, as shown in Figure 6-23. You could get away with a bit of 1-inch square tubing as well, but do not go any smaller than that, or the gusset will add no strength benefit to the rest of the frame.

If you leave the top line intact as shown in Figure 6-24, you can reduce the number of welds needed to enclose the ends of the gusset before it is installed under the seat. Think of it like forming a cardboard box from a single piece.

The top of the gusset can now be folded in place for welding, as shown in Figure 6-25, reducing the number

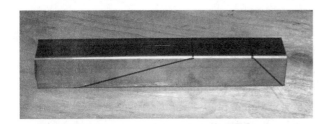

Figure 6-23 *Cutting the gusset from the square tubing*

Figure 6-24 *A neat cutting trick to reduce welding a bit*

Figure 6-25 *Clamp and weld*

of welds needed. This would have worked for both ends of the gusset, but the length of leftover frame tubing I was using was not long enough, so I will form the other end from the piece I cut from this end.

Just like you did earlier when capping the tubing ends, fully weld the corners of the gusset until it becomes a solid piece sealed from the outside environment. The gusset will be the lowest part of the frame, so you won't want it picking up moisture and crud along the ride. Figure 6-26 shows the gusset freshly welded at all the corners.

The gusset will be installed under the seat area, so the back 45-degree end of the gusset draws an invisible line along the back of the seat tube. Again, this position is not totally critical, as long as your installation looks like the one shown in Figure 6-27. If you had the strength to bend the frame, it would tend to fail ahead of the gusset, rather than directly under the rider at the seat tube and main boom joint, which is the weakest part of the frame.

Weld the front and rear of the gusset across the main boom, and then place a few inch-long beads of weld

Figure 6-26 *Fully welded*

Figure 6-27   *Installing the gusset*

Figure 6-28   *Only a few short welding beads are needed*

along the sides, as shown in Figure 6-28. There is no need to weld the entire length of the gusset, as this area will never be under any stress. You can also bypass the grinding process, as the gusset will remain hidden under the seat when the DeltaWolf is completed.

Now that the rear frame is joined to the main frame (Figure 6-29), you can start creating the rear hub/axles. The front part of the frame will be done after the rear wheels are installed because it is easier to set up the proper head tube angles and placement. Completing the front end at this time could leave your ground clearance too much or too little if the angle between the main boom and front tube are not done correctly. By having the rear of the bike standing on its wheels, this error will be easily avoided.

One of the most unique features of the DeltaWolf is that it has no rear hubs. The flanges that hold the spokes

Figure 6-29   *The rear and main frame completed*

are actually part of the rear axles, so this greatly simplifies the design, saves the builder a lot of money that would otherwise go to purchasing expensive aluminum hubs to be machined, and it also gives us the ability to widen the hub for use with a 26-inch rear rim rather than the typical 20-inch size. On a normal 26-inch rear wheel hub, the flanges are much closer together, so the wheel would not be able to withstand the great deal of side loading given to it on a trike, and may tend to bend or collapse. The wider the hub, the more lateral strength you put into the wheel, so this is how we are able to use a full-sized, 26-inch rear wheel on the DeltaWolf rather than being forced into the smaller size. A smaller drive wheel means less top speed and a much harsher ride, so this is a real plus.

I came up with the hub/axle combination when I was working out how to build a delta trike without being forced to use 20-inch rear wheels. There were a few readymade options for good strong 36-spoke hubs, but not only were they very expensive, they were designed for a 1-inch thick axle, which is way too much steel for a single-rider trike. Those hubs were made for pedicabs designed to carry two or more adults and the pilot, so that option was no good. I thought about having a mountain bike hub drilled to fit the $5/8$ axle I planned to use, but now I was into $150 per hub, plus the time it would take to have them machined. Even after that, the quest remained as to how to permanently affix the machined hubs to the axles—keyway? Ugly bolt or cotter pin? I thought there had to be a better way, and yes there was.

Why have a hub at all? An expensive machined hub bolted to an axle does nothing at all—it is wasted material. All that was needed was the flanges to hold the spokes, and even if I made them from steel, they would still end up weighing less than a complete hub machined and fit to the axle. Since a hub flange was nothing more than a washer, this seemed like a really good idea. As shown in Figure 6-30, I had the machine shop slice off four $1/8$-inch thick disc from a bit of 2.5-inch diameter steel stock, and then $5/8$-inch holes were drilled through the center. The discs are actually slightly less than 2.5 inches in diameter, but anything close to 2.5 inches, give or take $1/4$ inch either way, would be just fine. Although the four flanges only cost me $50 to have done at the machine shop, you could actually do this part without any machined parts at all, using a pair of hacked steel bicycle hubs, and I will show you this method

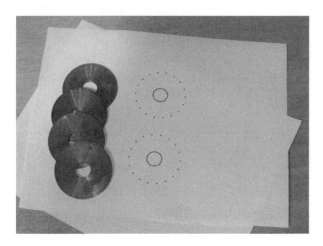

Figure 6-30 *Creating the DeltaWolf hub flanges*

as well. Personally, I think the machined flanges are the best way to go, since they will result in a nice, flat, outer flange that does not look like a hacked bicycle hub—something that seems to plague many delta trikes. Read the entire axle/hub section that follows, so you can see how this is done before choosing which method to use.

The four flange discs are made of mild steel $^1/_8$-inch thick and approximately 2.5 inches in diameter. The hole in the center is $^5/_8$ inch in diameter so that the discs will fit snugly onto the $^5/_8$ axles used on the DeltaWolf. If you decide to have these discs made at a machine shop, just print out this text and/or photo, or tell them you want four 2.5-inch diameter washers with a $^5/_8$ hole in the center of each with a thickness of $^1/_8$ inch. Also tell them that you plan to weld the discs, just in case they have some exotic or hardened steel; mild steel is what you want. The 18 spoke holes in each disc are drawn using a compass or by punching through a template, as shown in Figure 6-31. There should be no less than $^1/_8$ of an inch between the

edge of the disc and the closest edge of the spoke hole to make the disc as strong as possible. In other words, the meat between the edge of the flange and the hole should be about the same diameter as the spoke hole. Have a look at how a normal bicycle hub is made, and you can see this. If you are not a compass expert, or do not have a computer program that can make this template, then you can download the image from our website atwww.atomiczombie.com/support/dwhub.jpg—you can resize it in your favorite graphics program and print it out as a guide.

You will need to resize the pattern by trial and error since all printers will print using different sizes, and the original file will most likely be much too large. The circle drawn in the center of the pattern, as shown in Figure 6-31, is used as a guide to tape the pattern onto the disc for punching. I hold the pattern and disc up to a bright light so the light shines through the pattern, allowing me to tape the pattern directly to the center of the disc. The resulting punched disc, as shown on the right of Figure 6-31, is very accurate. If your machine shop buddy "owes you one," or you despise drilling holes, you could pay him or her to do this as well, but where's the fun in that?

The spoke holes will be drilled using a 7/64-inch drill bit. It may seem like a lot of work to drill 72 small holes, but in reality, this entire job only takes about an hour depending on the quality of your drill and bits. I am a minimalist and do not own a drill press, so I do this work using a hand drill, which is why the disc is held to a board in Figure 6-32 by a pair of wood screws. Drill and turn, drill and turn, drill and turn—I am truly thankful that I do not manufacture bicycles for a living or this

Figure 6-31 *Punching the spoke holes in the flange discs*

Figure 6-32 *Drilling the spoke holes*

hobby would quickly turn into a real chore! My hand drill was getting a little warm by hole 72, but I made it non-stop in less than an hour. If one of the spoke holes is slightly off center, do not fret, as you will be able to adjust this during the spoking process, and it will not affect the wheel in any way.

If you look closely at a standard bicycle hub, you will notice that the holes are very slightly beveled on the inside, so that the spoke bend is not subjected to a very sharp edge as it exits the spoke hole. This process can be recreated by simply running a $\frac{1}{8}$-inch drill bit lightly into each hole for about two seconds. The result will be a beveled hole, like the one shown in Figure 6-33 on the right disc. Do not put any pressure on your drill when doing this operation, or you may drill into the hole, which is why a hand drill may be the best tool to use when doing this operation. The goal is to smooth off the sharp edge without making the hole any wider. Do this to both sides of all the holes in each flange.

Now that your hub flanges are completed, you will need to create the two rear axles so they can be welded in place. The DeltaWolf has two rear axles made of $\frac{5}{8}$-inch diameter mild steel. You will need an axle at least 8 inches wider than the overall width of the trike's rear end, since we are going to work with a single axle at first. If you have a 24-inch wide trike, then ask the machine shop for a 32-inch mild steel $\frac{5}{8}$-inch rod. It is important to let them know you will be welding the rod, as there are different hardness rods available, and they may automatically assume you want a harder steel rod. If the steel is too hard, it may crack or weaken from the welding process.

Figure 6-34 shows the 31-inch axle (my trike is 23 inches wide), and a pair of $\frac{5}{8}$-inch pillow block

Figure 6-34 *One of the rear axles and a set of pillow block bearings*

bearings that will bolt onto the rear frame. You will need four of these pillow block bearings to hold the two axles to the trike, and these are available in many shapes and styles. The ones shown in Figure 6-34 are simple steel clamp housing pillow blocks, but they are available in cast iron, cast aluminum, and many other mounting styles. The important thing is that they do not weigh a ton and can be easily fastened to the 1.5-inch square tubing using a pair of bolts. The axle center should end up no more than 2 inches away from the end of the axle mounting tubes, or you will not have room for your disc brake.

As shown in Figure 6-35, the pillow block bearings are mounted to the axle mounting tubes so that the end of the pillow block is as close to the end of the tube as possible without hanging over. This places my axle center within 1.75 inches of the end, which gives plenty of room for clearance of the disc brake. If you are not sure if your bearing will allow clearance for the disc brake between the sides of the rear frame tube, just hold the disc roughly in place over the axle to check. Most pillow block

Figure 6-33 *Drilled (left) and beveled (right)*

Figure 6-35 *Mounting the pillow block bearings*

Figure 6-36 *The single axle held in place by the bearings*

Figure 6-37 *Marking the inner bearing holes*

bearing housings will be just fine, but it doesn't hurt to check before drilling the holes. To get the hole centered, simply drop the pillow block housing on the frame tube, mark the hole with a marker, and then punch the center. Do this on both sides of the frame, and then drill the appropriate sized hole for whatever bolt you need to fasten the pillow block to the frame. The pillow blocks are placed as far to the end of the tubing and as far to the outer edges as possible, as shown in Figure 6-35.

The reason I chose to work with a single axle at first is because any frame alignment problems will not translate into the two rear wheels being out of alignment. With a single axle held by the outer bearings, as shown in Figure 6-36, the other two pillow block bearings can be placed on the frame to mark their hole centers, while ensuring that all four bearings are in perfect alignment. If you tried to install the inner bearings in any other way, your rear wheels may have toe-in or toe-out misalignments that will cause excessive tire wear and rob you of your top speed. Why not leave the axle solid and just mount the freehub and brakes to it like it is, you ask? If you did that, you would create a trike that would not go around corners. Since the rear wheels turn at different speeds as you corner, the solid axle would try to make one wheel skid around the corners, and since so much of the weight is at the rear, this would most likely cause the front wheel to just skid along in a straight line, no matter how far you turned it. If you did manage to get the bike to turn a corner with a solid axle, the effort that this would take would make hill climbing seem like an easy task. To drive both rear wheels, you would need a geared differential, which is a complex setup that splits the difference between both wheels. This is unnecessary and

would increase weight and complexity to ridiculous proportions.

With the long axle securely held in place by the outer pillow block bearings, as shown in Figure 6-37, place the two inner bearings into the correct position on the drive and brake mounting tubes, and then mark the holes for drilling like you did before. Notice that the two inner bearings are as close to the inner ends of the tubing as possible, which is opposite to the outer bearings' position on the tube. Mark each site of the frame for drilling by bolting the axle onto each side of the frame to mark the center holes.

Figure 6-38 shows all of the pillow block bearings mounted to the frame using the appropriate sized bolts. I chose to use self-locking nuts just to be safe, but since the axle is going to be removed a few more times before permanent installation, I placed a few spaces on the bolt to avoid locking it on, as this would make removing the nut a little bit of a chore. With all the pillow block bearings held tight, you should be able to slide the axle right out by pulling at one end—this is proof that both rear wheels will be installed with almost perfect alignment. Notice that I have more than 4 inches of axle sticking out past the ends of the frame, which gives

Figure 6-38 *All of the bearings mounted to the frame*

Figure 6-39 *Marking the axle for cutting*

plenty of meat to work with when it comes time to create the built-in 3.5-inch wide hubs.

Now that you know your two axles will be perfectly aligned, you can cut them apart to create a left and right side. Since both axles have completely different lengths, you can't just find the center and cut; you have to find the center between the two inner bearings. As shown in Figure 6-39, I measured the distance between the two center bearings and drew a line by turning the axle around while holding a marker in place. Make sure that there is an equal length of axle sticking out past the outer bearings on the ends of each frame before you draw this line.

Figure 6-40 shows the freshly separated left and right axles after cutting with a cut-off disc. I like to cut the axles while they are held on by the pillow block bearings, as this makes the job easy. Just hold the cut-off disc in place and turn the axle as you cut through it. You will want to buff or slightly grind the rough ends off the freshly cut axles after cutting so that the bearings can slide on and off easily. The freshly cut ends are razor sharp and will make sliding the bearings onto the axle

Figure 6-40 *Cutting the axle into a left and right side*

Figure 6-41 *Tack welding one of the hub flanges in place*

difficult. Congratulations! You now have two perfectly aligned rear axles—a job that would have been very difficult to do properly using any other method. If there was any bowing on the main rear frame tube after welding the four small tubes in place, then this will no longer be an issue in rear axle alignment.

It's now time to integrate the flanges with the axle to create the axle/hub on both the drive- and brake-side axles. As shown in Figure 6-41, this is done using the frame as a stand, so that you can simply spin the axle in order to check flange alignment as you weld. Before you make the first tack weld, it is important to note that you should never place your welding ground clamp in any place that would cause the current to flow through the bearing into the weld area. If you clamped to the frame, the current would have to travel through the inner and outer bearing races, and through the actual ball bearings, in order to complete the circuit, which will damage the bearings. Since each ball bearing only contacts the inner and outer races by a very small surface area, the high welding current will create a pit in the bearing races, causing what would seem like years of wear on the bearing. Place the ground clamp on the axle, not the frame, and you will avoid this problem. Start with a small tack weld, as shown in Figure 6-41, leaving about $1/16$ of an inch of axle sticking out past the hub flange to make welding easier. Once the tack weld has cooled, remove the ground clamp and spin the axle to check alignment. It is very easy to see any misalignment of the flange as you turn the axle, and this is corrected using a few small taps of a hammer along the disc until it spins true. A very slight wobble is not a big deal in the end, since the spokes will correct this, but do try to get the disc spinning as true as you can while it is easy to manipulate.

Figure 6-42 *The hub flange is now aligned to the axle*

Figure 6-44 *Completely welded flange and axle*

Once you have tapped the flange into alignment from the previous step, add another tack weld directly opposite to the first tack weld, and then repeat the spinning and tapping process to true the disc back up. Repeat this process until your hub flange looks like the one shown in Figure 6-42, with four decent tack welds holding the flange to the axle in as perfect alignment as you can get it. Once this is done, you can remove the axle and weld the opposite side joint, with little fear of pulling the disc out of alignment with the axle.

The inner side joint can be welded around the entire joint, as shown in Figure 6-43, now that the four solid tack welds are securing the flange to the axle. I do this welding in four passes, switching to the opposite site of the axle just as a precaution against distortion due to heat, which is a good practice to get into when welding. Once you are done welding the entire distance, place the axle back into the pillow block bearings and give it a spin to check alignment. A very slight wobble is not a big

deal, but if you think there is too much misalignment, a few whacks of the hammer may still cure this. Any misalignment less than the thickness of the actual disc will be acceptable and will have no effect on the wheel-spoking process, nor will it show in the final build, so do not spend too much time and effort trying to get the disc spinning with "military accuracy."

Go ahead and weld completely around the front side of the joint now that the rear joint is solid. This can be done with no fear of heat distortion, since the disc cannot move at all now. As shown in Figure 6-44, a generous amount of weld is placed around the joint to ensure that there will be no pinholes after the face is ground flush in the next few steps. The final hubs will be very cool looking units once the ends are ground flush, leaving no visible line between the flange face and the axle. If you look at many production trikes, there is a huge, gangly looking bolt sticking out past the hub face, in order to secure the hub to the axle or, even worse, a locking pin. These integral hubs avoid that " I made this fit" look.

Before you start on the inner flanges, have a look at the completed hub in Figure 6-45, and you will notice that the holes on each opposing flange are offset by each other. If you place a rod or spoke through one of the holes, it will not enter a hole on the opposing flange; it will hit the space between the holes. Every bicycle hub is manufactured like this, even the cheap steel units, and this is an important thing to do for wheel strength once the spokes are in place. Keep this point in mind when you start tack welding the inner flanges to the axle. A spoke or small welding rod can be used to check this, like I have done in Figure 6-45.

Figure 6-43 *Inner joint fully welded*

Figure 6-45 *Holes in opposing flanges are offset by each other*

Figure 6-47 *A completed axle/hub unit with a flange width of 3.5 inches*

Figure 6-46 *Repeating the process for the inner flanges*

As shown in Figure 6-46, the same process used to mount the first flange to the axles is used to mount the inner flanges as well, but do not forget two important points: always place the ground clamp on the axle, not the frame, and make sure that each flange has the holes offset by one another. Other than that, tack, hammer, tack, hammer until you have completed all of the welding. The distance between the flanges will be 3.5 inches, as measured from the inside of each face—this makes for a much stronger hub that can take the heavier load demanded by a trike. A typical rear bicycle hub has the flanges at 2 inches apart, so our 3.5-inch wide hub will be much stronger.

Figure 6-47 shows one of the completely welded axle/hub units ready to have the face ground flush. These

pseudo-hubs will be much stronger than any "hacked to fit" aluminum hub would ever hope to be, and will certainly look much better once they are painted and spoked to the rims. When I show off the DeltaWolf, I get a lot of comments on the hubs from people who are into bicycle mechanics, asking where I purchased those cool trike hubs. I always tell them to take a closer look at the wheels and see that there are no hubs!

Once you have both axle/hub units completed, as shown in Figure 6-48, take your grinder and remove all weld material on the face of each hub until the disc is perfectly flat. Don't worry about the fact that most of the weld metal will be taken away, as the inner flange welds and weld on the other side of this flange will hold the wheel on to the axle. I like to start grinding with a heavy disc, then once almost to the face, place the axle back into the bearings and work it flush with a sander disc. You can have a helper turn the axle while you grind, or

Figure 6-48 *Completed pseudo-hubs on both axles*

Figure 6-49 *Hub flanges ground flush*

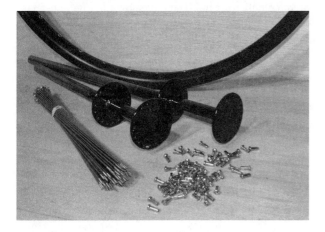

Figure 6-50 *Wheel building is well within anyone's grasp*

simply let it spin while you work, to create a face that looks as though it were turned down in a lathe. Yes, you could also turn the face down in a lathe if you are lucky enough to have such tools.

As shown in Figure 6-49, the hubs are a beautiful bit of work once ground flush and primed for painting. Normally, I save all of the painting for last, but you will need to paint the hubs before you spoke the rims, or this task will be very difficult with the spokes installed. Paint only the hub area, not the axle.

Figure 6-50 shows my primed and painted hubs ready to be spoked into my 26-inch, double-walled aluminum rims. A decent set of rims will cost you under $50 a piece, so if you plan on spending any money on the DeltaWolf, do it here. My goal was to lace the wheels myself, but since I am no veteran at the task, I brought one of the hubs and the rim to the local cycle shop and asked them to set me up with the 72 correct length spokes for the job. Cycle shops have a chart that takes into

account the hub flange widths and diameter in order to select the correct length spokes. I looked on the Internet for a simple wheel building tutorial, and printed out the simple instructions for building a 36-hole, 3-cross wheel.

Installing the spokes is an extremely easy task that requires no more than placing the spoke in the correct hole and giving the spoke nipple a few turns. It is the actual wheel truing operation that requires some skill or a bit of patience from those green to the art. I decided to install the spokes at this stage, and then continue building the rest of the DeltaWolf. I would come back and true the wheels once the paint was curing, just in case it took me a few days. I will show you how I made a simple wheel truing stand much later in this build—one that you can use to true the wheels yourself or bring to the bicycle shop to make their job easier. In true Atomic Zombie form, I will show you an alternative to using any machined parts on the DeltaWolf. Personally, I would go for the machine-cut hub flanges because they are a beautiful work of art once completed, but you could get away with simply butchering a cheap, steel bicycle hub to make the hub/axles, if no cost interests you.

If you take a close look at the garden variety steel bicycle hub, as shown in Figure 6-51, you will notice that this cheap, department store hub is simply made up of a few steel discs and tubes pressed together. The part that originally caught my eye when thinking of alternative hub designs was the tube that joins the two flanges together—it sure looks to be about the same size as the $5/8$ axle used on the DeltaWolf. I decided to rip one of these hubs apart and try the parts on the axle, and with only a small bit of effort, I was able to make an axle/hub using these parts. The completed axle/hub would be almost as strong as the machine's version, but somewhat less

Figure 6-51 *A department store rear bicycle hub*

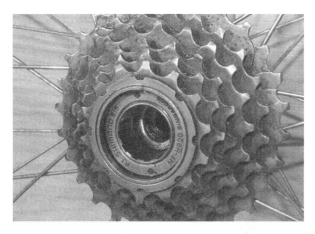

Figure 6-52 *A Shimano-style screw on a freehub*

Figure 6-53 *El cheapo rim busting tool*

visually appealing due to the bulk of the flange that would stick out past the wheels, giving that homebrew look I have been trying to avoid. Nevertheless, this system does work, and will cost you only a few parts from your junk pile, so here it is.

You will need a Shimano-type screw on the freehub, no matter which way you choose to build the hubs, so let me show you how to remove these from the hub in case you have a decent one lying around your scrap pile. The tell-tale sign that you have the correct freehub is the "spline" shown on the inside of the freehub hole in Figure 6-52. There is a tool you can purchase to fit inside this spline that will allow you to place an adjustable wrench on it to remove the freehub, but be warned, these tools break easily, especially when trying to remove a well-used freehub that has been torqued on by years of pedaling. Although I may be tooting my own horn, the tool presented next is a far better tool that the one you can purchase at a cycle shop in my opinion, because it will never break, and as usual, it will cost you nothing but your time to make.

Find a bolt that will fit snugly into the freehub spline, or grind one down that may be slightly too large, until it can be tapped into the spline with minimal force, creating a tight lock. This bolt is then welded onto an old steel crank arm or similar steel rod to create the indestructible freehub remover that will take any amount of hammer pounding you need to apply in order to remove the old freehub from the threaded hub. Believe me when I say that it may take extreme force to unscrew that freehub—usually by pounding on this tool with a large hammer while holding the rim on the ground. Oh, and you should always remove the hub with the wheel intact

or you will have no chance of holding it—a vice will not hold the hub, trust me. The tool shown in Figure 6-53 is three years old and has removed more old freehubs than I care to count. For the record, I broke two store-bought freehub removers in one month before building this tool.

Bang the bolt into the spline on the freehub, then grab the rim between your legs or by standing on a part of the rim with a few spokes removed. Most likely, you will not be able to crank out the freehub by hand unless you can bench-press a dump truck differential, so get out the largest hammer you can swing. Counter-clockwise is the direction of removal, just like any bolt, so make sure you are hitting the correct side of the tool. With a little patience, the freehub will unscrew out, as the tool always wins. A very long pipe over the handle is also good for increasing your leverage if you can find a way to secure the wheel in place.

Here is what the inside of the freehub looks like once it unscrews from the hub (Figure 6-55). Now you can remove the spokes by unscrewing them for later use, or

Figure 6-54 *Hammer away*

Figure 6-55 *Victory again*

Figure 6-56 *A steel hub can be taken apart*

by simply running your cut-off disc around the rim until they are all cut free. Only stainless steel spokes should be used for rim building, so if there are any signs of rust on the spokes, trash them, as they are much too cheap to be of any use.

Figure 6-56 shows the cheap steel hub separated after a few taps of the hammer while holding the center tube in a vice. There is not much holding the two flanges together because the tension of the spoked wheel tends to keep the two flanges pressed tightly to the center tube at all times. Knock the center tube away from both flanges. Do not worry about damaging the center tube in the vice; we will not need it for any part of this project.

If you are lucky, the large hole in the flange will fit snugly onto the DeltaWolf axles, as shown in Figure 6-57. If the hole is a bit small, a few minutes of hand filing with a round file will take care of that problem, as most of the hubs I took apart were already

Figure 6-57 *The flange fits directly on the axle*

the correct size or just a tiny bit too small. In Figure 6-57, the threaded part of the hub is placed in the end of the drive-side axle to hold the freehub in place. This is another part I chose to have machined, as you will see from reading ahead, but this method will work just fine to produce a solid mount for the thread on a Shimano-type freehub. In this photo, there is no welding done because I was only showing how to use these parts in place of the machined parts. If this is your chosen method, the joint at the axle and flange will need to be welded, as well as the joint at the edge of the flange and the center ring. In other words, every joint you see in Figure 6-57 would have to be welded completely around to create a solid mounting system for the freehub. Again, it's best to read ahead to see all of the alternative methods.

Figure 6-58 shows the alternative no-cost axle/hub completed and ready for spoking. Again, I did not use this system, which is why there is no actual welding done. To make the flanges secure, the joint at the axle and the outer ring will need to be welded all the way around on the inner sides of the flanges, in order to fully secure the

Figure 6-58 *The completed axle/hub*

Figure 6-59 *Spokes installed*

Figure 6-60 *Setting up the proper bottom bracket height*

parts together. Also, do not forget about the spoke hole offset, as explained earlier and shown in Figure 6-45. The flange faces are placed 3.5 inches apart from each other, just like the machined disc version shown earlier.

Whichever hub method you choose, you will need to paint the hubs and then install the rims in order to set up the frame for the rest of the building process. In Figure 6-59, the spokes are installed correctly, but they are not tight, and the wheels are not yet fully trued. I decided to leave that task until after the rest of the frame was painted and drying. Even though the spokes are fairly loose, the wheel is plenty strong enough to support my weight, although it would be a bad idea to test ride the trike before finishing up the wheels! You will also need the tires installed and inflated, since the goal is to set up the rear of the bike at the correct height, in order to build the front end correctly. I chose the largest rear tires I could find, because the roads around here contain some nasty pot holes and sewer grills, so this would take a bit of the road shock away. You do not kneed knobby tread tires, and should always find a tire that can be inflated to at least 75 PSI—the harder the tire, the better it will roll—it's all about the speed.

Install the axles into the pillow block bearings and place an object under the front of the frame, so that the clearance between the underside of the tube and the ground is 11 inches. As shown in Figure 6-60, a small plastic bucket does this job perfectly. Since the bottom bracket slides along the main boom for adjustability, there is plenty of boom length for a rider over six and a half feet tall, so you are free to make modifications to the frame if you are quite a bit shorter than that, and never plan on letting anyone taller than you borrow your

wheels. The goal is to have the pedals placed so that they have a 6-inch clearance from the ground, and if you plan to shorten the frame, the next few steps will ensure that the pedal clearance is maintained. In Figure 6-60, I have placed a temporary seat made of plywood and rigid foam onto the frame so I can measure the main rider's leg length for frame shortening. Making the frame shorter is a completely optional step, and only equates to a few inches of wheelbase, even if you are much shorter than six and a half feet tall.

Again, shortening the frame is a completely optional step, and not will not affect the overall ride ability or handling characteristics of the DeltaWolf—it's just a personal preference. If you choose to make the frame fit a single rider, then have that person sit on the temporary seat and extend his or her leg as far as it will go. With the rider's shoe placed on the pedal and crank arm, mark a point where the bottom bracket center would need to be. As shown in Figure 6-61, this is done by using the bottom bracket axle and a full length crank arm.

Figure 6-61 *Calculating the general bottom bracket location*

Figure 6-62   *Main boom shortening*

Once you have marked out the position of the bottom bracket according to the rider's full leg extension, you can safely shorten your frame at a point six inches ahead of this mark. In Figure 6-62, the frame will be cut to the same 24.5-degree angle, as originally shown in Figure 6-9 when you first cut the main boom. I was able to remove about 6 inches from the overall boom length, and I never plan on lending out my DeltaWolf, so taller riders are of no concern. Again, this step was purely optional.

The last part of the main frame to be made will be the front tube, the tube that will join the head tube to the main boom. This tube is a length of the same 1.5-inch square tubing as the rest of the frame, cut to a length of 12 inches with a 24.5-degree angle cut at each end, as shown in Figure 6-63. The 24.5-degree angle is the same as the one at the end of the main boom, so that when the two tubes are joined, there will be a 131-degree angle between them $(180 - (24.5 \times 2) = 131)$. The other end of the tube is also cut at 24.5 degrees, because this just happens to be the perfect angle to use when installing the head tube onto the frame. Precision is not required, as a few degrees either way will still work out in the end.

Once you have one of the angles cut, simply mark a line 12 inches away and trace the next cut using an angle

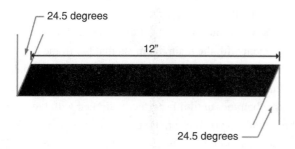

24.5 degrees

12"

24.5 degrees

Figure 6-63   *The front tube*

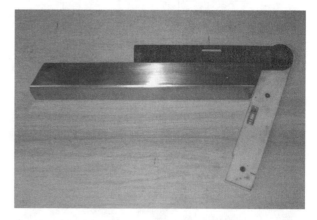

Figure 6-64   *Cutting the front tube*

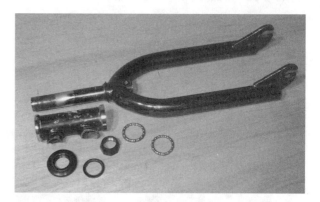

Figure 6-65   *Head tube, forks and hardware*

finder, as shown in Figure 6-64. You will have a rolling frame very soon!

The DeltaWolf has a 20-inch front wheel, so you are going to need a decent pair of 20-inch front forks and a matching head tube with all the hardware, as shown in Figure 6-65. I hacked the head tube from a mangled kids' bike and found a nice beefy looking pair of BMX forks, but feel free to install whatever forks suit your taste, including suspension, if you like. The key here is to make sure that the head tube is the correct length for the fork stem, so getting both from the same donor bicycle is the easiest way to go. There will not be a great deal of weight over the front wheel on this trike, so ultra heavy forks are not necessary. I just chose this style because they were in good shape, and I liked the shape of the large dropouts.

Grind away any excess "stubbage" metal from the heat tube, as shown in Figure 6-66. If there are a few small holes left over in the head tube after cutting it from the donor bicycle, you may want to fill these in, just in case they will not be covered once welded to the front tube.

Figure 6-66   *Grind the head tube clean*

Figure 6-68   *A good joint is easier to weld*

Figure 6-67   *Tracing the area to be ground out*

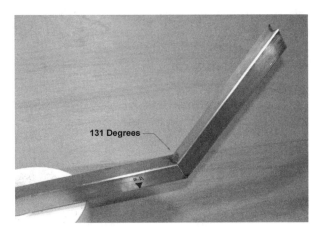

131 Degrees

Figure 6-69   *Front tube welded to the main boom*

Take this opportunity to grind or sand off any excess paint, so you can spare yourself that horrific smell when you weld the head tube in place.

Place the head tube onto either end of the front tube and draw a fish-mouth shape by tracing around its perimeter, as shown in Figure 6-67. This area will be ground from both the top and bottom of the front tube, in order to make welding the head tube to the front tube easier. The area can be ground out using an angle grinder, or hand file, if you like.

As shown in Figure 6-68, both the top and bottom of the front tube are ground to cradle the head tube, leaving almost no gap at all. This ground area may need to be tweaked a little bit later on when the head tube is finally installed, but at least most of the grinder work has been done before this tube is permanently welded to the main boom. The head tube should fit snugly in place, creating

a parallel line with the end of the tube. Do not weld the head tube in place yet.

The front tube will now be welded to the main boom, as shown in Figure 6-69, to form a 131-degree angle, almost the same as the angle formed between the seat tube and main boom, as presented earlier in the drawing shown in Figure 6-9. It is best to make two small tack welds on the top and bottom of the joint, so you can visually inspect alignment as if looking from the top of the frame. If your 131-degree angle is out a few degrees, do not worry, but do try to get the front tube in perfect alignment with the main boom, so that the frame forms one straight line as viewed from above. If needed, make slight alignment adjustments by tapping the front tube with a hammer until it is perfectly aligned with the main boom. When you are sure that alignment is good, weld the top and bottom of the joint, and then check it one last time.

Figure 6-70   *Front tube welding completed*

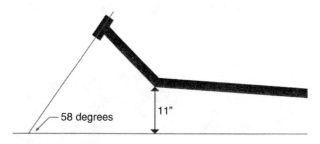

Figure 6-71   *Setting up the head tube angle*

The front tube and main boom are now permanently welded together, as shown in Figure 6-70, forming a perfectly straight frame as viewed from above. Avoid grinding away too much of the weld area, as this is a joint you do not want to weaken, since it will carry a fair bit of load.

With the front tube welded in place and the frame propped up so that there is at least 11 inches of clearance between the front end of the main boom and the ground, you can now set your head tube in place at the proper angle for welding. Look at the drawing in Figure 6-71 and you can see that the DeltaWolf has a head tube angle of 58 degrees as taken from the angle formed at the place where the head tube extends to the ground. This is a bit steeper than that of an upright bicycle, but this works out well on a trike and makes it easy to ride hands free. If you want to experiment with the head tube angle, feel free to do so, but be careful with the crank arm clearances if you bring the front wheel a lot closer to the frame. At 58 degrees, the DeltaWolf handles very well and can turn a tighter circle than an upright bicycle.

To get the head tube set in place, insert the forks, front wheel with inflated tire, and all fork hardware into the head tube, then place something in front of the wheel so it will not roll away. As shown in Figure 6-72, I have set

Figure 6-72   *Not bad for $10 worth of materials*

up the 58-degree head tube angle, and can now tack weld the top of the joint. When setting up the head tube for welding, make sure that the alignment of the head tube is correct as viewed from above, or your front forks will be off center. Also, you may need a bit of fine tuning at the end of the front tube where you ground out the area that cradles the head tube. Make these slight adjustments using a grinder, until the joint has very little gap on both sides as the head tube is sitting in the correct position.

Like all alignment critical welds, tack weld in a few places, then tap or move the part until alignment looks perfect from all angles. The easiest way to do the head tube is by first tack welding only the top, so you can simply push the wheel side to side to get the alignment as perfect as possible. When the alignment looks good as viewed from all angles (see next photo), then place a tack weld in each side, like I have done in Figure 6-73. Now the head tube should be secure enough to do the rest of the welding without heat distortion.

Figure 6-73   *Tack welding the head tube a little at a time*

Figure 6-74   *Checking head tube alignment*

Figure 6-75   *The head tube is fully welded*

Figure 6-76   *Finally, a rolling frame*

Set the front wheel as straight as possible, and take a look from the front of the bike to ensure that all the wheels are in perfect alignment, as shown in Figure 6-74. If your head tube is leaning in either direction, your bike will have a visible flaw as you head towards somebody looking at it from a distance. Although it is hard to tell from the camera lens distortion on the photo, my wheels are perfectly aligned, and the head tube is parallel with the seat tube.

When you are satisfied with the head tube alignment, finish all the welding by starting on the top and bottom joints, and then weld the sides, as shown in Figure 6-75. Once this welding is completed, you can insert the forks and hardware to get to the rolling frame stage, which is a milestone of progress! From this point on, the rest of the build will be smooth sailing, and you will be out riding the world's coolest looking delta trike in no time at all.

It always feels good to see a project progress to the rolling frame stage. Figure 6-76 shows the soon-to-be-completed DeltaWolf starting to look like a real vehicle. Now it's time to add the steering hardware.

The DeltaWolf has indirect steering, which connects the handlebars to the front forks via rod and ball joint.

Direct steering where the handlebars connect directly to a gooseneck placed in the front fork set would be impossible on a bike which puts the rider in a laidback position. For this reason, we will require a second head tube placed in the center of the frame, just ahead of the rider at the same approximate angle as the front head tube. To make sure you know which head tube I am referring to, from now on I will refer to the front head tube (the one with the front forks installed) as simply the "head tube," and this new head tube as the "steering head tube," since it is the one that connects the handlebars.

As shown in Figure 6-77, you will need any head tube and a set of forks with a stem of the appropriate length to fit together properly. Do not worry about the condition of the fork legs, as they are going to be removed anyway;

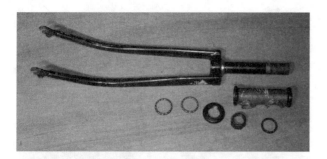

Figure 6-77 *Parts needed for the steering hardware*

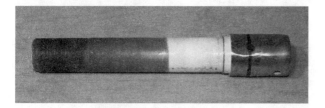

Figure 6-78 *Remove the fork legs and grind the area*

just make sure that if there is damage, it has not been done to the stem as well. One thing to note: you want to make sure the bearing races, head tube cups and bearings do work together properly, as there are several sizes used. Throw all the hardware together as a test—the fork stem should turn with almost no friction, even when all of the hardware is screwed on tight.

Start by cutting the fork legs away from the stem, and then grind off any excess material from both the fork crown and the head tube. What you should be left with will look like what I have shown in Figure 6-78. Before cutting the fork legs, remove the bearing race from the fork stem by tapping it off using a chipping hammer or chisel—this will ensure that it is not damaged by the grinder wheel when cutting the legs. It should be reinstalled once the fork stem is cut and ground clean. Draw a line around the base of the stem, as shown in Figure 6-78, leaving about half an inch of material just below the top ridge where the bearing race was located. This will remove some of the heavy tubing that was only needed when the fork legs were connecting and taking the weight of the rider. In this steering setup, there will be very little stress on any of the steering parts.

The large beefy chunk of fork crown material is really not necessary in our steering mechanism, since it will no longer be used to support the front end of a bicycle, only a small control arm, so it can be removed by cutting along the line you drew earlier. After you have cut the thick stub from the fork crown, a weld must be made all

Figure 6-79 *Cut and secure the crown together by welding*

the way around the underside, as shown in Figure 6-79. If you look closely at the area that you have just cut, you will notice that it is actually made up of two tubes that fit together, not a single beefy tube. The part you are throwing away would have a welded area just like the one you are about to make in order to ensure the two parts do not fall apart. This weld is very important, and if it is not done, your steering could completely fail, since the control arm is only going to be welded to the outer tube that is fit around the fork stem.

The DeltaWolf steering system will connect the front forks to the handlebars using a pair of control arms and two ball joints, so that the exact movements of the underseat handle bars will be transferred to the front forks, giving the feel as if you were holding on to handlebars directly connected to the front forks like that of a regular bicycle. The two control arms are just pieces of $1/8$-inch thick plate or flat bar cut to the size shown in Figure 6-80, and the ball joints are inexpensive steel units with a $1/4$-inch bolt hole. The ball joints can be found at many hardware or fastener stores and many automotive dealers—the exact type is not critical, and they do not have to be heavy duty since there is almost

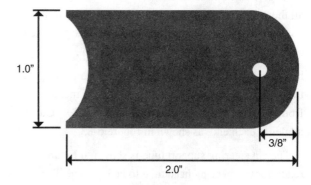

Figure 6-80 *You will need two control arms*

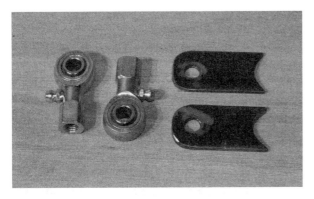

Figure 6-81 *The two ball joints and the control arms*

Figure 6-82 *Weld the control arm to the fork stem*

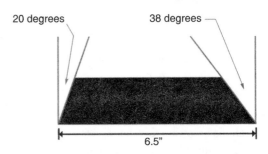

Figure 6-83 *The tube that connects the steering head tube to the main boom*

Figure 6-84 *The steering support tube*

no stress put on them in out application. The rounded area at the base of each control arm will form to the crown area on your front steering system and to one of the legs on the DeltaWolf's front forks. Look ahead a few steps to see how this works.

Figure 6-81 shows the two control arms after cutting, as well as the two ball joints I found at a local hardware store. The holes drilled in the control arms are the same size as the ball joints. My ball joints are designed to thread onto a bolt or rod with the appropriate threads, and other types have the bolt already installed on the body of the ball joint—either type will work just fine. The threaded ball joints allow for some fine tuning of the steering alignment, but once the rod is the correct length, this should never be necessary, so I always weld the ball joints directly to the rod, as you will soon see. Slight steering adjustments can be made by simply moving the gooseneck in the steering head tube if necessary.

Weld one of the control arms to the base of the fork stem, as shown in Figure 6-82. Do this with the bearing race removed from the fork stem (it is shown installed in Figure 6-82) to avoid weld spatter or accidental contact with the bearing race. The control arm will form a 90-degree angle with the fork stem, and should be welded all the way around to ensure strength. After welding is completed, tap the bearing race back onto the fork stem.

The steering head tube will be connected to the main boom, just ahead of the seat base by a 6.5-inch length of the same tubing that the rest of the frame is made from—let's call this the "steering support tube." Cut the angles at each end, as shown in Figure 6-83.

The cut steering support tube is shown in Figure 6-84, ready for the steering head tube to be installed. The 38-degree angle is the end of the tube that will connect to

the main boom, and the 20-degree end will be rounded out to cradle the steering head tube. These angles should place the two head tubes at approximately the same angles, as we shall soon see.

Place the steering head tube on each side of the 20-degree end and trace a line, as shown in Figure 6-85. This area will be ground out to cradle the steering head tube for welding by forming a good joint with little gap. A grinder disc works great to rough out the area so you can finish up with a hand file. If you accidentally take out too much metal, then just repeat the process on the other

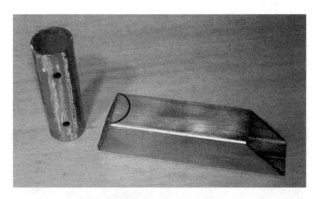

Figure 6-85 *Trace out the area to be cut*

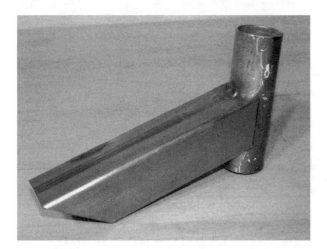

Figure 6-86 *Tack welding the steering head tube in position*

side, so the fish-mouth cut is the same on both sides of the tube.

When the steering head tube fits snugly in the cutout, forming a parallel line along the tubing end and the head tube shell, tack weld the top of the joint, as shown in Figure 6-86. Take a look at the tubing from all angles to ensure that there is no misalignment due to one of the cutouts being off center. Then, tack weld the opposite side of the joint to secure the head tube in place. If the alignment looks good, weld the top and bottom of the joint completely across to secure the part.

Now, weld the side of the joint until you have a good weld all the way around the entire joint, as shown in Figure 6-87. You must now install the bearing cups, the bearings, and the fork assembly with the control arm into this head tube, since it will be permanently installed in the final design. The reason it will be permanently installed is because there is not enough room to slide the fork stem in or out of this head tube once the steering support tube is welded to the main boom. This is done on

Figure 6-87 *Welding completed*

purpose, and is not a problem because you can still grease the fork hardware once installed. Since these bearings will never carry any load, the chance of ever needing to replace them is almost zero. If you do have to replace the bearings, only the lower one will be fixed, but you could simply grease the lower cup and feed the balls into place like they do on many expensive road bikes. Making the steering support tube longer in order to have the ability to remove the fork stem could certainly be done, but really would be a waste of extra tubing for no real reason, and it would place the steering rod higher up at an angle to the frame that would not look as good as it does this way.

The drawing in Figure 6-88 shows how the two head tubes should be placed at the same approximate angle. A few degrees either way is not a problem, but the closer the head tubes are, the more " balanced" the final trike will look. This angle also makes forming the steering hardware very simple to create a comfortable and well positioned handlebar. Figure 6-88 also shows the typical distance between the front of the steering support tube and the seat back to be 10 inches. This distance is good for a small to medium sized adult, whatever that means! Really, if you consider yourself to be a "big" person, then you may want to make more room by simply increasing this distance. There will be no effect on the final handling characteristics of the DeltaWolf if you play with the 10-inch distance, but you wouldn't want to move the handlebars too far ahead to be comfortable. If you think

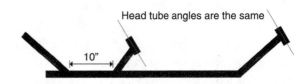

Figure 6-88 *The two head tubes will be at the same angle*

Figure 6-89 *Tack welding the steering support tube in place*

Figure 6-90 *Steering support tube completely welded*

you may need extra seat room, then only make a few tack welds in the next few steps, and save the final welding until after the handlebars are installed. As long as you read ahead to understand the rest of the building steps, you could easily do them in whichever order you like from here on.

If 10 inches of seat room will be enough, or if you just want to make your best guess, tack weld the steering support tube in place, as shown in Figure 6-89. The steering hardware is installed before you weld or you will not be able to insert it later. It is actually the lower bearing and cup that cannot be removed, but you might as well install all of the hardware now, since you will need it for setting up the steering in the next few steps anyhow. The support tube is angled away from the seat, not only for looks, but in the case of a severe front-end collision, you will not get wedged under the hardware as you would if it were reversed.

If you trust your seat base clearance, finish the rest of the welding between the steering support tube and main boom, as shown in Figure 6-90. Because I have built a few of these types of low racers, including the original Marauder (same steering design), I know that 10 inches of seat clearance is enough for myself. If you aren't sure, then just leave the tack welding for now and continue on, as you can finish this up after you create the handlebars and install them for a trial fitting. You do not have to weld the joint on the inside angle since it would be difficult to do without making several passes. The joint is plenty strong with the three other sides welded, so to avoid adding too much heat into the joint which may cause the boom to bow, just leave it.

We need a bottom bracket as well as a front derailleur tube, so to save a bit of time welding and worrying about finding the correct tube diameter for the derailleur clamp, I just hacked both from an old frame as one single piece. The front derailleur will only clamp to a seat tube on a bicycle, and I salvage all of my bottom brackets from old bicycle frames, so this method made the most sense and would spare some welding time. The size and condition of the frame is not important, as long as the seat tube is not bent out of alignment from the bottom bracket. Figure 6-91 shows the old 24-inch MTB frame about to be hacked. Goodbye old friend.

Before you select a frame for butchering, notice the two very different sizes of bottom brackets and crank styles commonly found: one-piece steel cranks with large bottom brackets, and the three-piece cranks sets with the smaller bottom brackets. Although both types will indeed work, choose the three-piece smaller bottom bracket as these are so much higher quality, and will yield a much better selection of decent crank arms and gear ratios.

Figure 6-91 *No, please don't cut me*

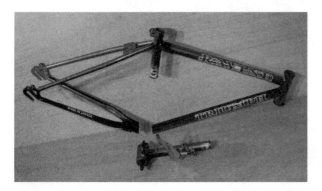

Figure 6-92 *Cut the bottom bracket and seat tube from the donor frame*

Figure 6-93 *Bottom bracket cut and ground clean*

The one-piece crank sets are very cheaply made, only come with heavy steel arms, and the crank rings are thin and prone to wear.

Install the front derailleur in whichever position it used to be, so that you can make sure to cut out enough of the seat tube in order to place the derailleur above the crank ring. If you are not sure, then cut way more than you need, like I did, as shown in Figure 6-92, as it can simply be cut off later. Just to be safe, I cut 3 inches above the derailleur because the largest crank ring on this bike might be smaller than the one I planned to install on the DeltaWolf.

If you are using a cut-off disc in your grinder to make the cuts, then beware when cutting either the seat tube or down tube, as the frame will tend to collapse on your disc as you get through the last bit of metal, which could cause your grinder to stall, or run out of your hand. If you have no way to secure the frame while cutting, then leave a sliver of metal during each cut rather than going all the way through the tubing, and you can simply bend the bottom bracket back and forth to release it after the cuts are made. If using a hacksaw, don't worry about it—just hack!

Once you remove the bottom bracket and derailleur tube from the donor frame, grind away excess metal and fill in any holes left by overlapping tubing or those drilled into the bracket shell. If you already had a bottom bracket for use in this project, then you will need to hack a length of seat tube from another frame and weld it in place to get the assembly shown in Figure 6-93. Make sure the tube forms a nice 90-degree angle with the bottom bracket if you have to weld the parts together.

The DeltaWolf has an adjustable bottom bracket that will allow riders of many different heights to adjust the trike to their inseam (leg length). This feature also makes it easier to build, since you don't have to worry about getting the bottom bracket installed on the main boom to within $1/4$ inch of your required leg length. On some recumbent bicycles, the seat is readjusted for leg length, but imagine how complicated it would be to design the DeltaWolf that way—you would have to move the steering control arm, as well as worry about the steering rod length. Moving the seat would really be impossible.

The bottom bracket and front derailleur tube move together along the main boom, so you do not have to worry about the position of the derailleur over the chain ring. The DeltaWolf also has an extra chain pickup idler under the seat, so large bottom bracket adjustment will not require you to add or remove chain links. Cut two $3/16$-inch thick pieces of plate or sheet to make the shapes, as shown in Figure 6-94. The round area is cut to conform to whatever size bottom bracket you decided to go with, and the two bolt holes will be drilled for $1/4$-inch bolts. The bolt holes are drilled at the edge of the plates so that there is $1/8$ of material between the hole and edge of the plate. These two plates will form a "vice" that will

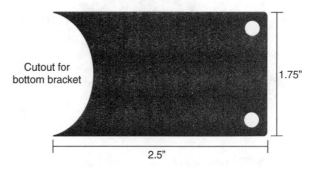

Figure 6-94 *The sliding bottom bracket mounting plate*

Figure 6-95 *Bottom bracket plates ready to be ground*

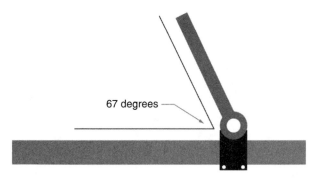

Figure 6-97 *The proper angle for the derailleur tube*

clamp the bottom bracket to whatever position on the main boom you want.

Once you have the two bottom bracket mounting plates cut, trace around the bottom bracket shell to draw a line, as shown in Figure 6-95. The marker line should meet the corners of the plate, in order to ensure that you don't cut out too much of the plate when grinding out the rounded area. If you take too much metal away from the plate, there will not be enough room between the bottom bracket and the clamping bolts for the main boom. When grinding out the traced circular area for the bottom bracket, place both plates in the vice so they can be done at the same time. This will ensure that both plates are exactly the same size.

Once both bottom bracket mounting plates are ground out to the same size, you can drill the bolt holes, as shown in Figure 6-96. Remember that the main boom

Figure 6-96 *Ground and drilled bottom bracket mounting plates*

tube must fit between the two bolts and the underside of the bottom bracket, so ensure that there is slightly more than 1.5 inches of room between the lowest part of the ground out area and the top of the bolt holes. The drawing in Figure 6-94 shows this. The two bolts that will tighten the plates like a vice along the main boom should have at least 1.5 inches of non-threaded area, as also shown in Figure 6-96, or you will scratch the paint off the underside of the frame when relocating the bottom bracket.

In order to allow the front derailleur to properly move the chain from one chain ring to the next, the angle of the derailleur tube must be set to 67 degrees to the main boom, as shown in Figure 6-97. It is actually the angle of the tube in relation to the angle of the chain that really counts, but it is easier to measure this angle from the frame once it is known. This angle will allow the use of a standard three-ring crank set as found on most mountain bikes. A degree or two in either direction will not have any impact on the derailleur operation.

In order to get the proper angle between the derailleur tube and the main boom, as shown in Figure 6-97, a bit of frame tubing can be clamped to the two plates (see Figure 6-98). This will allow you to set the bottom bracket into the correct position for tack welding, as well as ensure that both plates are aligned with each other, so that the bolt passes easily through both holes. I used an adjustable square set to the required 67 degrees, in order to set the bottom bracket and derailleur tube up for making the two small tack welds at the top of the joint. Besides the proper derailleur tube angle, the only other important factor is the position of the plates on the bottom bracket. With the piece of frame tubing as a guide, you will automatically have the correct distance between the plates, so the only other thing to ensure is

Figure 6-98 *Using a bit of frame tubing as a welding guide*

Figure 6-99 *The completed adjustable bottom bracket*

that there is approximately equal distance between each plate and the edges of the bottom bracket. In other words, the bottom bracket will be in the center of the main boom as viewed from above.

Figure 6-99 shows the completed adjustable bottom bracket and integrated derailleur tube. Since the mounting plates are fairly thick, there was no need to weld the inside of the joint, which helps to reduce alignment problems due to welding distortion. If you did weld the inside of the joint, the two plates would tend to move closer together, which would make it very difficult to slide the assembly onto the frame, or it would strip off the paint when moved.

With the adjustable bottom bracket and derailleur tube completed, you can mount the assembly to the main boom and tighten the two clamping bolts. Figure 6-100 shows the completed assembly mounted securely to the frame, just before cutting off the access tubing above the derailleur. When you trim the excess tubing, make sure that you have the actual chain ring you plan to use

Figure 6-100 *Bottom bracket installed into the frame*

installed, because there are many different diameter chain rings available with various diameters. In Figure 6-100, I have a 58-tooth chain ring installed, which is slightly larger than those found on a typical mountain bike. I installed the largest chain ring commonly available in order to trim above the derailleur so that I could swap them out later, yet still have the ability to run the large chain ring if I wanted. Trim the tubing above the derailleur so there is at least 1 inch of material above the mounting strap.

The DeltaWolf requires a very unique, long and straight gooseneck in order to mount the handlebars to the steering head tube. You will not find anything like it on a regular bicycle, so we will simply make our own by using a steel seat post to extend a commonly available steer bicycle gooseneck. Figure 6-101 shows the steel gooseneck I chose to cut—an old BMX seat post—as well as a set of steel mountain bike handlebars. Your goal is to place the gooseneck clamp in a position that will allow your handlebars to clear your legs and knees when pedaling, yet keep your arms and hands in a very relaxed and comfortable position. Any steel gooseneck that fits into the fork stem you used to make the steering head

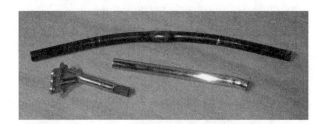

Figure 6-101 *An old steel gooseneck and a steel seat post*

Figure 6-102   *Extending the gooseneck*

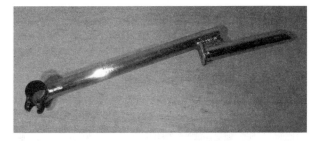

Figure 6-103   *The handlebar clamp is in the center of the gooseneck tube*

tube will work just fine, but be aware that there are two common sizes: BMX and mountain bike. BMX goosenecks have slightly smaller diameter tubing, and will create a very sloppy loose fit if placed into a mountain bike fork stem. A mountain bike gooseneck would not fit into a BMX fork stem. The seat post or mountain bike handlebar tubing shown in Figure 6-101 will be used to make the actual extension between the gooseneck stem and the clamp, so this tubing must be made of steel.

How tall to make your gooseneck is really a factor determined by your body size and the amount of padding you plan to use on your final seat base. Just for the record, the tube I used to extend the gooseneck, as shown in Figure 6-102, is 10.5 inches long and made from an old BMX seat post. Handlebar tubing, conduit, or any similar strength tubing will work just fine.

Cut the clamp from the gooseneck so there will be enough material to make a good joint between the extending tube and the base of the gooseneck. As shown in Figure 6-102, I cut the gooseneck an inch from the stem and ground the area to allow an easy weld of the two parts. The position and angle of the extending tube should place the top of the tube (where the handlebars will mount) directly over the end of the gooseneck tube. This placement of the handlebars will remove a tiller effect from the steering so that the handlebar turns directly on its center axis.

Before you make any cuts or welds, it's a good idea to sit on the bike while the bottom bracket and cranks are installed to see which handlebar position feels most comfortable. Do this by holding whichever handlebars you plan to use as you pedal so that you can find the optimal position. Make sure you have your seat base set up to the height it will be in the final design, including any padding. When you are holding the handlebars at what you think is the optimal position, have a helper measure the distance from the handlebars to the gooseneck and use that as a reference. In case you need

to make a slight adjustment later, you will be able to move the gooseneck up and down slightly just as it was done on a regular bicycle.

As shown in Figure 6-103, the handlebar clamp is positioned in such a way that if you drew a line extending from the center of the gooseneck tube, it would hit the center of the clamp; this is why the extender tube is welded in place at a slight angle. Feel free to use whichever gooseneck style you like to create this extended version; as long as all the parts are made of steel, you will be fine. In my version, I have a BMX gooseneck stem, a seat post tube as the extension, and a mountain bike handlebar clamp at the top. I needed the BMX gooseneck stem, since I made my steering system from a kids' BMX bicycle, but wanted the standard mountain bike clamp so I could use road bike handlebars.

Handlebar size and shape is totally up to you, but the ones I decided to use, shown in Figure 6-104, are really comfortable, and allow plenty of room to mount all the shifting and brake hardware. These "granny" style handlebars are common on cruiser-style bikes, and are often found on those older three-speed, 1970s ladies' bikes that often hide out at the local dump or scrap metal pile. The curved handlebars seem to make extra room for

Figure 6-104   *Handlebars installed*

your knees while placing the handgrips in a very reasonable position. You could also go for the "sawed-off" short racing style handlebars I used on a LWB recumbent called the Marauder, which can be seen on our website in the Builder's Gallery at www.atomiczombie.com.

Your handlebars should not only be comfortable and easy to reach, but should allow ample room to turn without jamming into your legs or body. You will never have to turn the handlebars more than 45 degrees, but should be able to do so without jamming yourself in the ribcage. In Figure 6-105, the handlebars easily make the 45-degree turn without interfering with the test pilot's legs while pedaling. When choosing your handlebars, don't forget about all those levers and shifters, as they take up a few inches of real estate along with the handlegrips.

Remember those two control arms that you made from the drawing way back in Figure 6-81? Well, now it's time to mount the second one to the front forks. The control arm will be placed on the left leg of the front forks just below the bearing ring. Because the fork leg will most likely meet this area, you will need to do a little extra grinding on the control arm, as shown in Figure 6-106. Try not to make the control arm too much shorter, but just grind enough material so that you can make a good weld around the joint.

There are a few things to note before welding the control arm to the front forks, once you are done grinding the part for a better fit. First, make sure that you tap the fork bearing ring before doing any welding, as you may accidentally hit the ring and it will need to be replaced. Second, remember that the control arm needs to be on the same side of the forks as the control arm on the steering

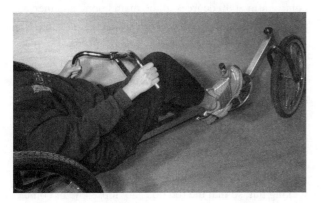

Figure 6-105 *Checking the handlebar fit and clearance*

Figure 6-106 *The front fork control arm*

Figure 6-107 *Make a tack weld at one end only*

system—the left side of the frame, opposite the chain. Third, the control arm makes a slight downwards angle to better support the ball joint, since the steering control arm is slightly lower than this one. Read ahead a bit and you will see what I mean. Tack weld the control arm at the rear corner of the joint, as shown in Figure 6-107. In the photo, the forks are placed frontside down, so the control arm is on the left side of the forks, and the tack weld shown is at the rear of the forks.

With the tack weld at the rear of the control arm, you can tap the front of the arm downwards in order to create a slight angle, as shown in Figure 6-108. A few degrees will be plenty. This is done to accommodate ball joints that do not have a lot of horizontal freedom. If you left the control arm perfectly straight and installed ball joints with very little horizontal ability, there would be rubbing between the ball joint shell and the control arm. This problem could easily be solved by placing washers

Figure 6-108 *The control arm is at a slight angle*

Figure 6-110 *Two ball joints and the connecting rod*

Figure 6-111 *Mounting the ball joints to the connecting rod*

between the control arm and the ball joint, but this solution is much tidier. In Figure 6-108, you are looking at the left side of the forks, and I have reinstalled the bearing ring by tapping it back into place. If you accidentally welded on the bearing ring area, then carefully grind or file it clean so the ring can be installed.

There is not a lot of room on the top side of the control arm for a heavy weld, so make the underside weld as solid as you can. You don't want your control arm to fall off while you are riding. Sure, the DeltaWolf rides perfectly hands free, but a steering failure is no way to test that point! Figure 6-109 shows the completely welded front fork control arm—note the ample weld metal on the underside of the joint.

The two ball joints will be used to connect the two control arms together via connecting rod, as shown in Figure 6-110, along with a light-gauge hollow steel tube. This hardware can be found at your local hardware store, and you are free to use whatever type of ball joints and connecting rod you can find that will suit this project. These ball joints do not need to be heavy duty like the type you would find on snow machine or offroad vehicle steering, since they are going to be taking a very light

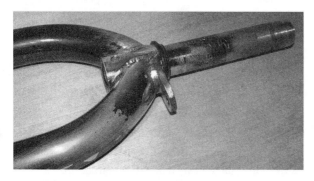

Figure 6-109 *Make this weld solid*

load. The heads on the ball joints shown in Figure 6-110 are about the size of a quarter, with a $3/8$-inch bolt hole. The connecting rod can be either a solid length of pencil thickness rod, or some type of light-gauge hollow tubing that feels fairly light, yet resists easy bending by hand.

The simplest and most reliable method of joining the ball joints to the connecting rod is to directly weld them to the ends of the rod as I have done in Figure 6-111. This method works as long as you know the exact length to make the connecting rod, and as long as you fully trust your welding. To calculate how far away the bolt centers of each ball joint should be, just place the steering control arm at 90 degrees to the main boom, and then place the front forks in their forward facing position. You can then use this as a reference to make the connecting rod the correct length. The other way to connect the ball joints to the control rod is by welding a bolt at each end of the rod which lets you screw the ball joints onto the rod and set them to the correct length. Some ball joints have bolt ends, and some have nut ends, so you will have to work this out for whichever type you have. Also, when using the screw-on method, make sure you include a locking nut, so that the rod does not loosen up or come out of alignment after time.

Figure 6-112 *Steering control arm in the center position*

The position of the handlebars is not important since the gooseneck is simply inserted into the steering fork stem and locked down. It is the control arms that must be aligned so that they are both forming a 90-degree angle to the frame when you are moving in the straight ahead direction. In Figure 6-112, the steering control arm is shown in the straight ahead position with the ball joint and control arm installed. When completed, the ball joint on the front fork will be installed on the top of the control arm, and the ball joint on the steering side will be installed on the bottom of the control arm. This positioning helps the ball joint body clear the control arm, and will keep the grease off your pant leg if you happen to come into contact with the control arm.

Figure 6-113 shows the steering side of the linkage. Notice that the ball joint is placed on the top of the control arm. It could not be placed on the underside of the control arm as it would come into contact with the fork leg after a few degrees of turning. When you put the DeltaWolf together for the final time, make sure to use

decent bolts with locking nuts on both ball joints, so that there will be no chance of bolts falling out after many miles of vibration and use.

Your steering linkage will easily allow more rotation than you would ever need on all but the sharpest of turns. At 45 degrees rotation, the DeltaWolf will turn a circle that would challenge a kid on a freestyle bike to copy. In typical riding, you only need a few degrees of rotation, but this steering system can really allow for some tight turning circles. Figure 6-114 shows the steering in the maximum left rotation, which would allow the DeltaWolf to turn around in a circle with a diameter of about 10 feet or so.

Figure 6-115 shows the clearance between the left crank arm and the steering control rod, so nothing to worry about here. As you can also see, it would not be possible to place the control rod on the right side of the frame since the chain ring and derailleur do not allow clearance. The control rod is closest to the crank arm in the straight ahead position, so the 1 inch of clearance shown in Figure 6-115 is more than enough for any style crank arm.

Figure 6-114 *Steering at full left rotation*

Figure 6-113 *Steering linkage is now completed*

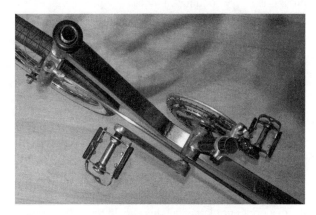

Figure 6-115 *The control rod has plenty of room*

On a very low recumbent bicycle or trike, the chain must pass under the pilot's seat in order to reach the rear freehub. This is rarely a simple path from one point to the other. Some recumbent cycles may have two, three, or even four chain guide pulleys in order to route the chain around all the obstacles from the front of the frame to the rear, and each time you add another pulley, you add friction and complexity to the design. The DeltaWolf has only a single drive-side pulley, and an optional but highly recommended return-side pickup idler to reduce chain flop over uneven terrain. Drive side refers to the side of the chain that leaves the top of the rear multispeed freehub to enter the top of the front chain ring. This is the only part of a bicycle chain that will ever be under any tension, since it does all the work of transferring your energy into the rear wheel. Return side refers to the side of the chain that exits the bottom of the front chain ring and enters the lower wheel of the rear derailleur. This part of the chain will always be slack as it never has any tension nor does it do any work.

We need to know which side of a chain is the drive side and which is the return side, so that we can choose the appropriate pulleys in order to route our chain under the pilot's seat. If you look at the two small plastic wheels that the return chain pass through on a rear derailleur, you will notice that they are nothing more than soft plastic wheels running on a steel pin. This works just fine due to the fact that the chain just lazily slips past the wheels with almost zero force. If you tried to use those plastic wheels on the drive side of the chain, they would disintegrate the instant you put any real energy into the cranks. For this reason, our drive-side chain needs a very large, low-friction guide pulley in order to keep friction to a minimum and not wear out after a few hard rides. Figure 6-116 shows such a pulley. These nylon idler pulleys are very common in fitness equipment that uses a wire or rope. They are a perfect size to carry a bicycle chain, have built in ball bearings, and will probably never wear out. The one I used is 6 inches in diameter, but any similar pulley from 3 inches and larger will work just fine.

I strongly recommend that you use a pulley of at least 3 or more inches in diameter, and it must have a ball bearing in order to work properly as a drive-side chain guide. Skate blade wheels have commonly been used for home-built chain idlers by carving out a slot for the chain using a grinder or lathe, but this should really be avoided,

Figure 6-116 *A nylon idler pulley with ball bearing*

as there will indeed be a power loss due to the small diameter of the pulley and the softness of the material. For $10, you can get these nylon idler pulleys online or at many fitness equipment stores, so why bother with a workaround? Once you have your idler pulley, place it near the end of the main boom where we left that 3-inch bit of tubing past the seat tube. The pulleys should easily fit in this space without rubbing on any part of the frame, although you may need a washer or two between the pulley's bearing and the frame. Once you know there will be no clearance issues between the pulley and the frame, drill the appropriate sized hole at the end of the tube, as shown in Figure 6-117. Even if you have plenty of clearance, try to keep this hole near the end of the tube in case you want to use a larger pulley later.

Once the hole is drilled for the drive-side idler pulley bolt, place the correct bolt into the hole, and place a little weld on the head side of the bolt and the frame to secure it in place. This welding of the bolt is done so that you

Figure 6-117 *Drill the idler pulley bolt hole*

Figure 6-118 *Securing the idler pulley bolt*

Figure 6-119 *Drive-side idler pulley installed*

can place a lock nut on the bolt, without needing to crank it so tight that you crush the walls of your tubing together. Since the bolt can never come free, you only need to place the lock nut on tight enough to secure the pulley to the bolt. Don't forget which side the bolt is facing—the pulley is on the right side of the frame, so Figure 6-118 shows the left side.

Install the drive-side chain guide pulley, as shown in Figure 6-119, using a lock nut to avoid losing parts on hostile terrain. My 6-inch diameter pulley has just enough room to clear all the frame tubing, which is why I left the 3-inch section of boom extending past the seat tube. Your pulley may be quite a bit smaller, but be aware of the energy loss placed on your transmission if you use a pulley less than 3 inches in diameter or something made of soft material. I have tried these, and learned that they just don't work very well, especially if you like to accelerate hard.

The last step before you can test ride your trike would be the installation of the freehub to the drive axle on the

Figure 6-120 *Machined version of the freehub mounting part*

left-side wheel. Figure 6-120 shows a nice, shiny machined part that has a hole drilled in it for the ⅝ rear axle, and a threaded outside that allows the installation of a standard Shimano-type freehub. This is a nice, clean method of installing the freehub to the axle, but it does require you to visit the local machine shop with fifty bucks in your hand, or to collect on that favor your machinist buddy owes you. I had several of these made, since I like to mess around with many different bike and trike designs, and this is the perfect part for mounting freehubs (multispeed and fixed gear) to an axle, so it was indeed worth the cost and wait. The more parts you have done at once, the cheaper it becomes for each one as well.

If you want to make the DeltaWolf "machine shop free" like I promised could be done, then refer back to Figure 6-51, and you will see that the dismembered steel hub not only makes the two hub/axle assemblies, but it also makes this exact part. The completed unit, shown in Figure 6-57, is what we are after—a method of installing a thread-on freehub to the end of the drive axle. If you do choose to have the part machined, do not look up the thread pitch for a standard freehub in your cycle book or on the Internet and attempt to bring a beautiful drawing to the machine shop—trust me on this one. Bring them the actual freehub, or even a threaded hub, as well as the axle or at least a part of the same rod you made the axle from, and explain to them exactly what you want, how it all fits together, and then leave the parts there so they can measure them and test fit everything. You should also print out Figure 6-120 and Figure 6-121 when you drop off the parts. By the time they get to actually machining the part, that drawing may tell them something different, and you may find yourself waiting another two weeks for the part to be fixed.

Figure 6-121 *The freehub threads onto the machined part*

As shown in Figure 6-121, the machined part will thread right into the freehub with little effort, and then slide onto the drive axle for a nice tight fit. Don't crank the freehub into the part just yet, or you may have a tough time removing it—just thread it on until it stops using little force.

With the left axle installed on the two pillow block bearings so that the hub is resting as close to the bearing as it can be installed, lock the little set screws that come with the bearings to hold the axle securely in place. Now, place your machined part or home-built part on the axle and then lightly thread on the freehub until it stops. Slide the freehub along the axle until the center of the freehub is lined up with the center of the drive idler pulley, as shown in Figure 6-122. Since the chain needs to travel from the largest gear to the smallest, you want the freehub to be centered with the idler pulley so the chain can reach all the gears without making too extreme of a bend. Bicycle chain is very flexible for this reason, but there are limits to everything. Once the center of the

freehub is in alignment with the center of the idler pulley, mark the axle so you can remove the freehub and place the mounting part back on the axle in the correct place. It will be welded directly to the axle.

I originally thought about adding a keyway to the axle and the freehub mount, then installing a bolt at the end of the axle so the part could be removed, but after realizing the amount of effort and cost added to making the part, I reevaluated this idea and decided against it. It's not that I decided to cut corners; I just realized that having the part removable would actually accomplish nothing, and would make the freehub less secure. With the freehub mount welded to the axle, the two axle bearings would no longer be easily removable, but why would I need them to be? These ⅝ bearings would never in a million years wear out with this kind of use, and I could still replace the freehub if I needed to, so the keyway idea was scrapped.

Figure 6-123 shows the freehub mount welded directly to the axle at the same position it was marked in the last photo. Make sure that both of your bearings are installed on the axle before you make this weld, or you will be spending some time with your grinder! Only the end of the axle needs to be welded, and as long as you remember to place the welder's ground clamp on the axle, not the frame, you can make this weld with the axle and bearing installed on the frame. I like to weld the entire joint while a helper turns the wheel slowly around in a full circle, as it makes a nice, continuous weld bead that way.

Once your freehub mounting part is securely welded to the drive axle, cut the leftover axle off just ahead of the weld and thread on the freehub as tight as you can by hand. The freehub will tighten up a lot more on the first few hard pedal strokes, so keep this in mind on your initial test ride as it may feel like something is slipping at first. Figure 6-124 shows the freehub mounted to the

Figure 6-122 *Checking the freehub to idler alignment*

Figure 6-123 *Weld the freehub mount to the left axle*

Figure 6-124 *Six-speed freehub threaded in place*

Figure 6-125 *Joining two bicycle chains*

drive axle in the optimal position for chain alignment with the large idler pulley.

The DeltaWolf is so reclined that you are going to need almost twice the length of bicycle chain that you would normally require if perched precariously on top a "wedgie" style bicycle. Shown in Figure 6-125 is a wad of chain, consisting of two new lengths of standard derailleur bicycle chain and an inexpensive tool that makes joining chains so much easier than the "smack it with a hammer and punch" system. Two chains may be a little too long, but it's a good start, so go ahead and join them into one long single chain. Look out for those brain-teasing loops!

Since there is no rear dropout, or threaded rear axle on the DeltaWolf, we will need to get creative in order to attach a standard bicycle derailleur to the correct position under the multispeed freehub. The easiest method I have

Figure 6-126 *A rear hub axle will make a derailleur mounting peg*

found is to simply emulate the original mounting method using an old rear axle from a mountain bike hub or a long bolt of similar size. Figure 6-126 shows a rear hub removed from a dead mountain bike hub—you will also need the two nuts and a few washers.

The plan is to cut the axle down and weld it to the rear frame tubing so that the rear derailleur will mount to the axle, just like it was on a regular bicycle. On a regular bicycle, the derailleur is held in place by the ride-side axle nut onto the outside of the right rear dropout. This placement of the derailleur puts it in the correct position for shifting the chain across the entire range of gears. Since we can't mount our derailleur in the exact center of the freehub in the DeltaWolf, we will place it slightly behind the freehub center, which is close enough for proper operation. Figure 6-127 shows the axle cut down to a length that will place the derailleur mount slightly past the smallest freehub gear, as shown next.

Figure 6-128 shows the derailleur mounting bolt welded to the frame at the bottom end of the tube, so that the derailleur will be placed slightly to the right of the smallest gear, and at the same height from the ground as the rear axles. Before you weld the bolt in place, make sure the chain can slip from one gear to the next without getting wedged between the end of the bolt and the

Figure 6-127 *Derailleur mounting peg*

Figure 6-128 *Derailleur mounted peg welding in place*

Figure 6-130 *Return chain management*

gear teeth. If the chain hits the bolt, then it is too long. If your rear derailleur is a very high quality unit, it may have a different mounting system rather than the simple dropout slide-in type that I have used. This will not present a problem, as long as you make sure that the derailleur is in a position to allow the full range of gear changes without the chain striking the mounting bolt, or whatever hardware you plan to use to secure the derailleur to the frame.

Figure 6-129 shows my derailleur installed on the mounting bolt that I made from a discarded rear bicycle axle. You will probably need to tweak the two adjustment screws on the derailleur to get it to travel to the far reach of the freehub without falling off the last chain ring. Start by loosening both adjustment screws out a few turns, then try moving the derailleur across the freehub range as a helper pedals the cranks. Your derailleur should be able to reach all of the gears without falling off the largest or

the smallest ring. There is usually a large margin of adjustment available on most derailleurs due to the multiple sizes of freehubs available.

At this point, the transmission is completed enough to test ride the DeltaWolf, although it would be a bit clunky without shifter cabling and that return chain banging around. The return chain has no tension, and is fairly loose, creating a sagging arc from the front of the trike to the rear due to its own weight. You will also notice that while shifted into the larger rear chain ring, the chain may actually rub against the large drive-side idler pulley a bit, causing a clicking sound. A very simple and effective return chain idler system can be built by cutting up an old rear derailleur, as shown in Figure 6-130. If you remove the part of the derailleur that connects the two idler wheels to the main body, you will be left with a small shaft that just happens to be the perfect length for an idler mount if welded to the frame in the position shown in Figure 6-130. There is no stress on the return chain, so this idler system will not add much friction into your drive system, but it will keep all chain flopping to an absolute minimum and stop the return chain from rubbing against the idler pulley when shifted to the larger rear chain ring.

With the return chain pickup mounted correctly, you will be able to lift the drive wheel and pedal the DeltaWolf without any chain rubbing or grinding noises throughout the entire gear range. With the little idler wheels cleaned up and slightly oiled, the DeltaWolf transmission will not make any more noise than that of a regular bicycle, and it should feel just as smooth.

Figure 6-129 *Rear derailleur installed*

Figure 6-131 *No more chain flop*

Figure 6-132 *Transmission top view*

Figure 6-131 shows the completed transmission working perfectly and silently.

Figure 6-132 shows the top view of the DeltaWolf transmission. Notice how the return chain passes alongside the drive chain in order to clear the large idler pulley. It's almost too bad that most of this clockwork will be covered up from view by the seat.

If you are not used to such a low, laidback, recumbent riding position, you may find yourself wanting to rest your head using some type of headrest. Everyone I know who has built an extreme low racer has stated that they need a headrest on the first few rides, but later decided that it was really not necessary, and became very comfortable without it. I guess you have to get used to sitting in positions that you are not accustomed to while using muscle groups that would not normally be used on a bicycle of this style. On my first extremely low recumbent, The Marauder, I added a headrest that would place a foam pad directly behind my head, but realized that this caused two new problems; I could never use a helmet, and the bumps from the road pounded the headrest on the back of my head. For this reason, I

almost never used the headrest unless I was cruising along on a perfectly fresh paved road.

On the DeltaWolf, I solved this problem by creating a neck rest instead that would allow the use of a helmet, and does not amplify road bumps nearly as bad as a headrest would. A nice, fat, round pad was also used to further pad the rider from vibration, and this did create a decent neck rest, although I still rarely use it unless I am on a leisurely cruise. If you choose to install the neck rest, an old set of road bike forks cut at the legs makes the perfect mounting system, and it looks great on the final design. Simply hack the legs from the fork crown at an angle of approximately 30 degrees, as shown in Figure 6-133.

The fork legs are then welded to the rear frame tube so they form a nice arc up to the position you want to install the neck rest pad. There is a lot of room for experimentation here, and you may even want to make your neck rest adjustable, spring loaded, or completely removable—an easy task that can be done with a handful of scrap bicycle parts and some imagination. I went for the fixed position headrest, as I thought the perfect

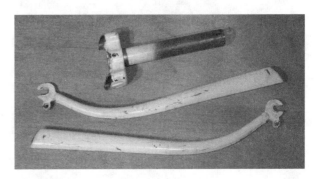

Figure 6-133 *Cut the fork legs at the crown*

Figure 6-134 *Fork legs are welded to the frame*

position for the neck rest pad would be directly above the rear frame tube where the fork legs are welded. The curved fork legs left plenty of room for my back so that only the neck rest pad comes in contact with my body. Feel free to experiment here; this is, after all, your own personal ride.

The actual neck rest pad is a round cylinder of black foam that came from some discarded fitness equipment I found at the dump. I believe you would ask for "leg extension pads" if you were looking to purchase these things new from a fitness store. The hole through the pad's center is the perfect size for the little bit of seat post tubing I cut off shown in Figure 6-135. The seat post has washers tack welded at each end so the long bolt will fit through it. The seat post is then slid into the foam pad so the fork legs can be pushed through the two small holes in the foam, in order to create a clamp that will hold the neck rest in between the fork dropouts.

Figure 6-136 shows the neck rest pad held in place by the bolt through both dropouts and the small bit of seat

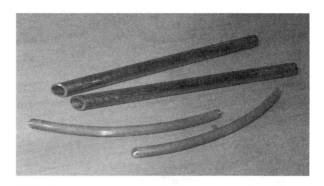

Figure 6-137   *Parts for the elbow guards*

post tubing. To plug the holes at each end of the foam pad and hide the bolts, I pushed a pair of plastic pop bottle caps that matched the color of my paint job. The completed neck rest is comfortable, not too bad on most roads, and allows the rider to wear a helmet.

If you couldn't resist the urge to test drive the almost completed DeltaWolf, then you probably realized how much you needed the brakes and how close your elbows are to the rear tires. On very uneven ground, or when driving like a maniac around sharp corners, you may end up rubbing your elbows on the rear tires, so some type of elbow guard will be necessary. The simple elbow guard I made from a few bits of thin scrap tubing and rod is very minimal, but works perfectly and makes it almost impossible to place your elbow against the tire, even if you wanted to. As shown in Figure 6-137, two small lengths of thin walled steel rod, about the same thickness and diameter as bicycle seat stay tubing, was used as a support arm to hold the two small curved pieces of steel rod that form the actual elbow guard. The two small curved rods are chinks of steel cut from a trashed shopping cart, and they were bent by hammering them in the vice to the approximate curvature of the rear wheel. Again, this part of the DeltaWolf leaves much room for improvisation, and as long as you find a way to keep your elbows out of the wheels, then feel free to experiment. You do not need a large elbow guard because your elbows really cannot move all that much when you are holding on to the handlebars, which is why my minimalist approach works perfectly.

The elbow guard is placed at a horizontal orientation towards the front of the rear wheel, as shown in Figure 6-138, as this area is where your elbow will be at rest. This simple elbow guard would also form a very sturdy mounting base for a pair of fenders or some type

Figure 6-135   *Neck rest mounting tube*

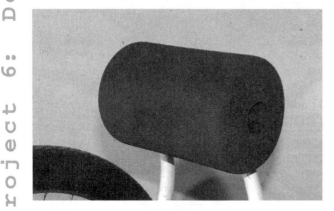

Figure 6-136   *Completed neck rest*

Figure 6-138 *Mounting the guard arms to the frame*

Figure 6-139 *Completed elbow guards*

of rear fairing, if you wanted to turn the DeltaWolf into more of a velomobile.

To strengthen the elbow guards and give them a touch of class, I added the two end rods and the small rod at the arm base that forms a gusset between the arm and the frame (Figure 6-139). This small steel rod is cut from a bucket handle—yes, garage hackers use whatever materials are on hand!

I hope you add at least a rear brake to your speed hungry trike before you take it out for any high speed testing! The DeltaWolf is designed with a rear disc brake in mind, and since this hardware has become very commonplace, it is inexpensive and easy to mount. The front brake is simply mounted to the front forks just like on a regular bicycle, so I won't cover it here, and if you have complete faith in the single rear disc brake, you could leave out the front brake, as most of the stopping power is at the rear wheel anyway. The disc brake will connect directly to the right axle by bolting it to a simple flange cut from a bit of steel with the approximate

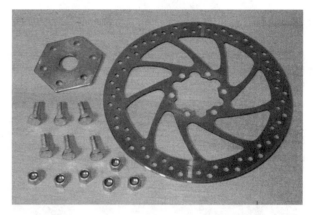

Figure 6-140 *The disc brake will mount to the flange*

thickness of the actual disc brake. This mounting flange can be round, square, or any shape you like, as long as the ⁵/₈ hole is directly through the center, and the disc brake mounting bolt holes form a perfect bolt circle. Figure 6-140 shows my standard six-bolt disc brake and the mounting flange that I hand cut from a bit of ³/₃₂ steel plate. Good quality, fine-thread bolts with lock nuts are also needed, and they should fit the disk brake holed perfectly with no play.

Install one of the bearings onto the axle, then weld the disc brake mounting flange to the axle, so that the disc brake will end up directly between the two frame tubes, as shown in the next photo. Use the same alignment techniques that you used when mounting the hub flanges to the axle to ensure that this disc brake mounting flange is aligned perfectly. When you place the disc brake on the flange and spin the wheel, there should be no side-to-side wobble in the disc, or it will make a rubbing sound when you are riding, or could create extra friction if it were too far out of alignment. A few tack welds and tapping of a hammer are usually enough to get the disc brake mounting flange installed so that the disc brake spins perfectly true. Do remember to place that bearing on the axle before you weld in the flange, or you will have to cut it off and reweld it. Figure 6-141 shows the flange welded to the right axle with bearing installed.

Figure 6-142 shows the correct position for your disc brake between the frame tubing. The disc is placed in the center so that there will be ample room to install whatever disc brake hardware you have.

I wanted to install a cable pull disc brake, as shown in Figure 6-143. This type of brake is simple, inexpensive compared to the hydraulic version, and works very well—much better than any pad brake. Since the

Figure 6-141   *Disc brake mounting flange*

Figure 6-144   *Setting up the brake mount positions*

Figure 6-142   *Disc brake mounted to the flange*

Figure 6-143   *Disc brake mounting hardware*

mounting system needed for your disc brake could be radically different from mine, there is no point in giving out dimensions for the two small steel mounting tabs I am using to secure the hardware to the frame. The goal is

to simply place the brake hardware over the disc so that both pads grip the disc in the appropriate place. The two small bits of steel shown in Figure 6-143 were cut from a bit of scrap angle iron and, as you will see in the next few steps, just about any method and placement of the disc brake hardware can be used, as long as the pads are in the correct position over the disc.

Most disc brakes have a generous amount of side-to-side adjustment for aligning the pads over the disc, so I placed the brake unit onto the disc, then tie wrapped the brake lever into the locked position to hold the hardware tightly on the disc. With the hardware held in place by the brake pads, you are certain to get a good alignment, and can simply tack weld your custom-made mounting hardware to the frame, like I have done in Figure 6-144. Be careful when welding near the brake hardware, and only use small tack welds to avoid damage to the brake hardware. I usually mask the parts with tape as well to keep any welding spatter away from the inside of the brake unit.

Once you tack weld the brake hardware mounting brackets to the frame, release the brake lever to free the pads from the disc so you can spin the wheel and check the alignment. You may need to slide the brake hardware a little bit either side to stop the disc from rubbing on the pads, but if all went well, this will be easy. Some rubbing may be normal according to your brake hardware instructions, but there should be no extra friction when you spin the wheel. If you are happy with the alignment, remove the brake hardware and weld the mounting brackets solidly to the frame, as shown in Figure 6-145.

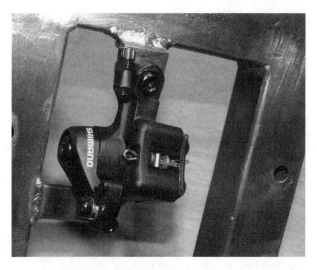

Figure 6-145 *Disc brake mounts securely welded*

Figure 6-146 *The seat mounts are cut and drilled*

Figure 6-147 *Seat mounts welded to the main boom*

Figure 6-148 *Trailer mounting tabs*

Fastening the seat to the frame will be done using two small brackets, as shown in Figure 6-146. Actual size and shape is not critical, as long as you can get a few holes for two or more woodscrews through each piece. I used a bit of ¹/₁₆ steel plate for my seat mounting brackets, which are about 2 inches square.

Weld the seat mounting brackets to the main boom between the rear seat tube and the steering tube, so the brackets will be placed directly under the center of our seat base, as shown in Figure 6-147. A board placed across the brackets will form a 90-degree angle with the side of the frame, and would touch both brackets and the top of the frame tubing at the same time since they are all flush. The rear of the seat will be held on by drilling a small hole through the top of the rear frame tube for a 2-inch long woodscrew, so no seat back mounting brackets are necessary.

You just never know when you might want to pull a trailer, or create some type of detachable tandem for the DeltaWolf, so a little engineering ahead of time will save you from having to repaint modified areas of the frame.

Figure 6-148 shows one of the tabs that I welded to the rear of the frame between the outermost bearing bolt holes, in order to attach some type of trailer later. Since the paint will be going on the frame very soon, this is the time to add the little parts you may need for mounting accessories such as lights, a horn, or whatever you may want on your final design.

Since most of the frame was made from new tubing with the exception of the head tubes, front forks and neck rest, a very light sanding with a few sheets of emery cloth was all that it took to get the metal ready for a coat of primer (Figure 6-149). I don't have a powder coating station, nor do I have any place in my garage to paint, so I simply hang my frames from an outdoor clothes line and go to town with a few cans of spray paint. If you take your time and follow the instructions on the can, a remarkably good paint job can be accomplished from a few cans of spray paint. It is true that spray paint will scratch easier than auto point, but to look at the final paint job, you would not be able to tell the difference.

Figure 6-149 *Sanding and cleaning the frame for painting*

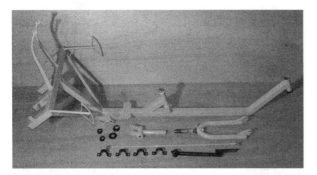

Figure 6-150 *Frame and parts painted*

The DeltaWolf was built to be my main mode of transportation, when not forced to take out the gas-guzzler, so I painted it yellow and black in honor of my last low racer, The Marauder, which recently retired from many years of faithful service. Figure 6-150 shows one of the most painful steps in building any new project: waiting for the paint to dry! It's too bad that this book is in black and white, as things always look better in color.

If you decided to tackle the wheel building yourself, then now would be a good time to tighten up all of those spokes to true up both rear wheels. If you are new to wheel building, then it may take you a day per wheel, and this is OK since the paint needs a few days of curing before you can start reassembling the DeltaWolf. As shown in Figure 6-151, an old bit of wood will make a perfect truing stand once you bolt the pillow block bearings to the wood. I am using a bicycle spoke as a guide when truing the rim, as this system works very well. If you plan on taking your wheels to the shop to have a bicycle wheel artist lace them up, then at least make some kind of truing stand like I did, as they will have no stand that can handle these alien hubless wheels. For the axle with the disc brake flange, you will need to cut a slot in the board.

Figure 6-151 *Something to do while the paint cures*

I cut my seat from a few pieces of ³⁄₄ plywood and some 1-inch thick, rigid, black foam. Seat construction is totally up to you, and there are hundreds of different materials you could use to make a great seat, depending on your personal preference. I like a hard seat that lets me easily shift my weight around, yet offers a little bit of padding from the rougher terrain. This hard, black foam needs no covering, will not soak up moisture and is very resistant to wear. It may seem that something like this would be uncomfortable, but to be honest, I can sit in this seat all day long with no pain at all, but can barely take two hours on an upright bicycle without a break. As shown in Figure 6-152, the two seat halves are held together by a pair of shelf brackets. The foam is glued directly to the wood using spray adhesive, and then trimmed by following the edge of the seat board with a sanding disc in the grinder.

Figure 6-152 *Gluing the foam to the wooden seat base*

Figure 6-153 *Finished seat*

Figure 6-155 *Rear transmission detail*

Figure 6-154 *The cockpit*

My completed seat is shown in Figure 6-153. Rounding off the corners on the front of the seat base is a good idea for safety reasons. The seat base is held to the frame by the six woodscrews through the bracket on the main boom, and the rear of the seat is fastened by a single bolt through the frame near the top of the seat.

The DeltaWolf goes together fast once you add all the components back onto the painted frame. Figure 6-154 shows the trike just about ready to ride as I finished installing all the shifter and brake cabling. Cabling a bike is a bit of a chore, but if you take your time and plan ahead, you will get away with minimal cable that will not rub on any of the moving parts. Notice the two yellow pop-bottle tops placed in the head rest foam—yes, the art of recycling!

Clean up all your transmission parts (Figure 6-155) and use a little bit of light oil on the freehub and

derailleur guide wheels to keep squeaking to a minimum. Tighten all the small set screws in each pillow block bearing so that the axles cannot slide side to side in the bearings. Your left axle is unable to slide side-to-side because it is held by the hub flange at one end, and the freehub mounting hardware at the other. The right axle could move a slight bit if the bearing set screw came loose, which will cause your disc brake to rub on one of the pads. The bearing set screws are usually enough to stop this from happening, but if you want to add redundancy, simply add a small hose clamp on the bit of axle that sticks out from the bearing to the left of the disc brake. Another way to secure the axle is to cut a small bit of tubing that will slide snugly over the axle, and then drill through both the tubing and axle to secure it in place directly against the bearing.

Figure 6-156 shows my new aluminum three-ring crank set with a nice large chain ring for top speed, as well as a very small hill-climbing chain ring. I am

Figure 6-156 *Good components make for a smooth ride*

Figure 6-157 *Ready to race, ready to cruise*

Figure 6-158 *Like clockwork*

Figure 6-159 *Comfortable and ultra-cool looking*

normally in high gear when riding in the city, but do find the lower gears to be a gift when hitting those seriously long and steep hills that would normally make you get off and walk your two wheeler up. You can move as slow as you need to on a trike, so there is no hill that the DeltaWolf will not climb.

Figure 6-157 shows a lean, mean trike with a speed hungry attitude, yet it also looks well-behaved and extremely comfortable, a fact which I can attest to. The DeltaWolf is like a fusion between your favorite recliner and a race car.

The first real ride with the DeltaWolf all painted up and finely tuned was a great experience. "I will be back in 15 minutes," I said, but three hours later I finally found my way home after blasting around town, passing uprights like they were standing still, and showing off my new ride to all those who flagged me down. What a perfect performer! I was seriously pleased with every aspect of the DeltaWolf's handling from acceleration to braking, and by the time I pried myself away from the pilot's seat, my legs were so worn out, I could barely make it up the stairs. Oh, don't get the wrong idea—the DeltaWolf is much easier to get up to speed on, compared with an upright bicycle, especially against the wind—it was my extreme desire to fly past every single bicycle I came up behind that wore me out. This thing just wants to race! Figure 6-158 shows a bird's eye view of my DeltaWolf engine—aka my legs.

When you just want to lay back and take in some great scenery without worrying about balance, or without having a 4-inch wide seat ruin your anatomy, the DeltaWolf fills this desire perfectly. The laidback position makes even my office chair feel like a medieval torture device, and I won't even attempt to compare the DeltaWolf seat with the killer upright bicycle seat! This

really is the most comfortable human-powered machine I have ever made, and I could easily spend all day on it, as long as my legs were up to the task. Having the ability to take in the view while looking straight ahead in a natural position is another serious benefit, when compared to how you are forced to lean over the handlebars and look up on many upright bicycles. Figure 6-159 shows a lean, mean trike cruising along with the Sleeping Giant in the background. The Sleeping Giant is a formation of mesas on Sibley Peninsula, which resembles a giant lying on its back when viewed from the north to west section of Thunder Bay, Ontario, Canada.

In Figure 6-160, I am hitting a high speed turn and enjoying the G-forces. Because of the extremely low center of gravity, the DeltaWolf takes corners at speeds that most delta-style trikes would have no chance at, no matter how far the pilot leaned to avoid a rollover. The full-sized, 26-inch rear wheels allow for some really great top speed, and the special hub/axles easily handle

Figure 6-160 *DeltaWolf hits the corners hard and fast*

Figure 6-161 *Yes, it rides as good as it looks*

Figure 6-162 *DeltaWolf brings new life to the trike world*

the extra demands that a non-leaning vehicle put on them. What a fun ride!

How far can you ride before you are chased down by a fellow cycle enthusiast to answer the same three questions: Where did you buy that awesome trike? How much did it cost? Can I try riding it? Hey, you can't blame them: look at this sweet machine! Figure 6-161 shows the DeltaWolf at rest, waiting for another ride.

So, that wraps things up! The DeltaWolf is ready to tackle any distance you feel like riding, and it will do it in pure style and at speeds that would make any upright cyclist nervous. It took a bit of work to get this far, but to be honest, the most difficult part of this project for me was not building it; it was sitting at my computer writing this text! Hey, I want to power this computer off and hit that open road on the World's Coolest Delta Trike, so I

will now bid you farewell! Figure 6-162 will give you one last look at the DeltaWolf—can you see yourself in the pilot's seat? Sure, you can!

Figure 7-1 *Two identical fronts, and one rear*

Since this section is dedicated to recumbent trikes, I thought it would only be fair to build a kids' trike so the younger bikers can also hit the streets in style and comfort. The Little Warrior is a nice simple trike that does not require any machined parts or wheel building, just a pair of kids' bikes and a few extra bits and pieces from your scrap pile. The Little Warrior has two front wheels, but uses the standard front forks and head tubes, so you do not have to worry about axle strength or custom brake parts. You can build this little trike using any size wheels you like, although I am not sure it would be well suited for an adult, so wheels of 20 inches in diameter or less would be most appropriate.

Let's start with what you will need in order to build the Little Warrior: two identically sized front wheels, forks, head tubes, and a similar sized frame for the rear that includes a crank set and rear wheel. You do not have to match the front wheels and rear wheel, but it does make for a better looking final product. Figure 7-1 shows the three identical 14-inch kids' bike frames I found at the city dump scrap pile. These department store starter bikes are so common at the dump and garage sales that I probably have at least 10 more like them in my huge scrap pile. Again, only the fronts of the bikes need to match, as the trike is basically a bicycle with "two heads," as you will soon see.

Remove the front fork hardware from the two matched frames so you can ensure that all parts are in good

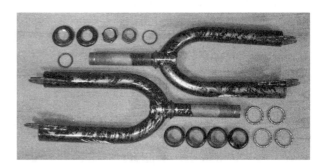

Figure 7-2 *Checking the front fork hardware*

working order, and that the forks are very close to the same size and shape. If they are identical, then great, but you could probably improvise if one of the fork stems is a bit longer than the other. The goal is to have the front of the trike balanced and symmetrical, so hunt around for the best matching set of forks and front wheels that you can scrounge. Figure 7-2 shows my two front forks, which just happen to be identical minus color, but you can't see that in black and white anyhow!

You will also need a crank set, bottom bracket and all the included hardware, as shown in Figure 7-3. Clean all grease and rust from the bearings to ensure that there is no damage, and replace any excessively worn parts.

Since the Little Warrior has two front ends, you will need to cut the head tube and down tube from both matching frames, as shown in Figure 7-4. The top tube is

Figure 7-3 *Cleaning up the crank set hardware*

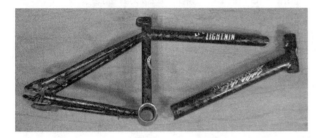

Figure 7-4 *The first cut made to both matching frames*

cut right at the head tube joint, and the down tube is cut right at the bottom bracket joint so you have the most material to work with. When making the first cut, be careful if using a grinder, as the frame may tend to collapse on the disc and pull the grinder from your hands. If you want to play it safe, cut the last little bit using a hacksaw to release any tension in the frame. Make the same cuts to both frames.

Figure 7-5 shows the result of butchering both frames: a pair of matched head tubes with an included length of down tubing attached. Again, both parts should be almost

Figure 7-5 *Two identical head tubes with included down tubes*

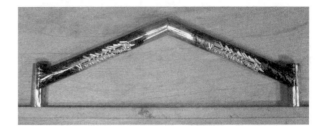

Figure 7-6 *Joining the head tubes*

identical so the angles on each side of the trike are the same when completed. You could cut one head tube to adjust it, but that is probably more work than simply sourcing the two identical donor frames. Grind away any excess material at the joint of the original top tube as well, filling the small holes in the head tube if you feel the need.

The two head tubes will be joined together so that they are perfectly parallel to each other, as shown in Figure 7-6. A length of angle iron is being used as a guide to ensure that the head tubes are perfectly aligned, and that the joint at the ends of the down tubes fit together with minimal gap. Don't worry about the width (track) of this trike; just take as little material away from each end of the down tubes as necessary, and the trike will be plenty wide enough for a safe ride.

Using a piece of angle iron or a flat surface as a guide, position both head tubes so they are parallel and tack weld your down tube joints, as shown in the top half of Figure 7-7. With the parts tack welded, do a visual inspection of the head tube alignment from all angles and then complete the entire weld, as shown in the lower half of Figure 7-7. You can also tack weld the head tubes to your guide metal to ensure that they remain aligned as you complete the welding.

Figure 7-7 *Welding the down tube joint*

Figure 7-8 *Cutting the rear of the frame*

Figure 7-10 *It's like a two-headed kids' bike*

Figure 7-8 shows how the rear of the frame must be cut, removing the head tube at the end of the down tube and the entire top tube. You can also clean up any excess top tube material from the seat tube joint as well at this point.

Since the end of the down tube on your rear frame must mate with the joint between the front parts of the trike, you will need to cut the appropriate fish mouth at the end of this tube, as shown in Figure 7-9. This process is a simple trial-and-error procedure, so take your time and rough it out with a grinder disc, finishing up with a round file. Take only as much material away as needed, or you will shorten the wheelbase of the trike. The next photo shows how the front and rear of the frame come together.

As you can see in Figure 7-10, the Little Warrior is nothing more than a standard kids' bike with two heads instead of one. All angles and clearances should remain the same, so you don't have to worry about head tube angles and bottom bracket clearances. Inflate all tires, and prop up the parts, as shown in Figure 7-10, so you can check the fish-mouth cut made between the rear of the frame and the front. It is also a good idea to make the

tack welds as the trike is sitting like this, to ensure that the angles are correct.

A few tack welds made as the parts are propped up should allow you to safely manipulate the parts if slight adjustments are needed. Check the trike from all angles to make sure the parts are in the correct place, making sure the angle between the front and rear tubing is at 90 degrees. What about head tube angles? Well, if you look directly at the side of the trike, it should look like a regular bicycle, with the head tube angle the way it was on the original bicycle. Of course, don't get too fussy about this angle; just make your best guess, or look at any other bicycle as a guide.

Figure 7-9 *Making the down tube fish-mouth cut*

Figure 7-11 *Tack welding the front and rear of the trike frame*

Figure 7-12 *Making use of the leftover frame tubing*

Once you are happy with the alignment between the front and rear of the frame, weld the entire joint right around the fish-mouth cut. At this stage, the frame is not ready for any weight, so do not let the eager young pilot jump on the frame, as it may bend the down tube where it joins with the bottom bracket. The frame needs a few more small tubes in order to form a triangle, the strongest of all structural shapes. By hacking up the leftover frame tubing, you will be able to make the two tubes needed to form a triangle between the front and rear of the frame. Figure 7-12 shows the seat stay tubing cut from the rear of the two frames that were used to make the front of the trike. Sure, you could use any scrap tubing with a diameter of ³/₄ inch, but recycling is the way to go.

There should be just enough tubing in those seat stays if you join them end to end to make two longer tubes, as shown in Figure 7-13. The top half of the photo shows how the tubes were welded together, and the bottom half shows the completed tubing after cleaning up the joints using a flap (sanding) disc.

The two tubes made from the recycled seat stays will create a triangle in your trike frame, as shown in Figure 7-14, giving it tremendous strength. I will refer to these tubes as "side rails," since the seat will be placed between them. Cut the side rails to whatever length you

Figure 7-14 *Adding the side rails*

need to make them run from the top of each head tube to the joint at the seat tube at the rear of the trike. Once welded, your frame can now take its full weight safely.

Although an adjustable bottom bracket is optional, it does make fitting your rider to the trike a heck of a lot easier, and since kids grow like weeds, you won't have to cut and reweld the frame every time they grow an extra inch. A simple adjustable bottom bracket can be made using a section of seat tube, a seat post or similar diameter tube, and the bottom bracket from one of the frame leftovers. Figure 7-15 shows the parts you will need in order to create a sliding boom adjustable bottom bracket.

The idea is very simple. The bottom bracket will be welded to the tube that fits into the seat tube and the original seat post clamp will lock it in place. You will need about 12 inches of seat tube, an 8-inch length of seat post or some diameter tubing, and the bottom bracket you plan to use for your trike. A fish-mouth cut needs to be made in the end of the tube that will slide into the seat tube, and then it is welded to the bottom bracket.

Figure 7-17 shows the completed adjustable bottom bracket after welding the seat post tubing to the bottom

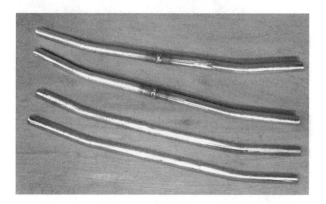

Figure 7-13 *Joining the seat stays to make two longer tubes*

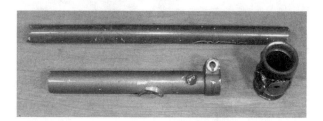

Figure 7-15 *Adjustable bottom bracket parts*

Figure 7-16   *Making an adjustable bottom bracket*

Figure 7-17   *The completed adjustable bottom bracket*

bracket. Now the bottom bracket can be adjusted either way to about 6 inches, to accommodate different rides, or small growth spurts. This assembly will now be called the bottom bracket boom tube.

The bottom bracket boom tube will be welded to the front of the trike frame, as shown in Figure 7-18, directly ahead of the rear frame joint and at an angle roughly parallel to the ground. Since chain routing is also a consideration, it might be best to read ahead and understand how the boom angle will affect chain routing.

The chain needs to pass under the front frame tubing on its way to the rear wheel, so some type of guide pulley

Figure 7-18   *Installing the bottom bracket boom*

Figure 7-19   *A chain idler pulley*

will be needed to do this job. Any half V-belt idler pulley with a center bearing will work for this purpose, as well as a grooved-out skate blade wheel, or any other idler you might find lying around your shop. A center bearing is a good idea, as a simple brass bushing may introduce extra friction into the transmission, robbing your young pilot of valuable top speed. Figure 7-19 shows my 2-inch diameter V-belt idler pulley and a bolt that will go through the hole in the center ball bearing.

Figure 7-20 shows the chain idler doing its job, as well as the general horizontal position of the bottom bracket boom tube. As long as your chain does not rub on any part of the frame, your idler pulley is installed properly, although just behind the front frame tubing seems to be a perfect place to weld the bolt. You will also need to adjust the length of the chain to suit the best position of the adjustable bottom bracket, using the rear dropouts to fine tune any slack. You will be able to adjust the rear wheel to account for about 1 inch of chain slack before needing to add or remove a chain link.

Figure 7-20   *Chain idler pulley installed*

Figure 7-21 *Steering control arms and bolts*

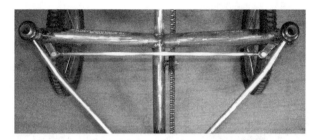

Figure 7-23 *Adding the steering rod*

The two front forks must turn together, taking into account Ackerman steering angles, so a pair of steering control arms is made from some 1-inch pieces of flat bar. The steering control arms are 2 inches long with a rounded end that will conform to the shape of the front forks, where they are to be welded. You will also need the appropriate sized bolts to pass though the ball joints and the control arms. These parts are shown in Figure 7-21.

Since this trike has such a short wheelbase, and will never see any high speeds, Ackerman steering angles are not as much of a concern as they were on the StreetFox described earlier in this section. Simply weld the steering control arms on a 45-degree angle to the front forks, as shown in Figure 7-22, and your steering will work just fine. The angle of the steering control arms should make them point towards the rear axle center, but 45 degrees will be close enough for this trike.

Figure 7-23 shows the steering control arms installed, as well as the steering rod and ball joints, which make both front wheels turn at the same time. The ball joints are small hardware store items which have been welded

directly to the half-inch rod that makes up the steering rod. You can add a bolt to make an adjustable steering rod like the one used on the StreetFox, but it may not be necessary. Ensure that both front wheels are tracking perfectly straight, and then measure the hole-to-hole distance between the steering control arms to find the optimal alignment position. Once the steering rod is at the correct length, nothing will ever go out of alignment.

The pilot's seat is a nice wide seat from an exercise bicycle, held in place by a small bit of seat post tubing that has been welded to the frame. Your seat should not interfere with the steering rod, and should be placed so that the rider's heels do not strike the frame tubing. Some adjustment of the bottom bracket may be needed to find the best seating position, which is why it was a good idea to make it adjustable. Figure 7-24 shows my seat installed in what seems to be the optimal position for balance and comfort.

Depending on the angle of your seat, you may want to create the optional backrest, as shown in Figure 7-25. Since the rear frame still has the seat tube and clamp installed, you can simply add a bit of wood and padding to a seat post and install it right into the frame. Figure 7-25 also shows how the seat post was cut and

Figure 7-22 *Steering control arm welded to the fork*

Figure 7-24 *Adding the pilot's seat*

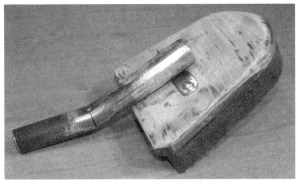

Figure 7-25 *Making a backrest*

Figure 7-27 *Trimming both goosenecks*

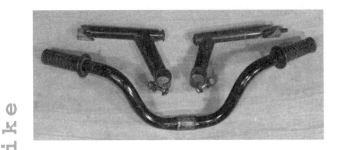

Figure 7-26 *Making the two handlebars*

Figure 7-28 *Making the handlebar halves*

rewelded, in order to put the backrest at a more reclined angle. Now the young pilot will not fall off the rear of the trike if he accidentally lets go of the handlebars during some vigorous riding.

The handlebars for this trike are made to install directly into the front forks, adding a handle to each side of the trike. Since the two front forks are forced to turn together by the steering rod, the two handlebars will also do the same. Figure 7-26 shows the bits that I will be using to make up the two handlebar halves; two equal length goosenecks and a mountain bike handlebar. The two goosenecks should have equal length stems, and fit properly in your front forks. There are two different diameter gooseneck stems, so make sure you choose the right ones.

The two goosenecks can be trimmed, as shown in Figure 7-27, removing the clamps to leave only the stems. You can also grind away any leftover material on the stems.

The position and angle of the handlebar halves should allow for easy reaching by your pilot, so tack weld them in place and see if they seem comfortable and functional before completing the welds. Figure 7-28 shows the position and angle of the handlebars which allowed an easy reach and full range of steering with no interference.

The goosenecks will be adjustable left to right, so you have the ability to set them up for best comfort. A slight upwards angle of each handlebar seemed to make the most sense ergonomically.

When you have the handlebar halves set up the way your pilot likes, complete the welds, as shown in Figure 7-29, making sure they are solid enough to take

Figure 7-29 *Completed handlebar halves*

Figure 7-30   *Accessorize and paint*

Figure 7-32   *Rear trike details*

the abuse they are most likely to soon see. Both parts should be the same size and angle.

Figure 7-30 shows the completed Little Warrior trike, complete with luggage rack (buddy carrier) and chain guard. The cool little trike required no machined or custom parts, yet looks as good as any adult-sized trike. The frame is strong enough to carry as many kids as can pile onto it, and it may even be able to handle the abuse of an adult rider. OK, fine, I rode it too!

It is a good idea to add the front chain guard to keep kids' new (and expensive) designer jeans from getting snagged in the chain ring. The chain guard shown in Figure 7-31 is an unmodified part from one of the donor bikes, requiring only a small tab to be welded to the boom in order to secure it in place.

Figure 7-31   *A front chain guard is a must*

Figure 7-33   *Riding the Little Warrior*

Figure 7-32 shows the rear of the completed Little Warrior trike, complete with buddy carrier and adjustable back rest. Since brakes are included in the rear coaster hub, there are absolutely no cables to worry about, and the trike has plenty of stopping power so your crazy young test pilots can fishtail around the corners.

Tanner puts the Little Warrior through its paces, handling the trike with the accuracy of a highly trained jet pilot! The completed trike (Figure 7-33) is so much "cooler" than the run of the mill kids' bike that every kid on the block will want one. Do you have enough time and spare parts to build a dozen more?

Hang on to your lug nuts! We have some wild and crazy rides coming up in the next section.

Wild and Crazy Wheels

The SpinCycle offers a non-stop, tricked outride that can only be described as a cross between go-cart racing and astronaut training! Seriously, this crazy machine has to be experienced in person to be fully appreciated. Before you dig right into the bike pile with your hacksaw in hand, let me tell you a little bit about the history of the SpinCycle and some theory on its basic handling characteristics.

The original SpinCycle was contrived in 2002 by my friend Troy Way and I in a small unheated garage during the coldest winter months using only an AC welder, angle grinder, and whatever scrap bicycle parts happened to be lying in the huge scrap metal pile. I was in the middle of writing the book, *Atomic Zombie's Bicycle Builder's Bonanza*, and Troy had the wonderful idea for some type of bike or trike that would send the pilot out of control as if riding on a sheet of ice. Since I had a chapter in the book entitled "Unclassified Rolling Objects," which included a bike that could fold in the middle, allowing it to turn on a dime, I thought that this idea would make an interesting bike for the book. Within a few hours, we threw together the first prototype, which was made from an old kids' BMX bike, a broken office chair, shopping cart wheels, and some recycled bits from an old robot frame.

We took the crazy contraption out to the gas station parking lot next door, and Troy sat in the nice comfy office chair pilot's seat and began to pedal the front wheel drive trike in a straight line. With one quick flip of

the wrist, he was spinning out of control as if on a patch of invisible ice. By the time he managed to get the trike back to the place he started, he was sick from the intense g-force and spinning action, and I was sore from all the laughing. It was an instant success! I took the trike out for a "spin," and loved every second of the intense, out-of-control ride. I went faster and faster, trying to flip the trike, bend the frame, or warp the front wheel, but to no avail, as the unit held together, which was surprising considering how fast it was put together. I did manage to flip the trike onto its side after some time, but only because I nailed the edge of the curb, causing the unit to stop dead and propel me into the neighbor's snow bank.

After a few hours of great fun on the new trike, we came up with the name SpinCycle, and I came back to the garage to build my version for the book. The second SpinCycle was made using the same basic principles: a 20-inch BMX bike was cut up to make a front wheel drive unit, which was then fastened to a basic T-shaped frame with shopping cart caster wheels at the rear. I made my trike as low as possible using light gauge square tubing, and built the seat from plywood to keep weight down to a minimum.

Once completed, we took both SpinCycles out for a test run and found that having two of them was even more fun than taking turns, especially since you could have crazy games of "chicken" and races that involved pushing your opponent's seat to throw him wildly out of control around corners. By the time the day was done, we had a synchronized SpinCycle routine that would leave onlookers in shock because of the amazing precision which we were able to achieve—missing each other while spinning in opposite directions by mere inches. I guarantee that you will not be disappointed by the incredible ride offered by this easy-to-build project, so get that hacksaw out and start digging in the scrap pile for the raw materials!

The first thing you will need is an old 20-inch kids' bike (Figure 8-1) with the front end in good working condition. We will be using the front forks, head tube, rear wheel, pedals, chain, and handlebars; the rest will not be needed. The front forks do not need to match the

Figure 8-1 *A 20-inch kids' bike with matching front fork*

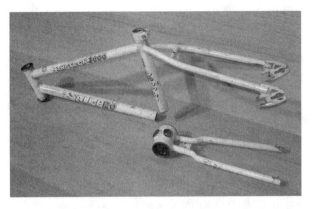

Figure 8-2 *Cut the chain stays and bottom bracket from the frame*

rest of the bicycle, as long as the fork stem is the correct length for the head tube on the original frame, since it will be used on the final design. If the forks you plan to use are from the original frame, then this is not a concern, but if not, it's a good idea to ensure that the fork stem is compatible with the head tube by placing the bearing cups, bearings and all fork hardware in place, just to double check. There is a 1- or 2-inch margin of fork thread either way, but it is always best to check this first. If the fork stem is too short, you will not be able to install all of the fork hardware, and your forks will be constantly coming loose. The problem is even worse if the fork stem is too long for the head tube, as you will not be able to screw down any of the fork hardware into the bearing cups at all.

Forks that are bent at the stem from extensive rider use should not be used, as this will compromise the SpinCycle's handling, making it impossible to drive in a straight line, or will induce unnecessary friction in the steering. Fork style is not really important, but if you have a choice, use the beefiest front forks in your stock pile, as the SpinCycle will tend to take a lot of abuse from the "big kids."

Now the fun part begins: hacking up the bike with your grinder or hacksaw. First, you will need to cut the bottom bracket and chain stays away from the main frame, as shown in Figure 8-2. Cut as close to the rear dropouts as you can, leaving as much length in the chain stay tubes as possible, and it may even be a good idea to leave a bit of the rear dropout metal as well (this can always be ground away later). The other two cuts will be made in the seat tube and down tube, and should be made as close to the bottom bracket as possible to avoid excessive grinding later. If you are using a cut-off disc in

your grinder to make the cuts, then beware when cutting either the seat tube or down tube, as the frame will tend to collapse on your disc as you get through the last bit of metal, which could cause your grinder to stall, or run out of your hand. If you have no way to secure the frame while cutting, then leave a sliver of metal during each cut rather than going all the way through the tubing, and you can simply bend the bottom bracket back and forth to release it after all four cuts are made. If using a hacksaw, don't worry about it—just hack!

The only other bit of the original bicycle frame we need is the head tube. This is cut in the same manner as the bottom bracket, trying to get the cleanest cut possible in order to reduce the time needed to grind away all the excess metal, as shown in Figure 8-3. There will be no risk of a grinder disc catching as you cut through the tubing, since the frame is no longer a closed shape. When you are done cutting the head tube and bottom bracket

Figure 8-3 *Removing the head tube from the rest of the frame*

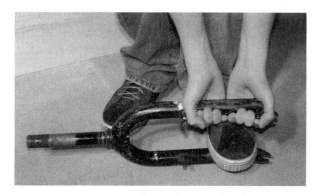

Figure 8-4 *Opening the front forks to take the rear wheel*

chain stays from the frame, toss the leftovers in your junk pile—they will probably be useful for some other bike hacking project down the road. When you take on this hobby, you can never have enough junk in the scrap pile—take it from me. My garage is filled with buckets and sorted piles of parts.

Are you feeling like a hero today? If so, then the next job is going to test your strength, especially if the front forks you plan to use are the beefy, thick-leg style like I have in Figure 8-4. After building several SpinCycles, I can assure you that not all forks are made with the same quality or thickness of tubing, and a simple visual inspection is no indicator of how easy it will be to spread the fork legs apart that magical extra inch. This fork spreading is necessary since the width of the front wheel hub is an inch less than the width of the rear hub, as you will see when you try to install it in the fork dropouts. There is no magical way to do this procedure, and I have seen all types of methods used to spread the forks, some of which include; prying them apart with a lever made of 2 × 4s, cranking them apart with a large bolt and nut, forcing them apart by sliding two large tubes over each fork leg to gain mechanical advantage, and, of course, my favorite method: grunting like a mad gorilla as you overexert yourself! Whatever method you do use, make sure that you don't pry the fork legs apart too much on the first try, or you will end up banging them back together, which may weaken the metal. I like to pull up on one of the legs as I stand on the other, trying to widen them up just a little bit at a time, until I can force the rear wheel axel into the dropouts. Each time I need a little more stretch, I pull up on the other leg—this seems to open both legs at an equal distance. Good luck!

So, how did the fork widening operation go? Can you still stand up straight? You will feel it tomorrow,

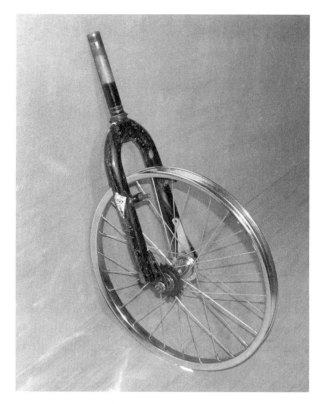

Figure 8-5 *Rear wheel installed in the front forks*

trust me! Anyhow, with the only real manual labor out of the way, we are free to get the rest of the SpinCycle together; it's all smooth sailing from here. As shown in Figure 8-5, the rear wheel should be easy to install in the front fork once the legs have been pulled apart a little bit. Ensure that the wheel can be centered, and that the drive sprocket does not rub on the fork leg or the dropout. If the fork dropouts are installed closer to the outer edges of the fork legs, you may need to install a few washers between the axle bolts to stop the drive sprocket from rubbing, but this is not going to pose a problem later.

The heart of the SpinCycle is the front-wheel drive system made from the original frame's bottom bracket, the front forks, and a small length of slightly bent tubing or electrical conduit. I like to lay out the various parts as I build, so I can get a visual picture of how everything will fit together before I begin the welding process. To create the front end, start by installing the front wheel and inflated tire (actually a rear wheel) into the forks. The bottom bracket that you cut from the original frame is going to be attached to the front forks right at the dropouts, so that the fork legs and chain stays form an approximate 95-degree angle. This is shown in

Figure 8-6 *Laying out the front end parts*

Figure 8-7 *Marking the steering boom to be cut*

Figure 8-6. Don't be too critical of the 95-degree angle, as it's just a guide, so you can set it up by eye with no protractor necessary! I chose the 95-degree angle because this will place the cranks at a comfortable level once the front end is installed in the main frame. If the cranks are too low, the pedals may hit the ground in turns, and if they are placed too high, your only view when riding will be your new running shoes.

Once you have the front wheel installed and the bottom bracket placed roughly in the correct position, you can now figure out the correct length and rough curvature for the steering boom (the bent tube that joins the bottom bracket to the fork crown). This tube is just a bit like bicycle type tubing or thin walled electrical conduit (EMT) bent in the center to allow clearance for the front wheel. Again, there is no critical measurement or system for bending this tube. I just take a longer than necessary bit of tubing, and then bend it over a car rim or some other round object a little bit at a time until it looks like it will fit. If you want to get fancy, you could make a nice curve around the front wheel but, in reality, any bend will do, even a cut and reweld if your tubing is too rigid to bend easily.

Once you have bent the steering boom so that there is no front tire rubbing, you can mark the area to be cut by drawing a line as the tubing is placed over the bottom bracket, as shown in Figure 8-7. Remember, the chain stays will be approximately at a 95-degree angle to the fork legs, and the steering boom should not rub on the front tire. If the boom looks like it might rub on the front tire, just bend it a little more until you reach a happy medium. To mark the steering boom for cutting, just draw a line at the approximate correct angle by visually "guesstimating" it, a method employed by all good

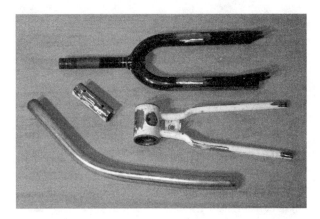

Figure 8-8 *Parts needed for the SpinCycle drive system*

garage hackers. Don't worry if your cut leaves a little gap after you place the parts back together; after all, that's what welding rod is for—filling gaps. You now have all the parts to complete the SpinCycle's drive system.

Figure 8-8 shows all the parts needed to create the SpinCycle's front end. I ground off any excess material from the bottom bracket, head tube and front forks. It is also a good idea to ensure that both legs of the chain stays are of equal length before moving to the next step, or your bottom bracket could end up off center. This is done by standing the bottom bracket up on the chain stays to see if it is level with the ground. Grind the end of whichever leg is longer, if necessary, until you are satisfied that they are of equal or close to equal length. Even if you cut both tubes at the exact same stop from the original frame, this does not guarantee equal length. After all, these are "department store" bikes we are dealing with, and exact quality is not something normally found in this variety of bicycle.

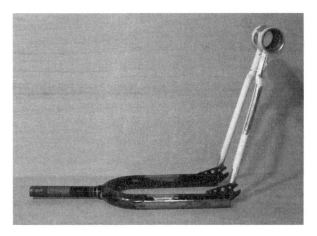

Figure 8-9 *A 95-degree angle is formed between the two parts*

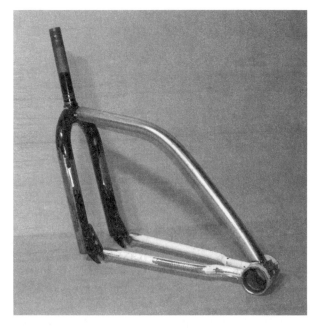

Figure 8-10 *The front end tack welded together*

As shown in Figure 8-9, the ends of the chain stays are tack welded to the front of the fork dropouts, so that the angle between the stays and fork legs is approximately 95 degrees. Only tack weld the lower front ends of the stays to the dropouts, so that you can bend the two pieces slightly in either direction to allow a good fit when you place the steering boom tube in place. With two decent tack welds holding the pieces together, you will be able to bend them several inches in either direction, which might come in handy if your angle was slightly off when you did the welding. At this point, you will want to examine the tack welded parts to make sure that everything looks aligned properly. It is also a good idea to carefully slip the front wheel in place as well, just to ensure there is no misalignment from either chain stay tube being too long or too short. If all goes well, the wheel should be aligned between both the fork legs and the chain stays. If this is not the case, just bend the parts back and forth a few times to break the tack weld and try again. If either chain stay needs a little grinding, then only remove what is necessary, as you do not want to make the tubes so short that the wheel will no longer clear the bottom bracket. When you are happy with the alignment, move on to the next step.

The steering boom tube will now be tack welded in place, as shown in Figure 8-10. Simply place the boom into position, then bend the bottom bracket in either direction to form a tight fit between all of the parts (this is the reason for the light tack weld earlier). The ends of the steer boom will need a little bit of grinding in order to conform to the curvature of the bottom bracket and fork crown for easier welding. Just take a little bit out of the

end of the tube with your grinder, and then do a test fit to see if the joint is properly fit for welding. Alignment of the steer boom for tack welding in place is no contest, as it is already made to fit perfectly in place, and as long as your bottom bracket and fork legs are aligned, everything will fit nicely together.

Before you commit to welding every joint, place the front wheel with the tire inflated into the fork dropouts to check for alignment and clearance just in case (see Figure 8-11). Most likely, everything will fit

Figure 8-11 *Test fitting the front wheel in place to check clearance*

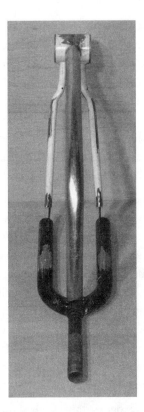

Figure 8-12 *Alignment checks out from every possible angle*

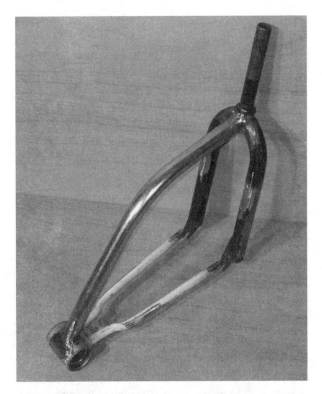

Figure 8-13 *The fully welded and completed front end*

perfectly, but if this is not the case, it is not a big deal to break a weld and make corrections at this point. Even with this fat 20-inch tire, I have a good half-inch clearance between the fork legs and the steering boom. A little clearance is a good thing, as it will allow for a slight adjustment of the wheel in the dropouts to ensure that the chain is tight later on.

Checking alignment from every possible angle is a good idea, since it's easy to break a tack weld if necessary to make corrections (see Figure 8-12). Once you weld the front end together, you will have no chance to repair any alignment errors later.

When you are finished test fitting the front wheel, and certain that the front end parts are aligned properly, go ahead and complete all of the welding, beginning with the joints between the chain stay ends and fork dropouts. By starting with those two welds, you will reduce the chance of any side-to-side distortion due to heat while you are welding the steering boom tube in place. In Figure 8-13, all of the welds are completed and ground for appearance. Before welding the joint at the steer boom and fork crown, you should remove the forks bearing ring by tapping it off with a chisel or welding hammer. The ring is only held on by friction, and should

come off easily. Removing the ring allows the weld to be completed without damage to the ring itself, which must be perfectly smooth and free of weld spatter in order to function. Be careful not to weld the area that seats the ring as well, or you will find yourself grinding the area in order to bang the ring back in place later.

With the front end completed, it's now time to begin building the SpinCycle's main frame. We will be starting from the back of the bike by creating the T-bar that will hold both caster wheels in place, preventing the trike from rolling over. In my original design, I bolted the caster wheels directly to the T-bar by drilling a hole in the underside of the square tubing and simply bolting each caster wheel in place. This method does indeed work, but only if the square tubing used to make the frame has at least a 1/8 wall thickness. After building a few SpinCycles, I found 1/8 wall tubing to be much too heavy, and went to 1/16 tubing or as close to .0625-inch wall thickness as possible to minimize weight and allow for better acceleration. The 1/16 wall tubing made a frame equally as rigid as the thicker tubing, but would not allow the caster bolts to be placed directly into the tubing ends.

The 1/16 wall tube would tend to bend if the caster wheels struck a curb during a spinout, and although this

Figure 8-14 *Two pieces of angle iron will secure the rear caster wheels*

practice should be avoided, I didn't want to release plans for a bike that could be so easily damaged. Another reason for not placing the bolts inside the T-bar was because this left the sharp open end of the tubing exposed, which could become a hazard under certain conditions. The new method of mounting the rear casters involves cutting two small pieces of 1/8-inch or 3/16-inch angle iron (Figure 8-14) to 1.5 inches in length, so that a solid wheel mount is formed which will also cover the exposed end of the tubing.

If you do not have the 3-inch length of angle iron needed to form the two caster mounts, just visit any machine shop or welding house, and they will gladly find you a bit of scrap from the pile; this stuff is extremely common. The thickness of the angle iron is not critical, but it should not be less than 1/8 of an inch, or strength may be an issue. As you can see in Figure 8-15, the two

Figure 8-15 *Rear caster mounts are cut, drilled, and rounded for safety*

pieces are cut at 1.5 inches and drilled so that they form a sturdy place to mount the caster wheel bolts, as well as plugs to cap the open end of the rear T-bar. The side of the angle iron with the hole drilled should also be rounded smooth, so that your SpinCycle is not a dangerous ankle-slicing death trap. Trust me; you don't want to be playing chicken with your buddy on his SpinCycle if the ends of the frame tubes are sharp. Remember the movie "Gladiator"? Yikes!

The key to the SpinCycle's wild and crazy maneuverability is the fact that all three wheels steer! When you turn the front wheel, the bike will move in that direction as expected, but if you lean into the turn, or exaggerate it, the rear casters will also begin to steer due to the momentum of the rear end. Remember what that Newton dude taught us: "every object in a state of uniform motion tends to remain in that state of motion unless an external force is applied to it." In other words, once you get that rear end swinging, you better just hang on and enjoy the ride!

I'm sure you recognize the wheels shown in Figure 8-16, and know exactly where to get them: from discarded shopping carts, of course. It's amazing how many shopping carts with perfectly new caster wheels end up at the city landfill or scrapyards, especially right after the snow clears. It's the front wheels you want from the shopping cart, and they can be removed simply by taking out the bolt that holds them in place. Try to find an identical pair of wheels so that the trike sits level.

The rear T-bar is made from a 2-foot long piece of 1.5-inch, 1/16 walled square steel tubing. You could use heavier gauge tube, but it is not necessary, and will only add weight to the trike. If you have no luck in finding the correct square tubing, you could always use the same

Figure 8-16 *Attention shoppers!*

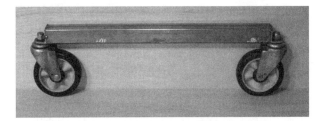

Figure 8-17 *Mounting the caster wheels to the rear T-bar*

length of 1.5-inch thin walled electrical conduit (EMT), as it has approximately the same wall thickness and strength as the square tubing. As shown in Figure 8-17, the two pieces of angle iron are tack welded to the ends of the T-bar, so that the caster wheels are both aligned at the same exact angle. If one of the caster wheels is misaligned, you may find your SpinCycle trying to turn in one direction like a badly aligned 1973 station wagon, or one of the casters may develop a seriously annoying flutter when you reach a decent speed. A nice trick I use to get the two pieces of angle iron aligned is to place the works on a level hard surface as I tack weld; this ensures perfect alignment every time.

It's always a good idea to install the part and check alignment before cranking up that welder (Figure 8-18). Bolt down both caster wheels and take a look at the T-bar from every direction until you are satisfied that both caster wheels are aligned at the same angle. If you leveled the T-bar and angle iron bits on a hard surface while tack welding, then you should have no problems at

Figure 8-18 *Check caster alignment before finalizing the welds*

Figure 8-19 *The completed T-bar fully welded and ground smooth*

all. You may also notice that there is a slight keyway thing on the caster wheel bolt, just below the threaded part. Do not bother trying to keyway the hole you drilled in the angle iron, and certainly do not make the hole bigger to accommodate this. Simply grind or file it off, as it has no purpose except to stop the wheel from turning as you install the nuts. All you need to do is hold onto the wheel as you tighten the nuts.

Once the rear casters are aligned, finish the welding of the angle iron wheel mounts to the rear T-bar and then grind away any sharp edges, as shown in Figure 8-19. Remember, when you pedal like mad and start pulling multiple 360s, the rear T-bar is going to be flying around like a lawnmower blade, so you certainly do not want any sharp edges. Really, this thing isn't anything like a plastic kids' toy, and it can move as fast as the bike you took apart to build it. I routinely pull five or more 360s in a row at a speed of at least 15 miles per hour!

The main frame is made from the same 1.5-inch, 1/16 walled square tubing as the rear T-bar. Again, if you have no luck finding decent square tubing, you could use the same length of 1.5-inch thin walled electrical conduit (EMT), as it has approximately the same wall thickness and strength but will be a little more difficult to cut at the proper angles. Have a look at Figure 8-20 to get a general idea of what you will need to cut in order to create the main frame. I say "general idea" because these angles and lengths are not critical at all, and you may want to experiment or alter the frame to suit your own needs. You could certainly make your SpinCycle using the same measurements and angles as mine, but before you start

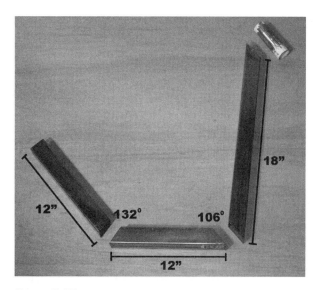

Figure 8-20 *Laying out the parts that make up the main frame*

Short riders might have to stretch a little, and tall riders may feel a bit cramped, but hey, it's not a touring bike, and either way the pilot is going to be having one heck of a good time. Again, it's a good idea to scan ahead a bit here before you start cutting, just so you understand why I am not dictating exact measurements for this part of the project. In Figure 8-21, I cut the top of the front tube to the correct angle for my front end and ground the joint to mate perfectly with the head tube. I will end up with a 1-inch clearance between my front tire and the front tube.

As you can see in Figure 8-22, the exact length of the main tube is dependent on the geometry of the front end, and your setup may be a little different due to the angle

cutting, read ahead and I will show you how and why I chose those angles and dimensions. Essentially, I will be working backwards through the process of building the main frame, starting with the mounting of the head tube.

Since the SpinCycle has little in common with the standard diamond frame bicycle, I am going to rename the three tubes that make up the frame, so you know which part I am referring to. Starting from the left in Figure 8-20, we have the rear tube (this is your backrest), the base tube (the bottom of the frame), and finally the front tube (joins the head tube to the base tube). From here on, I will be referring to the frame bits by those names. Begin by cutting a length of tube for the front tube. Cut this tube to a length of at least 20 inches, as this way it will not end up too short after you cut the ends to the proper angles and grind out the top for the head tube. This tube is like the "kingpin" for the entire frame, and will dictate how the rest of the tubes are cut and at what angles.

The reason the front tube is not cut to an exact length at this time is because it is really dependent on the front wheel, fork angle, and dropout position on the fork legs. The goal is to have the front tube pass by the back of the wheel with minimal clearance of an inch or less. By keeping the front tube as close to the front wheel as possible, it will allow riders with short legs to pilot the SpinCycle. Typically, the SpinCycle made with a 20-inch front wheel will fit riders ranging in height from 5'-5" to 6'-6" with an optimal height of 5'-9".

Figure 8-21 *Cutting the front tube to mate with the head tube*

Figure 8-22 *Front tube length is based on ground clearance and fork angle*

of the fork stem and placement of the fork dropouts. The goal is to allow the front tube to pass down past the front tire, leaving no more than an inch clearance. You should have your front tire fully inflated while you lay the parts down for marking the front tube to be cut. When the top of your front tube is cut to fit the head tube, as shown earlier in Figure 8-21, it can be tack welded in place, as shown in Figure 8-22, and you will now be able to calculate the correct angle and length of the tube, which is, again, based on whatever length is going to fit your SpinCycle best.

Take another look at Figure 8-22, and you will see that I have placed the front end (with front tube tack welded in place) at the same angle it would be if the entire trike were completed, which allowed me to cut the lower end of the front tube at the correct length and angle. The correct angle of the front end is derived by finding a good balance between pedal to ground clearance, head tube angle, and underseat clearance. Simply place the front end with front tube attached on the ground and draw an imaginary line for the ground. Your head tube angle should end up somewhere between 35 and 45 degrees to the ground, while allowing a clearance of about 6 inches between the lowest point of the crank arm and ground. You will have to measure the length of your crank arm or install it temporarily into the bottom bracket to get this distance. Again, none of this is too critical, and if your front end looks like what I have in Figure 8-22, then you will be OK.

When you are happy with the position and angle of the front end, you can then place the base tube over the bottom of the front tube to get the proper cut angle and position to join the two. Just make sure that there is approximately 5 to 6 inches of clearance between the bottom of the base tube and your imaginary ground line.

Figure 8-23 shows the completed welding between the head tube and front tube. It was easier to clean up this weld before any other frame tubing was welded, so the part could be placed at any position in my workbench vice.

Once you get the correct length and angle to cut and mate the front tube to the base tube, the rest of the frame is easy to finish. The rear tube (backrest) is simply a length of tubing approximately the same length as the base tube (about 12 inches) that forms a 130 to 135-degree reclined angle for your seat. In recumbent bicycle terms, your seat is at a 45-degree angle (the angle

Figure 8-23 *Welded and ground joint between front tube and head tube*

on the bottom side of the 135-degree angle). I chose this angle because it seemed to place the rider's weight in the perfect place for balance between front wheel traction and 360 ability. If too much weight ends up over the T-bar, you will have a tough time getting grip on the road, and if too much weight is at the front of the trike, you will never be able to pull the almighty 360 maneuver. So, if you have been reading ahead to understand the frame geometry, that was how I derived my original angles back in Figure 8-20. Your frame will look similar to the one in Figure 8-24, but will most likely have slightly different angles and lengths, and this is OK as long as your pedals are not going to hit the ground, and your butt will fit properly in the frame. If you are a "really big kid," then just add an inch or two to the base tube length, that's all.

Figure 8-24 *The main frame is completed*

The one last step to making the frame involves getting the proper placement of the T-bar in relation to the base frame. Again, there is no magic formula here, as it is all done based on what you have made so far and how you want your SpinCycle to behave. Before I explain the T-bar angle, install your head tube bearings and rings into the base frame and place the front end into the frame with tire fully inflated. We are going by ground clearance, so you will want the two parts set up just as they will be in the final design. You will now place the base tube into a vice or place a few blocks under it so that it is parallel to the ground, having the same clearance you planned for a few steps earlier. If your front end is flopping around like a fish in a boat, then just remove the top bearing and clamp the fork bolt down tight to add friction—this will not affect the position of the base frame. Once your SpinCycle is set up like the one in Figure 8-25, you can now place the T-bar up against the back of the rear tube with a pair of blocks behind the caster wheels to hold them in place.

The angle of the T-bar is very slight (between 3 and 5 degrees), and this is the main factor determining how your SpinCycle will behave. If you want to start spinning very easily, use 3 degrees or less, and if you want to get up to a higher speed before spinning, keep the angle at 5 degrees or slightly more. I always use approximately 4 degrees on my T-bar, as this allows a decent amount of control, yet gives me non-stop 360s on command. The T-bar angle is really setting the angle of the caster wheel bolts. The greater the angle, the more the casters will

Figure 8-26 *Marking the cutout for the rear tube and T-bar joint*

resist spinning around, thus keeping your SpinCycle traveling in a straight line until Newton's law takes over. I know, 4 degrees is pretty slight, and it would be difficult to get that right on my simply eyeing it up and marking it, like I did in Figure 8-26 but, in reality, your best guess will do just fine, as there is plenty of room for adjustment during the welding process, as you will soon see. Just place your T-bar against the rear tube at what looks like a 3 to 5-degree angle (clockwise rotation as looking at Figure 8-26), and make your mark on the rear tube. The cutout for the T-bar will be approximately an inch deep into the rear tube, or about halfway in.

Trace your marked area to both sides of the rear tube, and then cut the section out using a hacksaw or grinder disc (see Figure 8-27). Perfection is not necessary here, and if the hole is a little bigger than the T-bar tube, this will be fine, since you are going to be tinkering with the

Figure 8-25 *Determining the proper T-bar to frame placement*

Figure 8-27 *Cut out your marker area*

Figure 8-28 *Tack welded T-bar ready for adjustment*

Figure 8-29 *Make sure you have a 90-degree angle with the main frame*

angle to get it correct anyhow. Once the cutout mates with the T-bar tube, move ahead.

Do your best to tack weld the T-bar into the cutout so that it ends up as close to your desired angle as possible, as well as creating a 90-degree angle with the frame. The tack weld should only be done along the top rear end of the joint, as shown in Figure 8-28. This will allow for some fine tuning with your good old hammer to get that magical 4 to 5-degree angle. The tack weld will be strong enough to place the bike on the ground, but certainly not strong enough to carry any weight, so avoid the extreme urge to get into that pilot's seat—that time is just around the corner, promise! For alignment purposes, place both caster wheels facing the rear (this is their natural orientation).

To get the T-bar all lined up for welding, first make sure you have a good 90-degree angle with the frame, as shown in Figure 8-29, so you will not be leaning in the bike while traveling in a straight line. Check from the top of the bike and the rear while whacking the T-bar into position with your mallet. Do not worry about the 4 to 5-degree T-bar angle until your frame is sitting up straight. Remember to do this with the casters in their neutral position facing the rear. When you are satisfied with the frame-to-T-bar relationship, you can then adjust the T-bar angle by placing an adjustable angle square on the ground and against the rear of the T-bar. If you have a good eye and a lot of luck, that magical 4–5 degrees may be already there, but if not, just clamp onto the T-bar center with a pair of vice grips and force the tack weld to bend a bit until the angle is what you want. Make sure to recheck the frame alignment as well if a lot of bending

was necessary. Don't rush this step. Bend and check all angles until you are sure nothing is misaligned before any more welding is done. Once you are certain all looks well, tack weld the underside of the T-bar and rear tube, then repeat the checking process.

Continue welding in small steps, filling in the joint at the rear of the T-bar and under the frame first, making sure your angles stay the same along the way. Save the side welds for last, as you will still have the ability to flex the two parts into correct alignment with only the back and underside of the joints welded. When welding the sides, again do it in small steps, alternating sides often, as there will be a tendency for the T-bar to become out of alignment from heat distortion. Take your time on this part of the build, and don't be afraid to hack up a bed weld and start over, as it's worth it in the end. After this, the rest of the build is easy, and can be done with one hand in your pocket and blindfolded. OK, it's not that easy, but you have just finished all of the tough stuff. Figure 8-30 shows the completed weld.

It's time to get the pilot's seat mounts in place, and you can do this from bits of flat bar or the same angle iron that was used to make the caster wheel mounts. This metal does not need to be any heavier than the frame wall thickness, as its only duty is to bolt down the plywood seat and cap the sharp open end of the rear tube. Figure 8-31 shows the 1.5-inch bit of angle iron cut, rounded, and drilled for a bolt or large woodscrew.

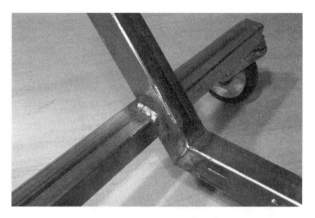

Figure 8-30 *Welding should be done in small steps, checking along the way*

Figure 8-31 *The top seat mount made of angle iron*

Figure 8-32 *The rear seat mount also caps the open end of the tube*

Figure 8-33 *Plywood and foam used to create the pilot's seat*

Figure 8-34 *The other seat mounts are welded to the frame*

There is nothing special about welding the seat mounts in place. Just weld around the edges, and then clean up the rough spots with your grinder (Figure 8-32). You will now need some 3/4-inch plywood to create the seat.

You can make your seat out of whatever material you like, but the old-fashioned 3/4 plywood with foam glued to the top method seems to be the easiest. Any width between 10 and 12 inches makes a fine seat, and the height will be dictated by the length of the rear and base tubing. I like rounding the top of the backrest part of the seat to avoid sharp edges and give it a nice look. Rigid foam can be glued directly to the wood, and may be fine without any covering at all. The six small plates shown in Figure 8-33 will be used for mounting the bottom of the backrest and the seat base to the frame. They are just pieces of thin flat bar or plate with holes drilled for a woodscrew, nothing critical.

Weld the other seat mounting plates to the frame, as shown in Figure 8-34. Yes, you could just drill holes right

Figure 8-35 *Underside shot of the seat-mounting process*

Figure 8-36 *Almost ready for flight*

into the base frame tubing and use bolts or long woodscrews into the plywood, but that would be an ugly shortcut, and not as strong as this simple method. The small plates are welded in place so that they are flush with the top of the frame where the plywood will be fastened.

A few good-sized woodscrews cranked into the plywood will be plenty of fastening strength for the two seat boards (see Figure 8-35). If the screws are a bit too long, just grind off the tips on the pilot's side of the wood with your grinder. You will also notice that I rounded the edges of the front of the seat board using the grinder disc, just to make sure that there will be no dangerous edges on the trike.

With your seat fastened to the frame, you can now install all the drive components to get your SpinCycle in a test-drive-ready state, as shown in Figure 8-36. You will certainly need to cut and join the drive chain to the correct length, and if all goes well, it should be nice and tight once you get it installed on both chain rings. There is not a lot of room for chain adjustment here, since we are not using rear dropouts to hold the front wheel, so you may need to do a small amount of filing or light grinding on the edges of the dropouts to accommodate your drive chain. If your chain is slightly too long, remove as many links as you can before the chain becomes too short, and then you may need to file the dropouts on the edge facing the cranks. Your goal should be a nice, tight chain, not one that is so loose that it flops

against the drive tire. A $10 chain link tool is really handy for this part of the build, and will save you a lot of time if you need to experiment with the chain links. Smacking the links with a punch and hammer is a wasted effort, and the cause of much unwarranted frustration.

The next step is optional, but highly recommended, as it will spare your pants from getting worn out as the front wheel rubs during extreme cornering. The original SpinCycles did not have fenders, but after many rides, I noticed that the slight rubbing of the front wheel against my legs during tight corners was wearing a hole in the material! On a few rides, this was not a big deal, but after a few hours, you will wish you had a fender installed. Can you just throw any old fender on the bike to fix the rubbing problem? Nope, it will not be strong enough, and will make your SpinCycle sound like a power saw when it rubs against the tire. The fender is made from 16 or 20 gauge sheet metal and, when welded directly to the fork legs, will become a very strong barrier between your legs and the hungry tire. Again, you could skip this point, but it is recommended, especially for shorter riders who are stretching to reach the pedals. Just take the SpinCycle out for a test ride like it is, and you will see what I mean. The fender plates are shown in Figure 8-37. Just use a few pieces cut to conform to the curvature of the inflated tire, so when the fender is installed, it reaches to the base of the seat in order to keep pants and tire separated. The center part of the fender is just bent by hand to form the curve.

The fender is welded together at the corners, as shown in Figure 8-38. To keep it straight, place a few tack welds at first, so you can bend it into alignment by hand. Once aligned, weld the entire length of the joint edges on the

Figure 8-37   *Making the fender to stop wheel rubbing*

Figure 8-39   *A little grinder magic does wonders*

Figure 8-38   *Welding the fender parts together*

Figure 8-40   *Optional but highly recommended fender installed*

outside of the fender. Welding on the inside is not necessary, and will only cause distortion.

Once fully welded, a sanding disc does wonders to make the fender edges clean and smooth, as shown in Figure 8-39. There is no point in making a clothing guard with sharp edges! Once welded, you will be amazed at how strong the fender becomes.

Test fit the fender on the fork legs, and grind the shape of the ends, if necessary. Once installed, it should allow a little clearance for the tire, but provide a tight fit, as shown in Figure 8-40. Only a few beads of weld may be needed to hold the fender in place, but I like to weld the entire joint, then grind it clean just for a nicer, professional look. I also like to tack weld the fender in place while the inflated tire is installed to ensure there is no rubbing. This is generally not a problem if you keep

the tack welds to a minimum. If you are worried about melting your tire, just throw a scrap wheel or tire on for this procedure.

With your fender installed and drive hardware mounted, you are ready for a test ride! Don't get too addicted to the SpinCycle just yet, or you will have a hard time tearing it down for painting. Just test the trike to make sure that nothing rubs, and then pull it all apart to clean up any rough welds or sharp corners. A light sanding with some emery cloth will remove any dirt or loose paint from the recycled bicycle tubing, so you can prime the metal parts for painting. Figure 8-41 shows the parts of the frame I decided to paint, including the caster wheel forks.

Although you can't see my cool, hot, green paintjob in Figure 8-42, due to the black and white photos, I can assure you that it is the greenest green I have ever seen. It's actually high-heat, brake caliper paint, but I liked the

Figure 8-41   *The trike is ready for painting*

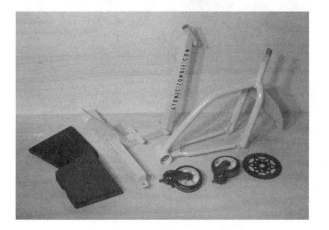

Figure 8-42   *A lovely hot green accented with a deep blue*

Figure 8-43   *High-quality blue vinyl will cover the seat*

Figure 8-44   *Stretch and staple*

color, and it seemed to make a nice durable finish for a trike that was about to take a beating. I also spray painted the plywood with flat black paint to get rid of that "el-cheapo" look that bare plywood would have given the final product. Speaking of seats, why not attempt some "poor-man's" upholstery while the paint is curing?

My original SpinCycle had only a bare plywood seat, and after a few hours of vigorous riding, this became a little uncomfortable. On this unit, I decided to pad the base of the seat, but leave the back unpadded so that there would be more room for riders of different heights and sizes. If you are a "big kid," the unpadded back will give you more leg room. For smaller riders, I had two foam pads cut out that I could slip in place to move them closer to the pedals. This worked like an adjustable seat. If you have a pair of scissors, a bit of vinyl and the ability to wrap a present, then you can make a seat. Just cut the vinyl so that its edges can wrap around the foam and wood, and then get ready to use an entire box of staples (see Figure 8-43).

Pull the vinyl over one side of the seat and staple it along the bottom, as shown in Figure 8-44. Now pull the other side as tight as possible and repeat the process. The other two sides are done basically the same way, but you will have to fold the corner of the material over as if you were wrapping a present. If you have a sewing machine and a little upholstery skill, you could make a fitted seat cover, but this system does the job, as, after all, this isn't a sports car we're building here.

With the SpinCycle completely assembled and all the bearings greased, you only have one task left—having fun! Tighten every bolt down, especially the wheel nuts and gooseneck, and then take the trike out for a test run. Oh, and don't forget that little strap that holds the coaster brake arm to the fork leg. Without it, your wheel will loosen and fall out of the dropouts once you hit the brakes. If you decided to use a free hub, then this is of no concern, but you may want to install a caliper brake. It is true that most of the time you will only want to go faster,

Figure 8-45 *A very clean-looking final product, ready for flight*

Figure 8-46 *Ready to roll, but don't forget the brakes*

but in the rare case when you need to stop, brakes do indeed come in handy! My completed SpinCycle is shown in Figure 8-45.

In Figure 8-46, you can see the coaster brake arm on the left fork leg. I had to make a new strap from a pipe clamp, since the original strap was only large enough for the much smaller chain stay tube that it once gripped. Other than that, I am ready to get into that pilot's seat and take orbit. I choose a nice, smooth, high-pressure, front tire as well, to maximize my ability to perform multiple 360s in a row. The balder the tire, the easier it will be to slide sideways.

The final product turned out so well that I had to take another glory shot just to show it off! The urge to burn the rubber off that front tire is killing me, so let's get on

Figure 8-47 *Another glory shot of the finished SpinCycle*

with this so I can get out and ride! The cool branding shown in Figure 8-47 was made by sticking on a bunch of letters, and then adding some clear coat to make them more durable.

The one last thing I want to mention is that you must have that gooseneck as tight as possible, in order to stop it from turning out of alignment as you put the SpinCycle through its paces. On a normal bicycle, there is almost no resistance at all when you turn the handlebars, but on this machine, there is extreme force as you counteract your pedaling force through the handlebars while traveling in a straight line or accelerating. Some goosenecks seem to hold better than others, and you may not have any problem at all, but if you just can't seem to get that bolt tight enough, then here is an alternative solution that works perfectly.

Remove the top fork bolt so that a small amount of the fork stem threads are exposed, as shown in Figure 8-48. You will still want to have the locking washer installed over the main bearing race and a thin lock bolt over it. Your fork hardware may or may not have this thin bolt in between the top bolt and lock washer, but these are easy to find on many scrap bikes. In Figure 8-48, I have the main bearing race installed over the bearing in the top cup, a locking washer, and the thin fork bolt (shown with the square notches). This configuration will keep the fork hardware tight, but allow me to weld the exposed part of

Figure 8-48 *The gooseneck bolt must be very tight*

Figure 8-49 *This gooseneck will never move out of alignment now*

the fork threads permanently to the gooseneck, as shown in the next photo.

As shown in Figure 8-49, a small bead of weld will permanently secure the gooseneck to the fork threads to withstand misalignment, no matter how evil your riding style may be. The downside to this system is that you will have to grind this weld away if you ever need to change the fork bearings but, in reality, have you ever had to change fork bearings on any bicycle? Just make sure that the handlebars are perfectly aligned, and that the bearings are well greased before you weld. Weld a little bit at a time, so you don't generate so much heat that all the bearing grease melts out.

Congratulations—you now have only the job of learning how to ride your new toy! Find a nice, smooth, flat parking lot (free of parked cars), and take the SpinCycle out for a test ride. As a newbie to the pilot's seat, you may find yourself spinning around backwards

Figure 8-50 *Starting one of many repetitive 360s*

quit a bit at first but, in a few minutes, you will learn that a combination of both subtle steering and body language can make you run a perfectly straight line at top speed, or send you into a spin that would make a trained astronaut barf. Pulling out of a fast 360 to regain control and continue in a straight line does take a bit of practice, but let me give you a hint: leaning back over the T-bar makes it much easier to slide that front tire back into place as you come around. I have managed to pull over six 360s in a row with only one pedal revolution between them to maintain speed, and the only thing that stopped me from pulling another was that very solid looking brick wall at the end of the lot (see Figure 8-50).

Some of the other cool maneuvers you will want to master are: T-sliding around the corners, "wagging" your tail as you ride in a somewhat straight line, and gliding in reverse at high speed after a good spin, then recovering back to the forward direction. There is no end to the amount of "impossible" tricks that the mighty SpinCycle can perform—the only question is: do you have the stomach for this kind of abuse? If you don't get motion sickness, you will certainly have sore legs in an hour or so, I guarantee that. Figure 8-51 shows another high speed T-slide, which is way too close to the wall again! Building a second SpinCycle is also worth the effort, as it multiplies the amount of fun you can have. We have played crazy games of chicken, practised synchronized stunt riding, and even invented a few games that involved batting a ball around with the rear of the bike as you spin around. Try racing around a track with your buddy and, as you are about to pass his SpinCycle, give a little tug on the back of his seat to send

Figure 8-51 *A photo just can't capture the fun of riding a SpinCycle*

Figure 8-52 *The SpinCycle is crazy, yet comfortable*

Figure 8-53 *You have to experience the SpinCycle to really appreciate it*

him wildly out of control. The amount of fun is really endless.

The next day after your first encounter with the SpinCycle will yield sore muscles that you didn't even know you had! Vigorous riding is truly a full body workout, considering you have to use your upper body as well as your legs to control the vehicle. I put so many miles on my first SpinCycle that I had to replace the caster wheels. I guess shopping carts were never intended for hitting the pavement at 20 miles per hour. There are pneumatic tire caster wheels available as well, and in the future I may make another SpinCycle or convert this one to test them out. Figure 8-52 shows a non-moving SpinCycle shot—something that is rare indeed.

By now you probably know what I mean when I say that the SpinCycle may be the most fun you will ever have on three wheels. Make sure to scoot on over to our website (that's the address shown on the back of the truck in Figure 8-53) when you have some photos of your SpinCycle, because I always enjoy seeing what other garage hackers have done. The next project is another fun ride that also has a practical purpose.

Project 9: UCan2 HandCycle

When you build creative bicycles and ride them on a regular basis, you will often find yourself pulled over by an interested cyclist or fellow garage hacker to answer some questions about your creation. Often I am asked if I built it, how much it costs, or if I would consider building a custom cycle for someone else who may need something a little different, due to physical limitations such as a bad back or some type of limited mobility. There seemed to be a large enough demand for upper-body-powered vehicles that I decided to search the Internet to see what was available, and why these people had just not purchased a ready-to-ride cycle to suit their needs. The answer was clearly cost.

For about $2,000 or more, you could indeed purchase a very basic hand-powered trike, but this seemed extremely steep for a trike that was made of little more than a few feet of steel tubing and some off-the-shelf bicycle parts. Not only was the price out of reach for many people, but the actual design of many of these trikes was better suited to the professional rider who wanted performance parts such as disc or drum brakes, an internally geared hub, and fancy, cambered rear wheels with the best hubs money could buy. Ironically, this also made for a bulky and heavy trike that looked more like a wheelchair with a bicycle front end thrown on the front, not something cool for younger riders.

I had the chance to closely examine a handcycle at a children's center, and could not understand how such a unit would cost $2,500. In fact, if you compared it to a bicycle with similar quality parts and engineering, it would be on a par with a $100 department store kids' bicycle. I understand that demand may be lower than that of a standard two-wheeled bike, but I felt that this still did not justify the inflated cost, especially when combined with the fact that the unit had obvious transmission flaws that should have been worked out. I decided that I could build a handcycle of better quality for a cost 10 times less than the unit I was examining, and the end result would be a stronger, sportier-looking trike that could be easily repaired using standard bicycle parts, and carried up and down stairs by any caregiver. The result was the creation of UCan2, a cross between a freestyle BMX bicycle and a handcycle that cost well under $500 with all new parts used.

Most handcycles use a pair of wheelchair type hubs on the two rear wheels, which are either spoked to bicycle rims or left as radial spoked wheelchair wheels. This causes the price of manufacturing the trike to skyrocket, since a decent pair of aluminum wheelchair hubs may cost $200 a piece. These wheels are often cambered (leaning inwards), due to the fact that they really are not the best type of wheels to use on a trike that may be moving fast enough around corners to take the side loading demands placed on them. After all, they were originally designed for hand-powered wheelchairs. Some manufacturers claim that these cambered wheels are a sign of high quality, and you should want them. Most people I show them to say they look like wheelchair wheels. The three 20-inch wheels shown in Figure 9-1 include two front wheels from a freestyle BMX bicycle which have 48 spokes and 14-mm-thick axles. The other wheel is a typical coaster brake hub wheel used on many kids' bicycles which allows the rider to simply pedal in reverse to activate the brake. The two front wheels which will be mounted as rear wheels on our trike are only slightly heavier than the expensive wheelchair wheels, but are many, many times stronger. Each wheel will probably cost you about $30 brand new from a bicycle shop.

There are two reasons for choosing the two freestyle rims as the rear wheels on our handcycle. First, the 14-mm axles are so tough that they can easily be held in place by one side without any possibility of ever bending

Figure 9-1 *You will need three wheels: two fronts and one coaster brake*

Figure 9-3 *A standard coaster brake hub*

them, even if your rider has a need for speed and high adrenaline riding. Second, the freestyle rim is very easily identified by the younger crowd as being "cool," so your handcycle pilot will look like they want to ride the UCan2, rather than look like they have to. When you're young, these things matter. Let's face it—these things always matter at any age!

As you can see in Figure 9-2, the 14-mm axle shown in the left dwarfs the standard thickness axle shown on the right. I can assure you that the standard axle on the right will not hold up if held by one side of the axle, even if your rider weighs less than 80 pounds, so don't even bother trying it, or your wheels will fold over within the first few rides. The 14-mm axles on the other hand are indestructible, and could easily take 400 pounds when used on a trike. These axles are commonly used on pedicabs that carry two or more

adults right between them, which is why the UCan2 has no problem with a friend hitching a ride on the rear of the frame.

The coaster brake wheel shown in Figure 9-3 will be used as our front drive wheel, and since it is held on by both ends of the axle, it does not require such a beefy axle as the rear wheels do. The coaster brake hub also allows very simple operation of the handcycle by allowing the rider to pedal forward to drive, and pedal in reverse to slow down or stop. Handcycles that have lever-operated brakes force the rider to remove one hand from the pedals and reach for the lever, which has to be mounted somewhere on the frame, which is a bad idea. Can you imagine steering and pedaling a cycle with only one hand as you precariously reach around for a brake lever? Try that in an emergency situation and things will likely end badly.

A coaster brake hub is easily identified by the large can-like hub and the steel arm opposite the drive sprocket. This steel arm is secured to the frame and acts as a transfer between your braking force and the frame. If you do a little digging like I did, you will be able to find three matching rims that give the handcycle a very professional and finished look. The black rims with brushed aluminum edges are the big style right now, and are instantly identified as being Freestyle wheels.

When choosing tires for your wheels, keep in mind that the heavier the tread, the more rolling resistance there will be, so find rear tires with as little tread as possible, and a front tire with minimal tread. Surprisingly enough, a tire with an aggressive tread will not give you any advantage over one with no tread at all, unless your

Figure 9-2 *14-mm axle on the left; standard rear axle on the right*

Figure 9-4 *Heavy tire tread is not necessary*

Figure 9-5 *Cut two plates from the flat bar*

Figure 9-6 *Round off the corners*

terrain is very rough such as gravel or dirt. I chose three similar tires, as shown in Figure 9-4, each with a rating of 75 psi, and minimal tread to keep the rolling resistance down to a minimum. If your rider is planning on hitting the skate parks, then spend the extra few bucks on brand name tires—these are way cool, so I am told!

Since the rear wheels on the UCan2 are held directly by one side of their axles, there is no need for any special bearings or rear transmission hardware, just a place to bolt the axles. Figure 9-5 shows two rectangular pieces cut to 2.75 × 1.5 inches from a length of 3/16-inch-thick flat bar. These will serve as mounting tabs for our rear axles, as will soon be shown. If you cannot find flat bar of exactly 3/16-inch thickness, then substitute the closest flat bar you can find without going under that. These two parts are so small that a little extra thickness will not have any impact on the final weight of the handcycle.

The corners of the axle mounting tabs are rounded off using a grinder so that there is no chance of being cut on the sharp edges (see Figure 9-6). Even though the tabs are well out of reach, who knows what kids may try to do

when out riding? Mark a point half an inch from each end of the tab and stamp a hole with a hammer and punch for drilling. A neat trick to make both tabs exactly the same size is to grind them into shape together in the vice at the same time.

14 mm is not exactly a standard size drill bit to have lying around the shop, so you may have to take the mounting tabs to a buddy or machine shop to be drilled. 14 mm is exactly .55 of an inch, so you could get away with a 9/16-inch drill bit to make the holes, as this will be pretty close. I did not have a 14-mm drill bit, and even if I did, it would not fit into the chuck on my little hand drill, so I purchased a more commonly available 9/16-inch drill bit and ground down the shank with a grinder so it would fit into my hand drill. I know that it's an ugly hack, but I have used it many times to drill holes for 14-mm axles, and it works perfectly. The drilled plates are shown in Figure 9-7.

If you do take the axle mounting tabs to be drilled, bring along one of the wheels just to make sure that the hole is not undersized. You could hand file the hole if it were just a bit too small, but this is annoying, since the 3/16-inch-thick plate is fairly thick for hand filing. If you are using a 9/16-inch drill bit, the axle will definitely fit through the hole. These 14-mm axles will have ample length sticking out from each side, since they are made to hold those freestyle stunt pegs in place. Shown in Figure 9-8 is the tab securely bolted to the 14-mm axle.

With the exception of a few standard bicycle parts, the entire trike frame is made from a few feet of 1.5-inch mild steel square tubing with a 1/16-inch wall thickness. This tubing can be ordered from any steel supplier and is

Figure 9-7 *Drill the 14-mm axle holes*

Figure 9-9 *The rear frame tube*

Figure 9-8 *Checking the axle holes*

Figure 9-10 *Weld the axle mounting tabs to the rear frame tube*

very easy to cut and weld. Steel suppliers like to rate tubing wall thickness using a gauge number, so 1/16-inch wall tubing will be called 16-gauge tubing, and although there is a slight difference between the two, your best bet is to tell them you want a length of 1.5-inch square tubing with a wall thickness as close to .0625 of an inch (1/16) as you can get. Round tubing should be avoided for this project, as it will be much more difficult to cut at the correct angles, and would most likely be out of alignment after all the welding is completed due to heat distortion. As shown in Figure 9-9, the rear part of the frame which will hold the two rear wheels is just a 22-inch long piece of tubing cut square at each end. This part of the frame will be called the "rear frame tube" from now on. Try to make the cuts at each end as square as possible, since this will affect the alignment of the rear wheels as you will soon see.

The axle mounting tabs are welded directly to each end of the rear frame tube, as shown in Figure 9-10. Notice how they cap off the open ends of the tube, as well as create a sturdy mounting point for the axles on the rear wheels. In Figure 9-10, I have welded on only three sides of the joint, leaving the joint just below the axle mounting hole untouched. It is important not to weld this part of the joint, as the extreme heat from the welding process will cause the axle mounting tab to bend slightly towards the weld area, throwing your two rear wheels out of parallel alignment. No amount of clamping will cure this warping, and there is no great strength benefit from welding this part of the joint anyhow, so make sure to leave it.

Once the three open sides of the joint are welded, grind the area clean and round off all of the sharp corners as best you can without taking off too much of the metal. The corners do not have to be rounded off too much to make them perfectly round, just enough so that there is no possible way to cut yourself on any part of the frame if you were to come in contact with it. Figure 9-11 shows my completed rear frame tube after finishing up with a sanding disc.

Figure 9-11 *Removing all sharp edges from the welded area*

Figure 9-12 *Checking rear wheel alignment*

Figure 9-13 *Both rear wheels should be parallel*

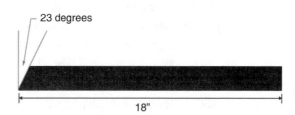

Figure 9-14 *The dimensions for the middle frame tube*

Once you have welded the axle mounting tabs to each end of the rear frame tube, bolt the rear wheels in place, as shown in Figure 9-12. If both the mounting tabs are parallel, the two wheels will roll in a nice straight line without any apparent wheel scrubbing. Test this by pushing down on the tube while you walk along the floor—there should be no rubbing sound.

Figure 9-13 shows the completed rear of the UCan2 HandCycle. Although there is little more than a few feet of tubing and a pair of wheels, this setup could easily carry 300 pounds or more without any problem at all. Yes, this would form the perfect backbone for a nice bicycle trailer or load carrying vehicle, but that's a project for another day.

To connect the rear frame tube to the head tube and front forks, we will need two more short lengths of the same 1.5-inch square tubing used for the rear of the frame. The tube that runs horizontally along the ground will be called the "middle frame tube," and the tube that connects the middle frame tube to the head tube will be called the "front frame tube." So basically the UCan2 is made from three lengths of 1.5-inch square tubing: the front tube, the middle tube, and the rear tube (already completed). The middle frame tube is shown in Figure 9-14 as an 18-inch-long square tube with a 23-degree angle cut at one end. Since this tube will be welded directly to the rear frame tube, the other end is cut off at a 90-degree angle.

Mark and cut the 23-degree angle using whatever method works best for you. As shown in Figure 9-15, I used an adjustable square and an angle finder to get this measurement. A chop saw is a better way to cut square

Figure 9-15 *Visualizing rear wheel placement*

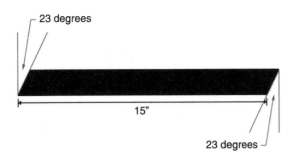

23 degrees

15"

23 degrees

Figure 9-16 *Dimensions for the front frame tube*

tubing at specific angles, but since I only use a handheld angle grinder, hand drill and basic welder to create everything I make, I had to mark the tubing by hand and then follow the marks using a cut-off disc. Do not worry about precision here, as even a few degrees either way will produce a perfectly working handcycle and, as always, feel free to modify the design for whatever reason you may have once you understand how all the parts fit together.

The front frame tube is another length of 1.5-inch square tubing with a total length of 15 inches and 23-degree cuts at each end, as shown in Figure 9-16. Notice that both angles are parallel to each other. Use the same method that you used to cut the 23-degree angle in the other tube and repeat the process for both ends of the tube. You now have three main frame tubes, so the rest of the handcycle will come together quickly from this point on.

The reason for the 23-degree cut at the ends of the middle frame tube and front frame tube is this. When welded together, they will form a 134-degree angle, as shown in Figure 9-17. It does not matter which end of the front tube meets the middle tube, since both ends have a 23-degree cut facing the same direction. When placed on a flat surface for tack welding, your frame tubes should

look like the ones shown in Figure 9-17, with an angle very close to 134 degrees between them. Clamp or weigh down both tubes on a flat surface, and then tack weld the two top corners of the weld and in the center of each visible joint. Once tack welded, pick up the two tubes and visually inspect them to make sure that they are aligned and at the correct angle. If you are satisfied with the tack welded parts, fully weld the top and bottom joints, and then do the two sides. This order of welding will help minimize side-to-side distortion of the frame.

Weld all the way around the joint, as shown in Figure 9-18, creating a solid and pinhole-free weld between the two tubes. The 134-degree angle formed between the two tubes may vary by a degree or two after welding, but this will not be a problem. If you look at the two parts from a bird's eye view, they should appear to form a straight line.

If you are a welder who likes to grind all welds flush for appearance, then do so sparingly here, as this needs to be the strongest weld on the entire handcycle because it joins the two main halves of the trike together. As shown in Figure 9-19, I did my best to make a nice-looking bead of weld so that no grinding would be necessary in order

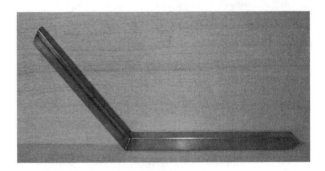

Figure 9-18 *Weld the entire joint*

Figure 9-19 *A good solid weld is needed here*

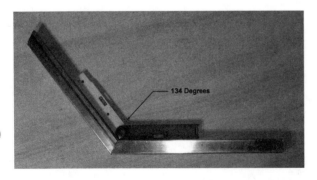

134 Degrees

Figure 9-17 *The front tube and middle tube join together*

Figure 9-20 *Joining the front and rear parts of the frame*

Figure 9-21 *Do not weld the sides of the joint*

to avoid weakening the joint. If you do need a little bit of grinding to make the weld look better, only take off as much material as needed, and weld any pinholes or open spots that may have appeared after grinding. If welding is done properly, the UCan2 frame will be amazingly tough. I have done things to this handcycle that would certainly void the warranty of any commercially produced handcycle, and doubt that I could intentionally break this cycle without first breaking myself!

Now that the front tube and middle tube have become one, the last main frame weld to be made will be the one that joins the rear tube to the back end of the middle tube, as shown in Figure 9-20. The rear frame tube and middle frame tube will form a T-shape, with a 90-degree angle between them. I simply laid the parts on a flat surface, marked the center point on each tube, and then placed a solid tack weld right at this point. The frame was then carefully flipped over so that another tack weld could be placed on the underside of the joint. Once the two tack welds that secure the frame tubes are in place, place a square along the two parts to make sure that there is a 90-degree angle between them. A small tap of a hammer will easily force the tubing into perfect alignment.

When you have a good 90-degree angle between the rear frame tube and middle frame tube, add two more tack welds to the joint at the top corners, and do one last alignment check with your square. If alignment is good, flip the frame over and weld the entire length of the bottom of the joint, and then do the same on the top, but do not weld the sides of the joint. The sides of the joint are not to be welded or your rear frame tube will be pulled out of 90 degrees, causing your handcycle to track in an awkward fashion. No amount of clamping will stop this distortion and, as you will soon see, the sides of the

frame will be getting gussets for added strength, so this weld is not needed. Figure 9-21 shows the completed weld between the rear frame tube and the rest of the frame tubing. Note the sides of the joint are not welded, and they will not be welded.

Although it would take a serious amount of force to damage the handcycle frame, even with the rear frame joint only welded in two places, you just never know what ambitious riders might try to do, so a few small gussets will make the frame virtually indestructible. The exact size, shape, and type of steel used for these rear gussets is not important, so dig into your scrap bin and use up some of those leftovers and cut-offs. For the record, the two gussets I have cut in Figure 9-22 are three inches long and are made from scraps of 1/8-inch-thick flat bar with a width of half an inch. Round tubing, square tubing, or even cut-off bolts will also work.

The rear frame gussets will install at the 90-degree corner between the rear frame tube and the middle frame tube, as shown in Figure 9-23. The size and shape of the

Figure 9-22 *Rear frame gussets*

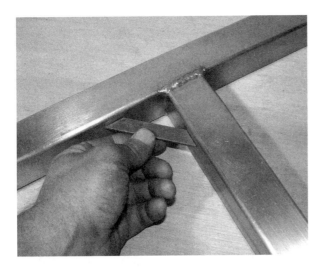

Figure 9-23 *Installing the rear gussets*

Figure 9-24 *Weld both rear frame gussets in place*

gusset are not critical, but do try to round off any sharp edges, and avoid creating tiny spaces that a finger could become lodged in.

There is no special procedure for welding the rear frame gussets in place; just try to make them even on both sides of the frame so that everything looks balanced. There is nothing worse than a frame that is non-symmetrical. Figure 9-24 shows my two small, rear frame gussets after welding both sides of the joint. The rear of the frame is now stronger than it would ever need to be for any type of riding your test pilot may want to do. Did I mention the time I took the UCan2 over a jump with a brave stunt rider standing on the rear of the frame in such a way that we landed on two wheels? Yes, this handcycle is really that durable! If you tried anything like this on some of the production models using radial-spoked wheelchair wheels on the rear, you would not have a good landing—that is for certain.

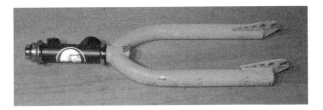

Figure 9-25 *Front forks and head tube*

You will need a pair of 20-inch front forks and a matching head tube and hardware to complete the front of the handcycle. Figure 9-25 shows a typical BMX-style fork set commonly found on any 20-inch kids' bicycle or BMX. The head tube and all of the hardware should fit properly onto the fork stem, and it is also cut from an old bicycle frame. My head tube is from a 1970s road bike, and the front forks are from a bent-up kids' BMX bike, but they work perfectly together. If your fork threads are too long for the head tube, just install all the hardware to make sure that everything fits, then trim off the excess fork thread with a hacksaw or cut-off disc. If you hold the head tube in your hand, the forks should spin around with practically no friction. If they do not, you may have the bearings installed upside down, or have the wrong sized fork bearings for the cups (there are two common sizes available).

Once you scrounge up a decent set of front forks and a compatible head tube, remove and clean all of the hardware for inspection. Badly rusted cups or bearings should be replaced, and forks with badly damaged threads should be avoided. To stay with the BMX look, I used the beefiest pair of forks I could find, as shown in Figure 9-26; they are not really any heavier, they just look stronger, which is why young people like them.

Clean up any excess stubbage on the dismembered head tube left over from the original frame, and grind away all the paint so you don't fill your garage with

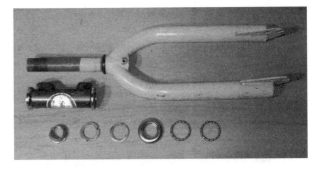

Figure 9-26 *Checking the steering hardware*

Figure 9-27 *Head tube ground clean*

Figure 9-29 *Testing the front wheel fit*

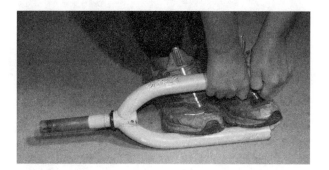

Figure 9-28 *Are you feeling strong today?*

horrible toxic smoke when you start welding. Clean metal is much easier to weld than rusty or painted metal, and it sure smells nicer! Notice the little bit of gold-colored brazing metal left around the welded area of my head tube, as shown in Figure 9-27—a sure sign that this frame was a museum piece before I hacked it.

If you are lucky, the front drive wheel may slide right into the fork dropouts with only a little effort but, most likely, you will need to do a little fork leg expanding, as shown in Figure 9-28. A rear hub is a bit wider than a front hub most of the time, so you will need to bend apart the fork legs ever so slightly to make installation of the drive wheel easier. Depending on the quality of the forks chosen and the condition of your back, you may be able to stand on one fork leg while pulling upwards on the other to get the extra quarter inch needed. Don't throw your back out trying to be a hero though; these forks are stronger than they look. I was almost at my physical limit when expanding them this way.

You could crank the legs apart with a long lever made with 2 × 4s, or place some type of screw between them and crank it with a wrench. There are many ways to get

this job done, but do be careful not to pull the fork legs apart too much in one mighty tug, or you will have to squash them back together, and this could weaken the welded area. I did this the manual way, pulling up on the forks little by little, and checking the distance with the actual wheel until it fit properly.

Once the fork legs are forced apart a little bit, the wheel will drop right in place, as shown in Figure 9-29. It's best to install the wheel by hand so you don't scratch up the forks once they are painted. This is why they are pulled apart permanently, rather than just forcing the wheel into the dropouts by prying about with a screwdriver. When installed, the drive sprocket will be on the right side of the frame, just like it was on a regular two-wheeled bicycle.

Once the front wheel has made friends with the front forks, reinstall the head tube and all the bearing hardware. At this point, you will also want the front tire fully inflated, since we will be using the front wheel and forks as a guide to set up the correct head tube angle. As shown in Figure 9-30, find an object to place the underside of the frame on so that the middle frame tube is roughly parallel to the ground. I found a 9.75-inch-tall bucket that was just perfect for this job. The middle frame tube does not have to be exactly parallel with the ground; just get it as close as possible. We may be lifting or dropping the frame to match the head tube in the next few steps anyhow. Place the head tube up against the top of the front frame tube and secure the front wheel from rolling away, as shown Figure 9-30.

Figure 9-30 *Setting up the head tube angle*

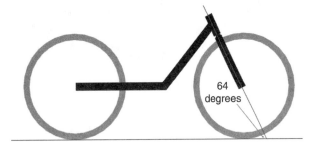

Figure 9-31 *Correct head tube angle*

The proper head tube angle will be approximately 64 degrees. This angle is measured from the angle formed between the ground and the head tube, as shown in Figure 9-31. This angle should be fairly close to the angle already cut at the end of the front frame tube, which is why 23 degrees was specified back in Figure 9-14. Don't worry about placing the head tube at the exact angle right now; this was just a trial mock-up. The actual installation will occur in the next few steps once the top of the front frame tube is ground properly.

The top of the front frame tube that must mate with the head tube will need to be ground out, as shown in Figure 9-32, in order to make a proper joint for welding.

Figure 9-32 *Grinding out the head tube joint*

Figure 9-33 *Checking head tube alignment*

Simply trace a circle the same diameter as the head tube shell on the top and bottom of the tube, then grind out the area using an angle grinder or hand file. Both cutouts should be the same depth on both sides of the tube, in order to keep the 64-degree angle constant. Make the cuts as equal as you can by eye, but don't worry about exact precision, as you will have lots of room to work with when it comes time to weld.

With the top of the front frame tube ground out to cradle the head tube properly, set the front end back in place, as shown in Figure 9-33. The three wheels should all be parallel when viewed directly from the front, and the head tube angle should be close to 64 degrees. If the top of the front tube is too low to meet up with the lower part of the head tube, just prop the frame up a little higher until it fits. It's not essential that the middle frame tube be exactly parallel to the ground, as this is just a guide.

The proper place to weld the top of the front frame tube to the head tube is a half inch from the bottom of the head tube, as shown in Figure 9-34. I like to make the first few tack welds while the front wheel is in place and the frame is supported by the underside to recheck alignment before any real welding is done. A few taps of the hammer will allow for slight alignment changes while there are only two or three small tack welds at the joint. Ensure that you look at the frame from all angles to ensure that the wheels are aligned correctly. All three wheels should be pointing in the same direction when viewed from top, front, and rear. When you are happy

Figure 9-34 *Tack weld the head tube*

with the alignment, do a little more welding until you can safely handle the frame to remove the forks from the head tube.

With the head tube properly aligned and securely tack welded to the front tube, remove the forks and finish up the welding by starting at the top and bottom of the joint, and then by welding both sides so it looks like the one shown in Figure 9-35. Make sure that all the welding is solid, and avoid grinding this area too much, just like you did at the weld between the rear frame tube and middle tube. This is another weld that needs to be strong. Now is also a good time to fill in any holes in the head

Figure 9-36 *You now have a rolling frame*

tube that may be left over from the removal of the original bicycle frame tubing. To fill the holes, make a single pass around the edge of the hole, chip the flux, then repeat until the entire hole is covered. Once covered, simply grind the area flush to remove all traces of the original hole.

Well done, as you now have a rolling frame like the one shown in Figure 9-36. This is a great place to be when building any wheeled vehicle. The rest of the UCan2 HandCycle will be a lot easier, since there are no critical welds or alignments to be done from here on.

The UCan2 places the pedals where the handle bars would be on a regular bicycle by connecting a bottom bracket (actually on the top here) to an extended gooseneck. I doubt that you will find a gooseneck with the required length to place the top 6.5 inches above the fork hardware, so we will make one from a standard bicycle gooseneck and an old seat post. Any old steel gooseneck can be used to create our new longer version, but do make sure that it fits properly into the fork stem you are using because there are two common sizes used. If your gooseneck feels loose in the fork stem, then it is the wrong size. There will be no mistake: it will either fit perfectly, be way too loose, or not fit at all. Figure 9-37 shows the old BMX-style gooseneck I found in my parts

Figure 9-35 *Completed head tube welding*

Figure 9-37 *Gooseneck to be cut*

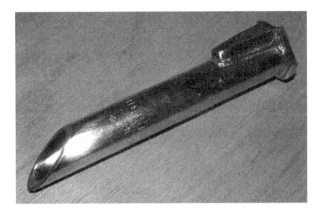

Figure 9-38 *Cut the top from the gooseneck*

Figure 9-40 *Grind the area to be welded first*

Figure 9-41 *Gooseneck extension fully welded*

bucket for chopping. Only the stem of the gooseneck will be used, so the condition of the clamp and hardware is not important. The stem needs to be straight, free of cracks, and made of steel.

Cut the clamp hardware away from the gooseneck stem, as shown in Figure 9-38, leaving just enough material to weld the extension tube to. If you are using a modern mountain-bike-style gooseneck, it will most likely be painted black and have a single clamp held to the stem by an arm. Either style will work, since it is only the stem we are after.

The actual gooseneck extension tube can be made from a 6.5-inch-long bit of steel seat post tubing, or any similar diameter and strength tubing. Handy alternative tubing can be found on mountain bike handlebars, bicycle frame seat tubes, or even some electrical conduit. As long as the tubing is not thinner than the gooseneck stem material, then it will be strong enough. As shown in Figure 9-39, cut the tube to a length of 6.5 inches.

The gooseneck extension tube is welded to the cut gooseneck stem, as shown in Figure 9-40. Grinding the leftover stem material to the shape of the extension tube before welding makes this task a lot easier, and will produce a much nicer-looking and distortion-free weld.

A tack weld at the top near the gooseneck bolt hole is a good start that will let you visually inspect and realign the two pieces so they are perfectly parallel before welding the entire joint.

When you have the gooseneck extension tube in line with the gooseneck stem, finish all the welding, making sure that the weld is strong and free of holes (Figure 9-41).

The bottom bracket will be placed on the end of the gooseneck extension tube in the UCan2 so the cranks can serve as both a source of motive power and steering at the same time. You will need to hack up another part of the bicycle frame you have been butchering to acquire the three-piece bottom bracket, as shown in Figure 9-42.

Figure 9-42 *A bottom bracket cut from an old frame*

Figure 9-39 *A seat post will be used to lengthen the neck*

A three-piece bottom bracket is easy to identify: the crank arms will be connected to the square axle, held on by a bolt at each end. The other type of bottom bracket, the one-piece bottom bracket, will have a single S-shaped crank arm, and for reasons that will become obvious very soon, only the three-piece bottom bracket will work for a handcycle. Cut the bottom bracket as close to the shell as you can to avoid excessive grinding later. If you are using a cut-off disc in your grinder to make the cuts, then beware when cutting either the seat tube or down tube, as the frame will tend to collapse on your disc as you get through the last bit of metal, which could cause your grinder to stall, or run out of your hand. If you have no way to secure the frame while cutting, then leave a sliver of metal during each cut rather than going all the way through the tubing, and you can simply bend the bottom bracket back and forth to release it after all four cuts are made. If using a hacksaw, don't worry about it—just hack!

A few minutes of grinding away the excess metal leaves the bottom bracket ready for welding (see Figure 9-43). You could probably order new bottom brackets from a bicycle frame building supplier but, in reality, recycling parts from damaged or old frames really isn't much work at all. Recycling parts also keeps your wallet happy and grinder skills finely tuned. Before you move on to the next step, install the bearing race on the right side (chain ring side) of the bottom bracket, so you know how to orient the bottom bracket when welding it to the front end of the bottom boom tube. There is nothing more annoying than building an entire frame, then realizing that your bottom bracket is installed backwards. Sure, you can just place the bearing races on

Figure 9-44 *Grind the top of the extension tube*

the other sides and get by, but there is a reason why one side has left-hand threads and the other does not—so the cups do not unscrew over time. I must admit, I have made this mistake more than once, and ended up putting the cups on the wrong sides without any noticeable bad results, but it really is easier to get it right the first time. As a general rule when building frames, always screw the right cup into the bottom bracket (counterclockwise for reversed threads), and make sure it is on the right side of the frame at all times—easy!

Mark the right side of the extended gooseneck, as shown in Figure 9-44, so you know which way to orient the bottom bracket when making the first tack welds. The gooseneck faces bolt forward, so this means the installed bottom bracket cup will also be on the right side. Grind the top of the gooseneck extension tube, as shown in Figure 9-44, so that it cradles the bottom bracket shell for an easy-to-weld, tight-fitting joint.

The bottom bracket is welded to the top of the gooseneck extension tube so that it forms a 90-degree angle between the shell and the extension tube. It is best to make a small tack weld on each side of the joint, as shown in Figure 9-45. This way, you can make any needed alignment adjustments with a small hammer. It is important to get the bottom bracket as aligned as you can, so that your chain does not run off either chain ring

Figure 9-43 *Bottom bracket cleaned up*

Figure 9-45 *Tack weld the front and rear of the joint*

Figure 9-46 *Checking bottom bracket alignment*

Figure 9-47 *Bottom bracket welding completed*

Figure 9-48 *Install the bottom bracket*

when pedaling. Don't forget which side of the gooseneck is the right side, so you can get the right side of the bottom bracket installed.

Do a close visual inspection of the bottom bracket once it has been tack welded to the gooseneck extension tube and make any necessary adjustments. From the top, the assembly should look like the one shown in Figure 9-46. Notice the good parallel alignment, as well as the position of the gooseneck bolt towards the front. With the two parts aligned properly, fill in the welds at the top and bottom of the joints before you do any of the welding along the sides, in order to reduce any distortion due to heat.

With the top and bottom of the joint securely welded and alignment looking good, you can proceed to weld the entire length of the joint, as shown in Figure 9-47. As always, make the weld strong, and be sure to fill in any gaps or pinholes when you are done.

The new extended gooseneck and bottom bracket combination will now be known as the "drive system."

Place the drive system in the fork stem and tighten it up so that it is aligned properly with the front wheel—bolt forward (see Figure 9-48). You can also install the crank axle and all the bearings back into the bottom bracket, making sure that one side of the axle is actually slightly longer then the other. The longer side of the axle is the drive side, so it will go in first. Bearings are also installed so that the balls face the cups. Do not bother to grease up the bearings yet because everything will have to come apart for painting soon.

To make the UCan2 as light and simple as possible to operate, a coaster wheel is used, which means there is no need for a multiple-speed crank set, or any type of shifting mechanism. The lack of shifters and levers is a good thing for those who are riding for the first time, or just want to get out and have fun on a tough, lightweight, easy-to-operate cycle. Just like many

Figure 9-49 *Reduce the crank set to a single gear*

Figure 9-50 *Crank arms are installed in the same direction*

single-speed kids' bicycles, a medium-sized front (drive) chain ring of between 38 and 42 teeth is used. Since most two- or three-speed chain rings are held together by the mid-sized chain ring as the main support, you can simply cut or unbolt the larger and smaller chain rings to leave only the one you want. A perfect style of chain ring to hack up is the old aluminum arm style found on old road bikes, as shown in Figure 9-49. You can simply unbolt or hacksaw the larger chain ring from the set to leave a nice light aluminum arm crank set with a single chain ring of 38 or 42 teeth. For an adult who wants a handcycle for some serious long-distance travel, you will want as many gears as found on a regular bicycle, but this added complexity is certainly not worth it for this project. From personal experience, this handcycle will travel as fast you will ever need for typical park or street riding. With this gearing suggested, the UCan2 will travel as fast as any single-speed kids' bicycle.

We needed the three-piece crank set and bottom bracket so that the crank arms can be placed on the square axle to point in the same direction, as shown in Figure 9-50. It may seem odd to place them this way, but if you think about how power is delivered to the drive wheel while at the same time needing decent control of the trike, it will seem more logical. You could certainly install the crank arms in whatever orientation you like and give it a try, but it is very obvious that there is less control over the steering and less available power to get moving if you have to push and pull in a swimming motion. Feel free to experiment, though.

The handcycle is ready for a transmission to bring that crank energy to the front wheel, but there are a few issues

that need to be resolved in order to get the chain to run properly past the frame tubing and fork leg tubing. If you simply installed a bicycle chain at this point, it would work very well until you tried to make a right-hand turn, which would place the drive side of the chain into the top frame tubing, creating a nasty rub or chain derailment. Some handcycles get around this problem by placing the front wheel and bottom bracket way ahead of the head tube, but this makes more of a chopper bicycle style than a handcycle. Although this system works fine on a fast and low frame, it just wouldn't suit the design of the UCan2.

"Drive side" refers to the side of the chain that leaves the rear of the front wheel sprocket entering the rear of the drive chain ring; in other words, it is the side of the chain that is closest to the rider. This is the only part of the chain that will ever be under any tension because it transfers your energy into the drive wheel. "Return side" refers to the side of the chain that exits the drive chain ring and enters the sprocket on the front wheel. This part of the chain will almost always be slack, as it never has any tension nor does it do any work unless the rider back pedals to engage the brake.

The reason we need to know which side of a chain is the drive side and which is the return side is so that we can choose the appropriate pulleys in order to route our chain around the frame. If you look at the two small plastic wheels that the return chain pass through on a typical rear derailleur, you will notice that they are nothing more than soft plastic wheels running on a steel pin. This works just fine because the chain just lazily slips past the wheels with almost zero force. If you tried to use those plastic wheels on the drive side of a chain,

Figure 9-51 *Drive-side idler pulley hardware*

Figure 9-52 *The chain should fit in the pulley groove*

Figure 9-53 *Idler pulley mounting tab and bolt*

they would disintegrate or be worn down in a hurry after using the handcycle for a while. For this reason, our drive-side chain needs a very large low-friction guide pulley in order to keep friction to a minimum and not wear out after a few hard rides.

Shown in Figure 9-51 is a commonly available idler pulley as used on many belt-driven machines powered by small gas engines. The pulley has a ball bearing built into its center, and a bicycle chain fits nicely in the V-groove so that it rides on the flat part of the pulley. Many hardware stores and small machinery stores have these. Also shown in Figure 9-51 is a cut-off bolt and nut that will fit through the hole in the pulley bearing, and a small bit of 1-inch-wide, 3/16-inch-thick flat bar, cut to a length of two inches to make a mounting tab for the pulley.

The pulley diameter should be at least 2 inches across, and the chain should ride on the flat part in the V-belt groove. My pulley is 3 inches in diameter and made for a half-inch V-belt. I purchased the pulley for a few bucks at a store that sells snow blowers, riding lawnmowers, and small garden machinery. If you're not sure if the chain will fit on the flat part of the pulley, just test it like I have shown in Figure 9-52.

The mounting tab is just a bit of flat bar with a bolt welded near the end that has been rounded off to remove any sharp edges. My mounting tab fits my 3-inch-diameter pulley so that, when bolted to the pulley,

there is about 1/4 of an inch of tab sticking out past the edge of the pulley. If you are using a smaller or larger pulley, cut the tab to the appropriate length. As shown in Figure 9-53, the pulley mounting bolt is welded directly to the top of the tab.

Figure 9-54 shows the correct mounting tab length for whatever diameter pulley you can scrounge up. The tab is sticking out past the edge of the pulley by about 1/4 of an inch so there will be no rubbing of the pulley on the fork leg tubing. It's also a good idea to use a nut with a round head to keep those sharp edges to a minimum.

Figure 9-55 shows the desired chain line, which will keep the drive side of the chain away from the frame when turning to the right, and will allow the chain to enter the drive sprocket through the front of the forks. You will also notice that a standard bicycle chain is too short to make the entire round trip. We will deal with that later.

Figure 9-54   *Pulley and mounting hardware*

Figure 9-56   *Pulley mounting tab tack welded in place*

Figure 9-55   *The chain will clear the fork legs*

Figure 9-57   *Mounting tab fully welded on both sides*

Test your chain run using a length of bicycle chain wrapped around the rear of each sprocket and over the pulley; this will represent the drive side of the chain. Look at the chain from the top so you can find the optimal place on the fork leg to align the pulley to be in the same line as the chain, as it travels in a straight line from one chain ring to the other. A couple of small tack welds can be used to secure the pulley mounting tab to the fork leg tubing, once you have found the optimal mounting position, as shown in Figure 9-56.

Once you have found the optimal mounting position for the drive pulley, weld both sides of the tab, as shown in Figure 9-57, to secure the tab in place. Because the bolt faces the front wheel, you will likely need to remove the

front wheel in order to remove the nut and the pulley, but it is best to have the tire away from any welding anyhow.

Now that the drive-side pulley is in place, we must deal with the return side of the chain, which will be under almost no tension. Because the drive side of the chain just idles along most of the time, a simple tensioner can be made by "slicing and dicing" an old rear derailleur from a road bike or mountain bike. As shown in Figure 9-58, a short arm rear derailleur is chosen for hacking, since it is no longer needed, and made mainly of steel. Practically any steel derailleur can be used to make this tensioner, as you will see by reading ahead a bit.

The only part of the derailleur we need is the longer arm with one of the plastic wheels attached and the spring-loaded pin. Start by drilling or filing off the two small rivet heads that hold the spring-loaded pin to the rest of the derailleur. A 1/4-inch drill bit works great for removing the pins, as shown in Figure 9-59.

Figure 9-58   *An old road bike rear derailleur*

Figure 9-59   *Removing the derailleur body*

Figure 9-60   *Derailleur wheels and spring post*

Figure 9-61   *Remove the rear wheel mount*

Figure 9-62   *Bolt the small derailleur wheel in place*

Once the pins are removed, you will be left with the two idler wheels and the spring-loaded pin, as shown in Figure 9-60. Now remove the bolts on each idler wheel to release both wheels and the backside of the derailleur arm.

With the idler wheels removed from the derailleur, cut the shorter arm off so that only the idler wheel mounting hole furthest away from the spring-loaded pin will remain. You should end up with something that looks similar to what I have in Figure 9-61, although your derailleur arm may be somewhat longer depending on what you started with.

The single idler wheel is now replaced using a nut and bolt that was the same diameter as the original bolt that held the two arm halves together. As shown in Figure 9-62, there is only a single arm and idler wheel remaining on the spring-loaded pin. We now have a simple spring-loaded chain tensioner. If you have a lot of small bits around your scrap pile, you could always make

your own spring-loaded chain tensioner, but this method is inexpensive and uses readily available parts.

Find a steel bolt of approximately 2.5 inches in length and weld the derailleur body to it, as shown in Figure 9-63. This bolt will serve as a mounting pin to

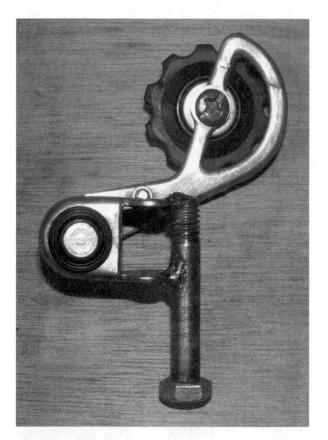

Figure 9-63 *Mounting post installed onto the derailleur*

Figure 9-64 *The return chain will be kept tight by the derailleur*

Figure 9-65 *Creating the correct length chain*

hold the chain tensioner to the front fork leg tubing, as shown next. You will probably have to crank the amps down on your welder for this job, as the derailleur body is only about 1/16-inch thick. Take your time and make very small spot welds and you will have no problem. I did this weld at approximately 50 amps using the same 3/32 6013 rod I use for everything. Three quick zaps and it was all done.

The chain tensioner mounting pin will be placed an inch or so up from the end of the fork leg, as shown in Figure 9-64, so the chain is held under tension and away from the drive-side pulley by the spring-loaded derailleur. You may have to fiddle around a bit for the optimal mounting position, and it may help if you cut a chain to the correct length before committing the weld at the mounting pin and fork leg. You will need just under two full bicycle chains to make the round trip.

On the UCan2, the bottom bracket is further away from the drive wheel than it would be on a normal bicycle, so you are going to need almost twice the length of bicycle chain that you would normally require. Shown in Figure 9-65 is a wad of chain, consisting of two new

lengths of standard, single-speed bicycle chain and an inexpensive tool that makes joining chains so much easier than the smack it with a hammer and punch system. Two chains may be a little too long, but it's a good start, so go ahead and join them into one long single chain. Look out for those brain-teasing loops!

When your chain is cut and joined to the optimal length, as shown in Figure 9-66, the chain spring-loaded tensioner will be pushing the return side of the chain nice and tight, and there will be no chain rubbing on any part of the frame or any other moving part. The return chain will pass directly over the drive chain pulley with as little room to spare as possible; this is the optimal configuration for the chain.

The UCan2 can make a very sharp turn without the chain rubbing on the frame. As shown in Figure 9-67, I can steer way beyond 45 degrees to the right with ample chain clearance. Under normal riding conditions, you would almost never crank the front wheel this far, so we have more than enough steering room.

Figure 9-66   *Chain installed and drive system complete*

Figure 9-68   *Coaster brake mounting tab*

Figure 9-67   *The chain will not rub on the frame during normal use*

Figure 9-69   *Brake mounting tab installed onto the fork leg*

Before you give in to the temptation of standing on the seatless frame for a test run, you must install the brake arm mounting tab. Without the brake arm held in place, your axle will turn in the dropouts upon application of the brake, which could possibly strip the axle threads, or make your front wheel pop right out—ouch! A simple brake arm mounting tab can be made from any thin bit of steel with a hole drilled through it. In Figure 9-68, I just cut the end of a reflector mounting arm and used that, since the end was already rounded off, and the correct size hole was drilled into it.

Bolt the brake arm mounting tab to the brake arm, and then rotate it counterclockwise until the tab comes into contact with the fork leg tubing, as shown in Figure 9-69. Once in the correct place, make two good tack welds,

then remove the front wheel so the tab can be fully welded to the fork tubing. Make sure that your front wheel is aligned properly between the fork legs before you mount the brake arm tab, or you will be cutting it off to fix it.

I will admit that the desire to test ride the UCan2 beat me at this stage, and I duct taped a small bucket to the back of the frame so I could hop on and take it out for a blast. What fun it was until the bucket fell off! Figure 9-70 shows the handcycle ready for a seat.

Depending on your intended rider's needs, you may want a simple, bicycle-style seat, or you may need something more elaborate, such as a seat with a full back and a seat belt. No matter what you plan to do for a pilot's seat, a simple way to start is by using an actual seat post mounting system as used in a standard bicycle. You can continue to rob your donor frame for parts, or cut the needed seat tube from an old bicycle frame, like the one about to die for my cause, as shown

Figure 9-70   *Transmission and brakes completed*

Figure 9-71   *Any old frame will do*

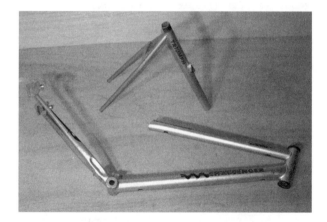

Figure 9-72   *Remove the seat tube and seat stays*

in Figure 9-71. Any frame will work for this part, but be aware that there are two sizes of seat posts, just in case you already have a bicycle-style seat picked out for use.

As shown in Figure 9-72, the seat tube and the seat stays are cut from the frame so they are left intact. Again, watch out when using a grinder disc for cutting,

Figure 9-73   *Welding the seat post to the frame*

especially on the last cut, which may tend to collapse on your disc, causing it to stick.

The hacked seat tube and seat stay unit are welded to the rear of the handcycle frame, as shown in Figure 9-73. The actual placement of the seat tube is totally up to you, and may vary from what I have, depending on the size of the rider and type of seat you plan to install. For reference, I cut the seat tube to a total height of 8.5 inches, and it is reclined at an angle of approximately 100 degrees, or about the same as it would be on a typical upright bicycle. Ensure that both seat stays are cut to the same length and angle before you weld, or your seat tube will be installed tilted to the side.

Complete the welding around all three of the tubes, as shown in Figure 9-74, to fully secure the seat tube setup. This simple system keeps the original bicycle look and

Figure 9-74   *A nice, big, padded seat to withstand road bumps*

allows a great deal of seat adjustment. It is also very easy to adapt the top of the seat post and rear of the frame for a large wheelchair-style seat if you need such a setup. A fishing boat seat and some kind of belt make a nice lightweight and sturdy alternative to a wheelchair seat that can be easily adapted to this setup using a little imagination and some scrap tubing.

Since the handcycle pilot may have limited or no leg use, the trike needs a method of keeping the feet secure as the unit is in motion. A few straight tubes welded to the front of the frame would do the trick, but this makes an ugly footrest and could become a hazard due to the sharp protruding ends of each tube. You can expect that the rider will eventually bump into something, or try to squeeze the handcycle through a space too narrow to fit into, so some type of sturdy rounded footrest should be considered. As shown in Figure 9-75, a few good candidates for a set of nicely rounded and sturdy footrests could be made from a pair of curved bicycle handlebars.

The placement of the footrest tubing is again a function of your rider's needs but, typically, a length of 10 inches for each footrest would be a decent start. As shown in Figure 9-76, I cut the old mountain bike handlebars in half and then cut an angle at each end that would place the rider's feet slightly below and in front of the middle frame tube. The rider's feet should not interfere with the front wheel when turning, and the rounded part of the footrest tubing should be placed in a position that will not snag on any objects that may rub past them.

These simple footrest tubes can easily be adapted for foot straps if needed using a pair of belts with buckles, Velcro strips, a pair of loose bungee cords, or even some straps made from old laptop carrying straps. As shown in

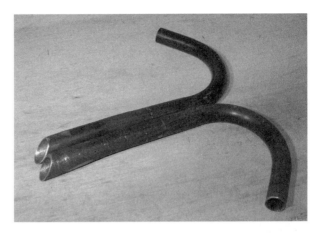

Figure 9-76 *Leg mounting tubes cut to the same angle*

Figure 9-77 *Leg mount tubes tack welded in place*

Figure 9-77, you will weld the foot-rest tubes at the front of the middle frame tube where it meets the lower part of the front tube.

The two footrest tubes will need a bit of help to survive a good smack against any solid object, so a set of simple gussets are formed using whatever materials you have lying around, just like the two small gussets at the rear of the frame were. Thin flat bar, tubing, or even some leftover bits from the frames you have been cutting up will do the trick. Figure 9-78 shows the two 8-inch-long bits of half-inch flat bar I used to strengthen the two footrest tubes. Again, keep an eye out for sharp edges, and round them off, if necessary.

Although it is my opinion that an unoiled chain is far superior to an oiled chain (this is largely debated), you will still want some type of chain guard to keep fingers and clothing bits from easily getting snagged in the chain. Some handcycles have huge, fully enclosed chain guards over the entire top-drive chain ring, and although

Figure 9-75 *Possible leg mount tubing ideas*

Figure 9-78   *Leg mount support gussets*

Figure 9-80   *Three arms will hold the chain guard in place*

this is certainly effective, it does look a bit gaudy, and makes it very annoying to pop the chain back in if it becomes derailed. Some handcycles just use the chain guard that was connected directly to the actual chain ring, but this does not stop things from entering the chain on the drive side. My system is a cross between both styles of chain guards that will create a minimal-looking chain guard that is quite effective in keeping things out of the chain, yet allows easy access to the chain ring in the event of a derailment.

As shown in Figure 9-79, my chain guard is just a bit of thin, half-inch-thick flat bar bent into a circle that approximates the diameter of the chain ring plus a quarter inch. Again, I used the same 1/8-inch-thick, half-inch-wide flat bar as used on all the gussets earlier. The three small bits will form mounting fingers that will hold the guard directly to the bottom bracket.

Bend the main guard in shape by hand so there is about a 1/4 inch of distance between the top of the chain

and the inside of the guard. The three mounting fingers are then welded at a slight upwards angle, so that they can be welded to the bottom bracket and will position the guard ring right over the chain. In Figure 9-80, I welded the fingers in place at a slight upwards angle, and at this point they are a bit too long, since I did not know the exact length to cut them. I will grind them to the correct length by trial and error so I can get the chain guard exactly where I want it.

Figure 9-81 shows the completed chain guard welded directly to the bottom bracket by the three small mounting fingers. The mounting fingers were ground to the correct length by trial and error in order to keep the space between the chain and the guard to a minimum. This chain guard is very sturdy and resistant to bending. I often use it as a handle to carry the handcycle around.

The excess rear axle threads can now be trimmed off flush with the axle bolts, in order to keep the total handcycle width to a minimum so that it will easily fit

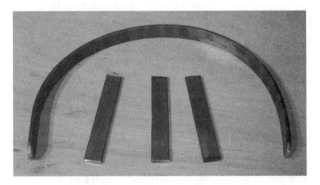

Figure 9-79   *Some half-inch flat bar bent around the front chain ring*

Figure 9-81   *A simple, yet strong chain guard*

Figure 9-82 *Trim away the outer axle stubs*

Figure 9-84 *Perfect ergonomic handgrips are made*

between standard door frames. As shown in Figure 9-82, a quick zip with a thin cut-off disc will do a perfect job of removing the extra axle stub. Once cut, you should paint the bare exposed metal area so that it does not rust, and recheck the nuts to make sure they are nice and tight using the proper cone wrench.

A decent set of "proper" handcycle grips may cost you as much as you put into the UCan2 so far, so let's just make our own set using a pair of commonly available plastic bicycle pedals. Not only will these homebrew handcycle grips feel great, but they will take a standard set of bicycle handgrips perfectly. As shown in Figure 9-83, simply hacksaw the plastic pedal wings away from the core, which will leave you with the perfect sized ergonomically correct plastic handle, ready to take any rubber handgrip. Not bad for a few bucks!

Once you stretch the rubber grips over the plastic pedal body, you will be left with a perfect, ergonomically correct handcycle grip, complete with high-quality, low-friction bearings. As shown in Figure 9-84, the hourglass shape is perfect for power and a solid grip when burning up the track on the UCan2 HandCycle. Don't forget that there is a right and left side pedal, and the one on the left has reverse threads. There is an R or L stamped on the end of the threaded part, in case you are having trouble getting them reinstalled in the crank arms.

Now that the UCan2 is completely rideable, the hardest part of any project is upon you: waiting for paint to dry! Since most of the frame was made from new tubing with the exception of the seat mounting tubing and front forks, a very light sanding with a few sheets of emery cloth was all that it took to get the metal ready for a coat of primer and paint.

I don't have a powder coating station, nor do I have any place in my garage to paint, so I simply hang my frames from an outdoor clothes line and go to town with a few cans of spray paint. If you take your time and follow the instructions on the can, a remarkably good paint job can be accomplished from a few cans of spray paint (see Figure 9-85). It is true that spray paint will scratch easier than auto paint, but to look at the final paint job, you would not be able to tell the difference.

When the paint is dry and cured, reassemble the handcycle and repack all the bearings with fresh grease. You may also want to install some grip tape on the footrest tubing, and the rear of the frame where passengers are very likely to hitch a ride. How about a

Figure 9-83 *Plastic pedals cut down to the shell*

Figure 9-85 *A good paint job makes your ride look sweet*

Figure 9-87 *The first test ride*

Figure 9-86 *The UCan2 can turn a very small circle*

Figure 9-88 *The UCan2 laughs at this kind of abuse!*

horn, or some cool flame stickers? Accessorize! Figure 9-86 shows my fully assembled UCan2, about to get some grip tape and a few cool stickers. Notice the small, black, plastic plugs placed at the ends of the footrest tubes. It's always a good idea to plug all open-ended tube to keep moisture and critters out of your frame and to remove all sharp edges.

Well, what are you waiting for—get over to that skate park and see what this little beast will do! At first, your handcycle pilot may find getting started a bit awkward, but after an hour or so, the trike will feel fast, stable, and get moving with the flick of a wrist. The UCan2 makes the rider work muscles that normally don't get much action, so as time goes on, top speed and climbing ability will also increase. Oh, and don't forget to practise using

the brakes; we all have to slow down once in a while! Figure 9-87 shows the mighty Brittany blasting those upper-body muscles on the UCan2.

Sure, we can give you a lift—try this on your spindly, pedal-powered wheelchair, pal! The UCan2 will handle as much weight as you can sensibly pack onto it, as long as your passenger can handle the extremely fun G-forces on every corner. As shown in Figure 9-88, I am getting one serious upper-body workout as I race towards a nice steep hill to see how agile my passenger really is. For the record, it is the passenger who really needs the helmet (sorry for running you over, Brittany)! So there you have it—the UCan2 freestyle handcycle looking cool and ready to tackle any terrain that its two-wheeled relative can. This handcycle is fast, agile, and extremely durable and looks like it wants to be ridden, not like it *has* to be ridden. Not bad for a couple of weekends of work and a handful of standard bicycle parts!

Figure 10-1 *The donor bike ready to be chopped*

Now what would an Evil Genius vehicle book be without a tallbike? A tallbike gives you the sensation that you are hang gliding without a hang glider! You get to face your fear of heights, confuse your onlookers, and peer into the neighbor's second storey window all at the same time! I have been a fan of tallbikes ever since I figured out how to hack bicycles, and have even had some tallbike fame by getting my photo in the 2005 gold edition of the Guinness Book of World Records. Yep, that dude on page 134 slinging the Canadian flag would be me—cool bike, eh? Anyhow, you have not earned your bicycle hacking stripes until you have ridden a tallbike, so let's get building.

This tallbike is called the SkyStyle because it is based on a freestyle BMX bike, and would be perfect for jousting competitions and many other foolish tallbike tricks. The overall height is high enough that you have a good time defying gravity, yet not so high that you can't mount the bike without a ladder. Because we will keep the wheelbase the same as the original BMX bike, you can play fair at jousting competitions, pull wheelies if you dare, and many of the "flatland" stunts you might try on an earthbound BMX. For starters, you will need to dig in the back of your tool shed for that 1980s BMX that your mom bought you after you watched Eliot fly through the air with E.T. in the basket. Actually, any BMX bike with 20-inch wheels will do the trick, just like the one shown in Figure 10-1, before getting the axe.

Figure 10-2 *The parts ready for hacking*

You need two decent 20-inch wheels, a sturdy frame, front forks, and all the crank set hardware in order to make the tallbike. Figure 10-2 shows all the parts disassembled, cleaned, and ready for severe modification.

The overall height of your final tallbike will depend on the length of the head tube plus the diameter of the front wheel. Since this tallbike is designed so that you can step up from the ground using foot pegs, you will want to keep the total height from the handlebars to the ground at reaching distance. The total height of my tallbike from the ground to the handlebars is approximately 7 feet, so the head tube was made to a length of about 5 and a half feet (7 feet minus the height of the front wheel to the top of the fork hardware). You will need to find a tube with

Figure 10-3 *Making the ultra-long head tube*

Figure 10-4 *Extending the fork stem to match*

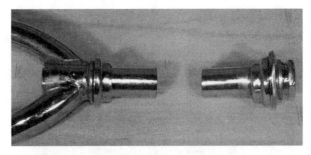

Figure 10-5 *Figuring out the extension length*

Figure 10-6 *Extending the fork stem*

about the same inside diameter as the original head tube so that the bearing cups will fit inside, as shown in Figure 10-3. A piece of electrical conduit or fence tubing may be just the thing you need.

Once you have chosen an overall tallbike height and cut the tube to make the head tube, you will have to cut and extend the fork stem to match the new head tube. Start by cutting the fork stem in the middle of its length, as shown in Figure 10-4. A tube cutter does a nice job here if you have one.

To calculate the correct fork stem extension length, install the baring hardware, as shown in Figure 10-5, then lay the cups along the extended head tube so you can measure the missing section of fork stem. This will help ensure that all the fork hardware fits correctly once you have extended the fork stem to match the new head tube.

To extend the fork stem, find some tubing that is fairly close to the diameter of the fork stem, cut it to the required length, and then weld the parts together. Figure 10-6 shows the extension tubing I have found to make my fork stem longer. This tubing was cut from an old barbell, which had the exact diameter of the fork stem tubing.

Make the fork stem extension welds as strong as you can, ensuring that there are no cracks or holes. Remember, if these welds let go, you are going to learn the meaning of free fall as you make an instant flight towards the pavement! Figure 10-7 shows the completed weld on one end of the extending fork stem. Also, ensure that the bearing cups fit over the weld, and only grind as much material as necessary if there is a little too much weld buildup around the joint.

Figure 10-7 *Welding the fork stem extension*

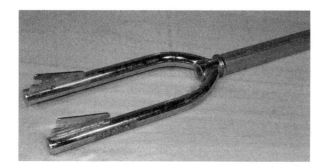

Figure 10-8 *Completed fork set and head tube*

Figure 10-10 *Installing the new tall head tube*

When you are done extending the head tube and fork stem, the two parts should fit back together just like they did on the original bicycle they were taken from. Figure 10-8 shows the completed fork set and head tube with all bearing hardware installed. The forks should turn freely and without friction once everything is installed.

Now that your extended head tube is completed, you need to hack away the original head tube from the BMX frame so you can replace it with the new one. As shown in Figure 10-9, make the cuts as close to the head tube as possible to keep the frame tubing as long as possible. You don't want to shorten the wheelbase any more than necessary, or you will be doing wheelies all the time! You could also slice the head tube in half lengthwise and weld it directly to the new head tube if you wanted to.

The new extended head tube will be installed onto the BMX frame, as shown in Figure 10-10, just like the original head tube once was. The one thing to keep in mind is head tube angle versus handlebar-to-seat distance. As you add more head tube angle, you make the

Figure 10-9 *Removing the original head tube*

distance between the seat post and handlebars decrease, so you will have to find an angle that gives you the best of both worlds. Draw an imaginary line straight up from the original seat post tube, and then try to install your new head tube so that the distance between this imaginary line and the top of the head tube is no less than half of what it is at the bottom of the frame. Dude, does that make sense to you? If not, let me explain it another way. If the distance between the original seat post and head tube of the BMX frame was 28 inches, then the distance from the top of the new head tube to the new seat post should not be less than 14 inches.

The seat tube must now be extended upwards vertically as high as the top of the new head tube, as shown in Figure 10-11. Any length of 1 inch or 1.25 inch tubing will do. I used a bit of electrical conduit that was lying around the shop. The new seat tube should extend upwards from the frame at an almost vertical angle, keeping in mind that the more it leans back, the easier it will be to wheelie your tallbike. Maybe you want this instability for a jousting tournament? For the best stability, try to plant your butt ahead of the front wheel, not over the top of it.

The new top tube is installed between the top of the new head tube and seat tube, as shown in Figure 10-12. The length of this tube is determined by the angle of your head tube and seat tube, and it should be at least 16 inches long if you plan to ride this bike without having your knees banging off the handlebars. I just found an old tube hacked from some other frame that was used for another evil vehicle. Recycle, reuse!

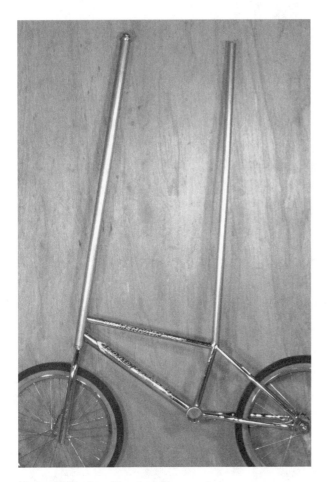

Figure 10-11 *Installing the long seat tube*

Figure 10-12 *Adding the new top tube*

Your tallbike is almost ready, but the frame will need some reinforcement if it is to withstand the crashes that will surely be coming in the near future. With only the head tube and seat tube extending the frame, the ride would be seriously wobbly, and the frame would easily warp if you were to mow into a curb or smack head on into "Sir Bleedsalot's" tallbike during the jousting finals.

Figure 10-13 *Adding some strength to the frame*

By adding a simple truss into the open frame, you create a triangle, which will make the frame many times more rigid and stable. This tube can be any appropriate length of 1 inch or even ¾ tubing, and it is installed as shown in Figure 10-13.

To step up to the cranks during a mounting procedure, you will need a lower set of foot pegs unless you have abnormally long legs. You can purchase some nice foot pegs at the bike store and weld a bolt to the frame to hold them, or build the poor man's foot pegs, as shown in Figure 10-14, from some scrap tubing and a few washers. The washers are a safety precaution that will save your hide from impalement during one of those wonderful moments when you are lying on the road and your tallbike is falling over on you. Oh yes, you will understand this one day, I promise!

As shown in Figure 10-15, the washers are welded and then ground smooth so there is no sharp edge to your foot pegs. Sharp edges are not a good thing on a tall bike, especially one designed for battle.

Figure 10-14 *Poor man's foot pegs*

Figure 10-15   *Completed foot pegs*

Figure 10-16   *Foot peg position*

Figure 10-17   *Adding the bottom bracket*

Figure 10-18   *Installing the bottom bracket*

The lower foot pegs are installed on the lower part of the extended seat tube, as shown in Figure 10-16. This position makes it "easy" to step up to the pedals as you launch the tallbike from a standing start. If you plan on being a total maniac of the sky, then add more pairs of foot pegs all over the bike, so you can astound your buddies on the ground with your fancy footwork.

You will need to hack a bottom bracket from another frame in order to install the cranks onto your tallbike. The best way to do this is to cut it from the donor frame so an inch or so of the original tubing is still there, as shown in Figure 10-17. This way, you can simply weld the tubing stubs directly to the new frame.

The new bottom bracket is installed in an appropriate place on the new seat tube, as shown in Figure 10-18. Where you put the bottom bracket will depend on how long your legs are (your inseam), but you could simply measure this from a bicycle that fits you.

If you get lucky, the distance between your crank set and rear wheel sprocket will be just perfect so that you

can join together your new chain and have it nice and tight. You can move the rear wheel slightly in the dropouts to pick up some of the tension, but there is a chance that your chain might be way too slack once it is made. A chain tensioner is needed when you add one more link and your chain is too long, or if you remove a link and the chain is too damn short. Any half-inch V-belt idler pulley can be used to pick up the chain slack by simply welding it in the correct place somewhere on the frame. An example pulley is shown in Figure 10-19.

Figure 10-20 shows how the idler pulley can be used to pick up any chain slack that cannot be taken up by simply moving the rear wheel in the dropouts. Join your chain until you have as many links as necessary,

Figure 10-19 *You may need a chain tensioner*

Figure 10-20 *Installing the chain tensioner*

Figure 10-21 *Dude, let's ride!*

Figure 10-22 *Step one of the launch procedure*

and then hold the pulley somewhere on the frame so you can mark the best place to weld the bolt that holds it there. A tallbike chain should be fairly tight, as a derailment will mean a dismount—an ugly dismount if the chain happens to get tangled in the rear wheel!

Figure 10-21 shows the finished SkyStyle tallbike, complete with a sloppy, black, battle-ready paintjob. Hey, why worry about paint quality on a bike that is certain to end up in a tangled mess at least a dozen times during a jousting competition? The thick clothing helps me avoid road rash, as well as the harsh subzero climate of northern Ontario!

Mounting the SkyStyle tallbike is a process that is easily learned. As shown in Figure 10-22, step one involves gripping the handlebar with one hand as you kick off from the ground to get the bike rolling, while one foot is on the lower foot peg. It may seem odd to steer a bike by reaching 7 feet into the air, but if a

monkey can learn to ride a unicycle, then you certainly have no excuse!

As shown in Figure 10-23, step two of the launch procedure is a little more precarious. At this point, you must not only steer the tallbike from one side as it leans, but also climb from the lower foot peg up to the pedal on the opposite side of the bike. Hopefully, you had the pedals in the correct position before you started. Oh, did I forget to mention that? Oops! Now you know.

Figure 10-23 *Step two of the launch procedure*

Figure 10-24 *Step three of the launch procedure*

If you are actually standing on one of the pedals with both hands on the handlebars, as shown in Figure 10-24, then you are past the hardest part of the launch procedure and home free. If your launch has not gone well according to plan, then you are lying on your back on one side of the street, with your tallbike on the other side of the street, as the neighbors point you out to their kids through their picture window. "Now, Billy, this is why I won't let you play with that Graham boy—he is a bad example." You know the routine by now. Well, we hope you have fun with your SkyStyle tallbike. Watch out for low-hanging power lines, and remember to put your hands out in front of you on the way down during battle! Figure 10-25 shows the SkyStyle pilot out for a leisurely ride, free of ankle-biting dog attacks, and highly visible in traffic. Oh, and yes the tallbike does have brakes; they are built into the rear wheel in the form of a coaster brake which engages when you pedal in reverse. Have fun, and respect the laws of gravity!

Figure 10-25 *Off to battle we go!*

Next up, an entire section devoted to electric vehicles, from a cool mini bike to an electric trike suited to the younger crowd.

Electric Power

The goal of this project is to arm you with the necessary information and skills to create your own fun electric vehicle (EV) from scratch. There are many inexpensive electric scooters and pocket bikes available, but let me assure you that the acceleration, hill climbing ability, top speed and overall durability of these units cannot match the quality of a home-built unit such as Sparky. Of course, these traits are dependent on your choice of drive motor, batteries and overall gear ratios, but even with a small investment in time and money, you will be able to build a one of a kind EV that can climb hills with ease, operate for well over an hour on a single battery charge, accelerate with enthusiasm, and be able to take on most terrain without falling apart. Unlike petrol-burning mini bikes, the electric motor and transmission system used in Sparky are so quiet that the only sound made by the unit is the sound of the wheels hitting the pavement, even at full speed. This silent operation allows the unit to be ridden practically anywhere without polluting the area with noise. Your neighbors and the general public will likely not be annoyed with this type of quiet EV.

Unlike many of our do-it-yourself plans, this project does not require a rigid set of materials and measurements, since the goal is to build the electric vehicle to your standards using the parts you have available. Sure, you could build an exact replica of Sparky by following the instructions, but with the great availability of various small electric motors, batteries and wheels, why limit yourself to a single design? You may want to build a smaller EV for youngsters to ride at camp, convert an old gas mini bike to electric for hunting or trail riding, or go all the way and build a fully street legal motorcycle—this is all possible by choosing the appropriate parts. When I built Sparky, my goal was to pack as much power, range, and ruggedness as I could into the smallest possible size, and with a top speed of approximately 31 mph (50 km/h), streetbike-like acceleration, and a full hour run time on a charge. To add icing to the cake, I accessorized Sparky to resemble a pint-sized Harley fat boy, complete with tree-trunk front forks, large fenders, and an old-school-style seat.

Electric motors have nothing in common with their gas-powered counterparts, even though they both may be rated in horsepower (HP) and rotations per minute (RPM). At only 2.5 HP, the coffee-can-sized motor used on Sparky will generally outperform a 5 HP gas engine (commonly used on go-carts), especially during acceleration. Motor controllers are another alien technology, especially if you have worked on gas-powered vehicles before, but don't worry, it will all be explained in this section, which is why the best way to approach this plan is by reading the entire text all the way through in order to understand the technologies used, and to see why I choose the parts used to build Sparky. Once you understand the core technologies and components needed, you can build a small electric mini bike just like Sparky, or create your very own custom electric monster from scratch.

The batteries like the ones shown in Figure 11-1 are probably the most important component in any electric vehicle, large or small. If your batteries cannot deliver

Figure 11-1 *Two 12-volt GEL batteries*

the needed amperage, then it really does not matter how powerful your drive motor is. One should not compare battery size to gas tank capacity on a gas-powered vehicle, since a teaspoon of fuel will run a gas-powered engine at full power, even if for a limited amount of time, whereas a tiny battery will simply fail to turn an electric motor at all. Batteries are rated in voltage and amp hours, and these two variables will determine how fast your motor will turn, how much horsepower it will deliver (more on this later), and for how long. OK, get out your pocket protector; things are going to enter the "nerd zone" for a while!

Although the "fuel" for an electric motor is a function of the amperage and voltage of a battery, there is no way of comparing this to a simple tank of gas. If I had to guess, then I would estimate that amperage is the quantity of fuel per hour delivered to a gas engine, and the voltage is the octane, or burn quality, of the gas. The more amperage a battery can deliver, the more torque the motor shaft will deliver, and the more voltage this amperage is delivered with, the faster the motor shaft will spin. Although this is an oversimplified explanation, it does hold true for many of the small electric DC motors you will have to choose from when sourcing parts for your mini bike. Before you can choose a motor, or the appropriate batteries, you should know the basic facts and formulas dealing with amps, watts, volts, and horsepower.

Watts

Most small, electric, direct current (DC) motors are rated in watts, and wattage is derived by multiplying voltage by amperage (volts × amps). So, if a motor claims that it will deliver 900 watts at 36 volts, then you know it will take 25 amps from your battery pack (900 watts divided by 36 volts). This simple formula works, as long as you know any two pieces of the watts, volts and amps puzzle. Remember that 250-watt light bulb in your garage? Sure, just divide 250 watts by 120 volts, and you will then know that your light bulb needs just over 2 amps to operate. The wattage ratings of a DC motor is the most important of all ratings, and this value will determine the size and type of battery and motor controller needed, as well as the overall performance of the vehicle. The motor I used on Sparky is rated at 2000 watts continuous.

Horsepower

This rating isn't really very useful in the electric motor world, since an electric motor's horsepower is rated at zero RPM under load, whereas a gas engine's horsepower is rated at its maximum RPM under no load. What this means is that an electric motor rated at the same horsepower as a gas engine will put the gas engine to shame when it comes to acceleration and torque. To make things even more confusing, gas-powered engines are rated at their peak horsepower, while electric motors are rated at their continuous rated horsepower. The peak horsepower of an electric motor is usually 8 or 10 times its continuous rating, so this makes an electric motor of an equal horsepower seem all the more powerful. If you really want to rate your motor in horsepower, or find one that is rated in this way, then you can convert by using the ratio of 1 horsepower equals approximately 746 watts. To compare this to a gas engine, I like to assume that a 1-horsepower electric motor will feel like a 3-horsepower gas engine. The motor used on Sparky is 2000 watts, so this works out at 2.68 horsepower (2000 watts/746), which should seem like a gas engine of approximately 8 horsepower.

Amperage ratings

Battery amperage may be stated in various ways depending on the intended use for that particular battery. You will see things like CCA (cold cranking amps), MCA (marine cranking amps), RC (reserve capacity), and AH (amp hours). It is the last rating, AH (amp hours), that we are really interested in because this is the maximum energy that can be delivered from the battery every hour. By knowing the power rating of the motor, we can figure out the minimum run time for your vehicle before it needs a recharge. I say minimum run time because your motor will not always need maximum amperage, and if you are not climbing large hills, or driving like a stunt person all of the time, you may only be using a small percentage of the motor's rated current. Using a few of the above formulas on Sparky's 2000-watt motor, I can determine that it will draw a maximum of 83.33 amps at its rated 24 volts (2000 watts 24 volts), and since my battery pack can deliver 32 amp hours, I would expect a run time of only .38 hours, or about 20 minutes (32 AH/83.33 amps). This assumes that I am forcing the motor to work at its full rated capacity at all times, and this is never the case; in reality, I typically get 1.5 hours or more between charges, even when spinning up dirt and pulling wheelies. A realistic measure of run time between charges will be something you will have to find out from actual use, as it will certainly vary.

Voltage

Each battery has a rated voltage, with 12 volts being the most common for batteries typically needed for small electric vehicles. Since most small electric motors run at voltages between 24 and 48 volts, you will have to connect two or more batteries in series to achieve the required voltage rating for the battery pack. A series connection means that the positive terminal of battery #1 connects to the negative terminal of battery #2, and power is taken from the free terminal of each battery (one positive, and one negative). In series, a battery cannot deliver any more amp hours than the rated capacity of a single battery, but it does double the wattage available because it also doubles the voltage. My two batteries shown in Figure 11-1 are rated at 12 volts, and 32 amp hours each, so in series they deliver 32 amp hours at 24 volts, or 768 watts (24 volts × 32 amps). A single battery would only deliver 384 watts (12 volts × 32 amps). If wired in parallel (not recommended), the batteries would double their amp hours to 64, but the voltage would remain at 12 volts, so they would again deliver 768 watts (12 volts × 64 amps). Parallel wiring of batteries should never be done, since one battery may try to charge the other if they are not equal, which could result in damage, heat, or fire. To sum this up, you will need to connect as many 12-volt batteries in series as necessary, in order to achieve the desired voltage, but your overall amp hour capacity will be the same as one single battery can deliver.

Battery type

Not all 12-volt batteries are the same, and this fact is largely reflected in the price. Typical car batteries are not well-suited for electric vehicle use, even though their price may be inviting. A car battery is made to deliver an enormous amount of power in a very short time and then become fully charged shortly after. As many of us know, a car battery will be destroyed if run completely down more than a few times in a row, which is why it will not be a very good candidate for our purpose. Car batteries are also very messy, require some fluid maintenance, and do not like vibration or uneven mounting. A much better choice for your battery needs is the non-spillable, sealed, GEL deep cycle battery. These batteries have no removable tops since they contain no liquid (it is in a gelled state)—they can be mounted in every orientation except upside down, they can withstand moderate vibration, and they can be depleted completely many times without damage. These batteries are commonly rated as marine or wheelchair-use batteries, and the smaller units may be rated for security or battery backup use. They are more expensive than conventional car batteries, but will outlast them many times over, especially in our electric vehicle. Multiple batteries should always be the same type, and approximately the same age, especially when charging them together as a 24 or higher voltage pack.

Charger

The best way to charge your batteries is together as a single pack, so if your battery pack voltage is 24 volts, you will need a 24-volt charger. These are readily available, and often you can find older ones at secondhand stores, or from shops that repair or sell wheelchair parts. A common 12-volt charger used for a car battery will also work, but you will either have to charge one battery at a time (very slow), or purchase two chargers and run them both, one on each battery. Whichever you choose, the charger should be fully automatic, shutting off when the battery is at peak capacity in order to avoid damage from overcharging. Only the worst chargers available will not have a current meter or automatic shutoff, and these should be avoided. Also, make sure that the charger you have is rated for the type of battery you are using, such as wet lead acid, or sealed gel.

I hope that wasn't too much nerd jargon in such a small space! It really does help to know what to expect from batteries and motors before you start searching for parts, and the above information is the basic lingo you will have to know when searching the numerous sources for motors and batteries for your vehicle. I still recommend that you read this entire plan before sourcing any parts, as there are more tips and facts further ahead that will help you beat the pitfalls of part sourcing, and avoid many common mistakes while hunting for parts. As the build progresses, each major component will be described in more detail. Hey, buddy, put that furnace motor down and read the rest! It wouldn't work for this project anyhow, trust me.

Once you have read through the entire plan, and started from the beginning, you will probably have chosen a motor and the appropriate batteries to achieve the desired acceleration and run time for your mini bike. The batteries make up the bulk of the weight and size, so we will start by building a strong cage or battery box to hold them together as a single unit. A battery box can be easily removed from the rest of the frame for charging, transporting the mini bike, and allowing you to have multiple packs for continuous non-stop fun. A nice battery box can be made by forming a base, like the one shown in Figure 11-2, from some 1-inch, 1/16-inch-thick angle iron. You will want to cut the pieces, as shown in Figure 11-2, so that the battery sits into the base of the box, but not too tight that it becomes stuck. The inside measurement of the box should be at least 1/8-inch larger than the battery.

The four pieces of angle iron that make up the battery box base are shown in Figure 11-3, tack welded together at the corners. The angle iron could have also been formed by cutting the ends at 45 degrees, but I find this way to be a bit more tolerant of small errors that could make the box too small for the batteries to slide in easily. With only the corners tack welded together, it

Figure 11-3 *Welding the battery box base pieces*

is easy to force the box into a square by tapping it with a hammer.

Before you completely weld the battery box base together, test fit one of your batteries to make sure that it is snug, but is not wedged in too tight. As shown in Figure 11-4, my 32-amp-hour GEL battery drops right into the base with approximately 1/8-inch clearance between the battery edges and the angle iron. Once tested, you can weld the outside corner and underside of all joints to complete the base of the battery box. Do not weld the inside joints, or your battery may not fit. For reference, my battery measures 7.75 by 6.5 inches, and is 6 inches tall to the top of the terminals.

As stated earlier, my battery pack will be made from two series-connected, 32-amp-hour, deep-cycle,

Figure 11-2 *The base for the battery box*

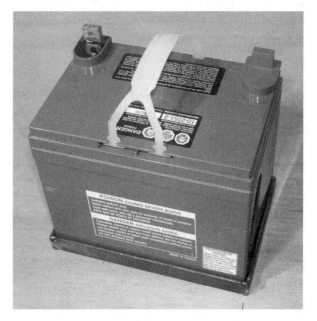

Figure 11-4 *Testing the battery fit*

GEL-type batteries at 12 volts each. The resulting battery pack will be able to deliver 32 amps at 24 volts for one hour, or any ratio of 32:1 up to the battery's maximum discharge ability, for example, 64 amps for 30 minutes, or 16 amps for two hours. The current draw from the motor during typical riding will vary quite a bit, with acceleration from a dead stop requiring the most current, and cruising on the flats without any headwind requiring very little current. Your ultimate acceleration will depend on your battery pack's maximum current discharge, and the maximum current that will flow through your motor controller if you decide to use one.

On a small electric vehicle like Sparky, the maximum discharge of most deep cycle batteries will be more than enough for powerful acceleration, bordering on the wheelie zone, but don't expect to throw on a huge amp-hungry motor and a pair of tiny batteries and have them deliver 900 amps in one minute just because they can deliver 15 amps in one hour—there is a limit. If you really want to know what the maximum discharge capacity for your battery is, then you will probably have to spend some time digging around on the manufacturer's website.

The sides of the battery box are made from the same 1-inch angle iron as the base, and are welded at each corner to continue the box upward along the corners of both batteries, as shown in Figure 11-5. The top of the box should be at least half an inch higher than the top of the battery terminal to protect the terminals from the rest of the frame. I made these four side pieces 14 inches in length for my battery pack. The four side pieces are first tack welded in place, so that you can do another test fit with the batteries before welding the outside of the joints. There should be a bit of wood or rigid foam between the two batteries, so that the top battery is not resting directly on top of the lower battery's terminal, as excessive vibration may damage the top battery. Do not attempt to make any welds with the batteries installed or they will melt!

The top of the battery box is made by welding flat bar of the same thickness as the angle iron across the top, as shown in Figure 11-6, so that it completes the box. Flat bar is needed since angle iron would make it impossible to install the batteries due to the lip it creates. If you do not have any scrap bits of flat bar to use, you can cut the angle iron along the edge with a grinder disc to make two pieces of flat bar from it. Again, it's a good

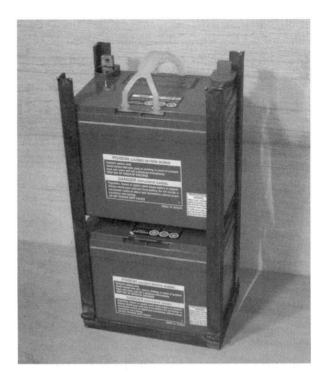

Figure 11-5 *Installing the battery box sides*

Figure 11-6 *Finishing the battery box*

idea to tack weld and test fit before committing any welds, just in case you need to make a few adjustments. Once everything is looking good, make the final welds and clean up the joints with a sanding disc, as shown in Figure 11-6, to complete the battery box.

Once you have the battery box completed and have chosen a drive motor (or at least know the measurements of the drive motor), you can begin working on the frame. My goal was to build the smallest frame possible around my battery pack, drive motor and motor controller without an inch to spare. I thought it would be fun to have a cute-looking "mini hog" that would look as though it was a kids' toy, but have more than enough power to leave any department store scooter or pocket bike whimpering in the dust during a race.

My original electric mini bike had a frame made of square tubing, but this time I wanted a little more style, so I decided to go with a nice curved "spine" over the top that would define the general proportions of the rest of the frame. To achieve this without a pipe bender, I bought two 1.5-inch thin walled electrical conduit (EMT) elbows from a hardware store. In Figure 11-7, I marked the area of each elbow that I planned to cut, in order to weld them together to form the appropriate sized spine for my frame. If for some reason you cannot find EMT where you live, then any 1/16-inch walled mild steel tubing (even square) would work. You could even butcher a mountain bike frame for the large tubing used in the frame.

Feel free to experiment with your own frame designs. As long as you build the frame around your key components in such a way that everything fits together, and so that your belt or chain will have adequate clearance, your imagination can run wild. As you read on, you will see why I chose the size and style of materials I did, but you could use the same techniques to create a variation of Sparky, a full-size motorcycle, a chopper, or even a four-wheel go-cart if you wanted to.

Since I was going to build the frame spine over the main components, then build the rest of the frame around

this configuration, I only tack welded the two elbows together at this point, just in case I decided to cut them shorter. As shown in Figure 11-8, there would be plenty of length for the battery box, drive motor and rear wheel under this smooth-flowing spine. It's always easier to cut the excess away after you have a plan, rather than being stuck with a piece that's too short.

Figure 11-9 shows visually how I attack a frame when I have no plan, but do have most of the components. Since the battery box would require the most space, and dictate the overall height of the frame, it was important to finish it first so that the other components could be placed around it. My goal was to pack as much "stuff" into the smallest area possible, so it was very important to lay out the frame like this, just in case something unexpected came up. The space between the battery box and the rear wheel is the amount of space the motor will occupy, and although my motor was still on order from the supplier at the time, I had detailed drawings, and knew the exact size and shape. With the motor installed directly between the rear wheel and the battery box, there would be just enough room above the motor for either a large motor controller, or an automotive solenoid, depending on which way you plan to control the current

Figure 11-8 *Elbows tack welded together*

Figure 11-7 *A pair of 1.5-inch conduit elbows*

Figure 11-9 *Visualizing the frame layout*

(more on this later). This curved spine was a perfect fit because it placed the rider low over the rear wheel, and then made a nice flowing curve over the battery box, so that the final mini bike would be very strong yet stylish.

Even though I was satisfied with the main spine, I decided to leave it uncut for now. You just never know how things are going to progress when you tackle a project like this without a rigid plan, and sometimes the best ideas present themselves randomly as your build progresses. This ability to build a custom project, just like an artist paints a picture from scratch, is the very skill that will set you apart from those that can only build from a strict set of plans dictating every single cut and exact parts. Trust your instincts, and don't be afraid to try new things!

The main spine is the kingpin tube for the entire frame, and all the other tubing simply flows around the main components, basically holding the guts in place like a ribcage. The two diagonal tubes shown in Figure 11-10 are called the "down tubes," and they will connect to the front of the main spine just behind the head tube, and run at an angle to the bottom sides of the battery box. If you look at most motorcycle frames, you will see a very similar geometry to many frames, but the gas engine is where the battery box in our vehicle will be. These tubes do not have to be nearly as strong as the main spine since they are mainly there to "hang" the battery box, so I made them out of two lengths of ¾-inch thin walled electrical conduit. Again, if you have no source for EMT, then any thin walled mild steel tubing will do the trick, and a perfect candidate would be the tubing used for the seat tube in a bicycle frame.

The lengths of both tubes will depend on the height of your battery box, the goal being an inch or so longer than the diagonal distance across the top and bottom, as shown in Figure 11-10. For the record, I cut my two tubes at 16 inches each to clear my battery box and allow a bit of space to work with. Both tubes will be fish mouthed to form a joint with the other frame tubes at a later time, so a little bit of length will be lost.

The down tubes will connect to the bottom tubes, which will travel under the battery box and continue to the rear of the frame, where they will hold the rear wheel bearing mounting plates. For this reason, the bottom tubes are made from slightly heavier tubing than the down tubes, but do not need to be as large as the main spine tube. As shown in Figure 11-11, I cut two lengths

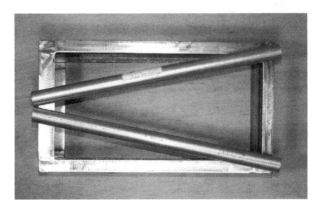

Figure 11-10 *These will be the down tubes*

Figure 11-11 *Preparing the down tubes and bottom tubes*

of 1-inch thin walled electrical conduit to be used as the bottom tubes, and then I made the appropriate fish-mouth cuts in the down tubes to form a proper joint for welding.

The bottom tubes need to be long enough to make the distance from the front edge of the down tubes, under the battery box and drive motor, and then past the area where you plan to mount the rear wheel bearings. It's best to read ahead again before you choose a length to cut these tubes, and when you do have an idea as to how long they need to be, add at least 6 inches extra just in case (you can always chop it off later). By the time you get to Figure 11-16, the function and length of the bottom tubes will be clear. My bottom tubes were cut to a length of 20 inches.

Nothing screams out homemade more than an unpainted frame or a frame with open-ended tubing. Open-ended tubing is sharp, prone to filling with dirt or water, and not as strong as capped tubing. It is really easy to plug the end of a round tube by simply welding an appropriate sized washer over the end. As shown in Figure 11-12, I tack welded a pair of washers over each end of the bottom tubes. It is much easier to get this done ahead of time, since you can work with the tubing on your bench as a single unit. Once welded to the frame, it will be difficult to get your grinder into position. Get 'er done!

The tubing caps are welded all the way around then ground down, as shown in Figure 11-13. The completed product looks very professional, and will resist filling with debris when you ride, and will not be a source of danger if for some reason the frame ever came in contact with somebody. You could also fill the washer holes if you like, but I decided to leave them as they are, just in case I want to add a few accessories, such as reflectors or lights to the frame.

As stated earlier, the bottom tubes will connect to the base of the down tubes and travel under the battery box to the rear of the frame, so they need to be welded, as shown in Figure 11-14. I chose an angle of 90 degrees so that the down tubes run directly along the front of the battery box up to the main spine tube. This allows the most compact configuration possible, placing the front wheel very close to the front of the battery box. On many motorcycle frames, the down tubes travel from the base of the engine towards the head tube at a slight angle, and the front wheel has several inches of clearance between the motor and frame due to the rake of the front forks. More front fork rake generally means a more stable bike at higher speeds, but my goal for design was size, and I did not plan to hit the highways or break any speed records, so minimal rake would be fine. Again, feel free to experiment with your own design, and look at other mini bike and motorcycle frames for ideas. Wouldn't a chopper be cool? You know it!

Once the down tubes and bottom tubes are welded completely around and ground clean, they will be installed, as shown in Figure 11-15, so that the bottom tubes continue to the rear of the frame, and the down tubes travel along the front of the battery box at a diagonal angle to meet just under the main spine tube. You may want to clamp the down tubes to the bottom of the battery box, just so they stay in place while you test fit the rear wheel and make the other tubes for welding. You may also want to grind the tops of the down tubes to form a good joint for welding with the main spine tube,

Figure 11-12   *Plugging up open tubing ends*

Figure 11-13   *Washers welded and ground*

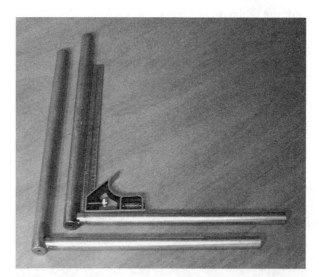

Figure 11-14   *Joining the down tubes and bottom tubes at 90 degrees*

Figure 11-15 *Laying out the frame tubing*

but don't do this until you visually lay out the rest of the frame, as will be shown next.

The next weld you make will pretty much dictate the final shape of the frame, position of the rear wheel and maximum motor size that you can install into the frame. As shown in Figure 11-16, it's important to visually check your layout before making any final welds, just in case your drawing or rough plan is not as accurate as you might have thought. There are some potential problems that may not be obvious in a drawing. Will the rear wheel slide out of the frame? Will the inflated tire rub on the

bottom tubes with pressure applied? Can the battery box slide out of the frame easily? Once the rear bearings are installed, will the rear tire clear the back of the main spine tube? As you can see, the only way to feel confident about committing to a final frame design is to try to answer all of these questions visually.

You should try to set up the mini bike using as many of the main components as you can, such as the rear wheel, axle, bearings, and motor before welding the top of the down tubes to the main spine tube. Don't forget about fender, sprocket or pulley clearances, and brake drum arms if your wheel has one.

Figure 11-17 shows the clearance above the battery box and main spine tube after tack welding the frame parts together, based on the visual layout in Figure 11-16. I ended up with a quarter inch just above the top of the battery box, which would be plenty of room to allow the quick removal of the pack for charging or swapping. Make a few good tack welds between the down tubes and the spine tube to hold it in place and allow for any alignment fine tuning. Alignment should be checked from directly above the frame as well to make sure that both bottom tubes are parallel to each other, and that the spine tube runs right down the middle of the bottom tubes. At this stage, it is easy enough to manipulate the frame to get it back into alignment, but if it is too far out, you are better off breaking the tack welds and redoing them. The best welders are not those who get things perfect on the first try, but those who fix all their mistakes.

To add a bit of style to the little frame, and make good use of the mountain bicycle frame to be used for the head tube, I decided to recycle the fork legs for the last part of the frame, which I will refer to as the "rear tubes" from now on. The rear tubes will complete the frame by

Figure 11-16 *Installing one of the truss tubes*

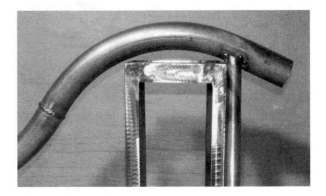

Figure 11-17 *Tack welding the frame together*

fastening the rear of the bottom tubes to the rear of the main spine tube and allowing a place for the bearing mounting plates to connect, if your design calls for them. Although you could have simply used a few lengths of 1-inch or 3/4 EMT for this part of the frame, the fork legs offer a nice tapered and flowing look to the overall design that perfectly complements the curve of the main spine. Remember that building a custom is an art, not a rigid set of rules!

When cutting fork legs, try to get as close to the stem as possible without cutting into the stem, because we will later use that stem to create the triple-tree fork. You can also cut off the fork dropouts as close to the end as possible, and add them to your ever growing pile of parts for some other project. Figure 11-18 shows the legs cut away from the stem with minimal waste. I make all my cuts with a 3/32 cut-off grinder disc, but you could also do this with a hacksaw if you are worried about slicing into the stem.

As shown in Figure 11-19, the fork legs will form the rear tubes by completing the main frame. There is no

magic measurement dictating where on the main spine tube the fork legs should be placed, but it is important that when welded to the bottom tubes, the axle will end up in the correct position to allow for adequate clearance between the motor and rear tire. I have built enough custom designs from drawings and sketches to know that you can never fully trust your two-dimensional drawings for accuracy, especially when working with such a compact design. The only way to be certain that all the components will live together in harmony is to lay out the frame visually again. It is a bit annoying to get out all the parts and prop them all up again, but this sure beats the disappointment that you will feel when your only option is to cut fully welded tubing later (my first-hand experience).

As shown in Figure 11-19, I decided to place the fork legs (rear tubes) up as far as length would allow along the main spine tube, so that it would look like the legs were flowing directly from the top curve of the frame. There is adequate clearance between the tire rubber and the tubing, and enough room for my belt drive to be installed on the axle without rubbing on any of the tubing. Things are very close, but by visual inspection, I expected to have no problems with this design.

Even though the mock layout seemed perfect, I only tack welded the rear tubes in place, as shown in Figure 11-20. The tack welds are strong enough to handle the frame without fear of misalignment, but allow for easy adjustment or removal if something unexpected

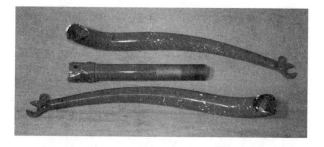

Figure 11-18 *A de-limbed mountain bike front fork*

Figure 11-19 *Positioning the rear tubes*

Figure 11-20 *Tack welding the frame tubing*

217

happens. At this point, I cut some of the excess rear spine tubing away, but left all the bottom tube length extending past the rear axle area, as I thought it would look cool in the final design, almost like a pseudo exhaust pipe. My goal was a small yet powerful mini bike that would resemble a Harley, so this was a bonus.

Figure 11-21 shows the head tube and fork stem taken from the unfortunate mountain bicycle that I mercilessly butchered for parts. The length of the head tube is not important, but it is important that the fork stem and head tube form a matching pair. If you have taken both parts from the same bicycle, then you have no worries, but if not, install the bearing cups and all the hardware to ensure that the fork stem isn't too short. In Figure 11-21, I already ground off excess tubing from both the head tube and the fork stem. The little ring just above the fork stem (bearing race) has been tapped off with a hammer in order to avoid damage when grinding.

There are numerous ways to create the forks for your electric mini bike. You could simply use a set of bicycle forks as they are, as long as the front wheel will fit. You could bend a few tubes and weld them back onto the fork stem to create a wider bicycle-type set of forks, or even install an actual motorcycle front fork. I decided to go with the conventional triple-tree-type fork that you see on almost all motorcycles and even some hardcore mountain bikes. A triple-tree fork is very strong, looks like it means business, and is really not all that hard to make, requiring only the fork stem you already have and a few plates cut from some scrap 1/8 sheet metal.

As shown in Figure 11-22, the tree plates are drawn out on the plate using your fork leg material and the fork stem as a guide. There is no need to give you a drawing, since the exact size of the plate is based on the width that the fork legs need to be so that they are positioned on

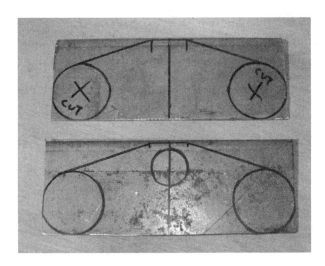

Figure 11-22 *Cutting the triple-tree plates*

each side of the front wheel with adequate clearance. Determine the height of the plate so that the distance between the center of the fork stem circle is one inch from the center of the fork leg circles, as shown in Figure 11-23. There is a quarter inch of material left around the fork stem hole, since it will be drilled out of the plate.

The triple-tree plates are cut using either a jigsaw or zip disc, but before you start cutting, read ahead so you can see why the plates are made this way.

The top-tree plate is cut in a certain way so that it will be placed over each fork leg, while the bottom tree plate is cut so that it can be welded directly to the inside of the fork legs. The top plate also needs to slide over the fork stem at the threaded area, so the hole needs to be cut out. Because I used the same 1.5-inch EMT for the main

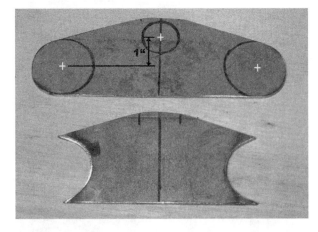

Figure 11-23 *Triple-tree and top-tree plates cut from sheet metal*

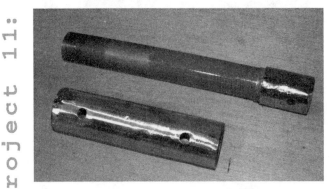

Figure 11-21 *Fork stem and compatible head tube*

spine, this is how I cut and ground both plates. With a front wheel width of approximately 4 inches, and a fork leg width of 1.5 inches, my top plate had an overall dimension of 7.5 inches in width and 3 inches in height.

You may want to use thinner tubing for the front forks, and may have a completely different size and style of front wheel in mind, so your tree plates may be a lot different to mine. Again, the best way to test your dimensions before cutting is by visually laying out the wheel and fork legs to see what type of distance you need between the front fork legs. Shoot for at least a quarter inch of clearance between the edge of the front tire and the fork tubing. This way, you can add fenders later.

If you do not have a hole saw of the correct size for the fork stem hole, drill a series of small holes around the inside of the traced area, and then tap the hole out with a hammer. In Figure 11-24, I punched several points along the inside of the line to drill them out with a 1/8-inch drill bit.

This crude hole cutting method works well when you aren't in the mood to spend the day driving around looking for the perfect hole saw, which will usually wear out after three uses anyhow. Once the area is drilled, punch out the center, as shown in Figure 11-25, and then file off the teeth with a round file. Building a custom can be a lot of work, but like most artistic projects, if it was too easy, then everyone would be doing it. All this tedious work will eventually pay off.

Figure 11-26 shows how the triple-tree plates will be connected to the fork stem. The top plate is held in place by the fork stem hardware, and the bottom plate is welded directly to the base of the fork stem where you removed the original fork legs. If viewed from the top, both plates would be in perfect alignment, since the center of the fork stem passes through them at the same

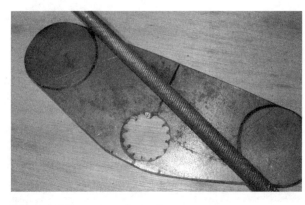

Figure 11-25 *Hole ready for filing*

Figure 11-26 *Tack welded triple tree*

point. As you can see, if the top plate covers the top of the fork legs, then the fork legs continue to the bottom plate, cradled by the cut-out area so that all three tubes (fork legs and fork stem) remain parallel to each other.

Test the position of the bottom tree plate by placing both fork legs in position to check parallel alignment. Once you have the bottom plate in the correct position, tack weld it in place, as shown in Figure 11-26.

Once you have the bottom plate tack welded in position, and have checked the alignment of all three tubes, weld the entire joint, as shown in Figure 11-27. There may be a small amount of distortion to the bottom plate from the welding heat, but this will not present a problem, since the distortion will be the same on both sides, and have almost no effect on the alignment of the forks.

I decided to use the same 1.5-inch thin walled electrical conduit tubing as used for the top spine tube

Figure 11-24 *Cutting the fork stem hole from the top-tree plate*

Figure 11-27 *Bottom tree plate welded in place*

Figure 11-29 *Half-inch bolts and 1.5-inch washers*

when making my front forks, in order to get the thick "fat boy" look in my final design. On a mini bike of this size, 1-inch conduit would have been strong enough, but the difference in weight between the two sizes of tubing was miniscule. Rather than trying to calculate the exact length for the fork tubes at this point, I simply cut them longer than they would need to be by placing the frame on blocks to estimate ground clearance, and measuring from the top of the front spine tube where the head tube would be installed. Although my final fork length was 17.5 inches, I gave myself plenty of room to work with and cut the two tubes, as shown in Figure 11-28, to 24 inches long each.

Since the front fork tubes need to be held in place by the top-tree plate, we will need to figure out a way to achieve this, while still allowing the assembly to come apart for removal. Weld a pair of nuts to a pair of washers that will be welded to the top of the fork tubes. As shown in Figure 11-29, a pair of half-inch nuts and bolts and some 1.5-inch washers will do the trick. The washers do not have to be exactly 1.5 inches in diameter, just something close enough so that you can weld them to the top of the fork tubes, or directly inside them. If you are

using other fork tubing, then choose appropriate width washers.

The first step is to weld the nuts to the washers so that the bolts can be placed into the nuts through the washer holes. A good way to make sure the nut is in the correct position is to hold it in place by the bolt while you are welding it. The final product will look like Figure 11-30.

The washers will now be welded to the top of the fork tubing, as shown in Figure 11-31, so that the nuts are inside the tubes (right in Figure 11-31). Weld the joint all the way around and then grind the area clean, so that it looks presentable and creates a flush top surface to mate with the triple-tree top plate (left in Figure 11-31).

The fork tubes will be held in place by the top-tree plate, as shown in Figure 11-32. Not much to say here: if all went well, the bolts will thread right in. If the bolt seizes, remove it and look at the nut threads for any welding spatter that may have found its way inside and remove it with a pick.

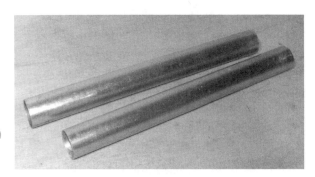

Figure 11-28 *Fork leg tubing*

Figure 11-30 *Weld the nuts to the washers*

Figure 11-31 *Welding the washers to the top of the fork tubing*

Figure 11-32 *Test fitting the fork tubes*

Now install all the fork hardware and let the lower tree plate cradle the fork legs, as shown in Figure 11-33. Both fork legs and the head tube should be in perfect alignment with each other, once the lower plate is fit snuggly against the fork tubing. You can now weld the lower tree plate in place while the fork hardware is holding everything in place. A light bead on the underside of the plate is all that is necessary for strength, and you should avoid welding the top of the joint to reduce the chance of warping the fork leg tubing. If you are worried about

warping the fork legs, tack weld a bit of plate or scrap tubing across the ends before you weld; this also helps hold the assembly together prior to welding.

Now that your triple-tree assembly is completed, the head tube needs to be welded to the front of the main spine tube. Head tube angle can have a large effect on the overall steering characteristics of a two wheeler, so some thought should be put into this cut. The more angles you have, the more stable the bike will feel at higher speeds, but the less responsive the steering will be to small adjustments of the handlebars. With very minimal head angle, the steering will react instantly to any small changes in the handlebars, so high-speed riding may feel "twitchy," and take some getting used to. I let the front wheel diameter dictate my head tube angle by leaving as little space between the tire and frame as possible. My resulting head tube angle is fairly steep, as you can see in the final design photos, and this resulted in a fairly stable ride that felt a little "nervous" at high speed for the first few rides. Overall, I am happy with the steering characteristics.

If you want to choose a head tube angle that will offer a great ride at all speeds, simply look at the head tube angle of a typical bicycle, and shoot for that. Extend an imaginary line from the head tube to the ground, and use the angle between the ground and that formed by the head tube as your guide. For reference, this angle would typically be between 65–70 degrees on a bicycle, and on my Sparky it is 80 degrees.

As shown in Figure 11-34, I placed the head tube at the angle I desired, and then marked a line along the end of

Figure 11-33 *Setting up the fork set for welding*

Figure 11-34 *Marking the head tube angle cut*

the top spine tube showing the angle that I needed to cut with my grinder disc.

Once you cut the spine tube at the correct head tube angle, you will need to cut the fish mouth in order to make a correct joint for welding. As shown in Figure 11-35, the same amount of material should be ground from both the top and bottom of the joint in order to maintain the proper head tube angle. You should place the triple-tree fork assembly into the joint often as you grind, to make sure that you are maintaining the proper head tube angle.

Once the end of the spine tube has been ground to mate perfectly with the head tube, install the head tube and tack weld it in place only at the top of the joint, as shown in Figure 11-36. Ensure that the head tube angle ends up being what you planned for, as well as the alignment as viewed from the front of the frame. The head tube and fork legs should look perfectly aligned

with the center line of the frame, or perpendicular to the ground. With only a single tack weld at the top of the joint, manipulation will be easy if alignment is visually off.

Although it is hard to tell from Figure 11-37 due to lens distortion, my fork legs and head tube are perfectly aligned as viewed from the front of the frame. When you are satisfied with both the alignment of the head tube with the rest of the frame and your head tube angle, add two heavy tack welds on both sides of the lower part of the joint and then recheck the works. At this point, you will still be able to give the head tube an "attitude adjustment" with a rubber mallet if need be.

Continue to weld the head tube to the main spine tube by completing the top and bottom areas of the joint first, in order to minimize side-to-side distortion due to

Figure 11-35 *Head tube joint ground*

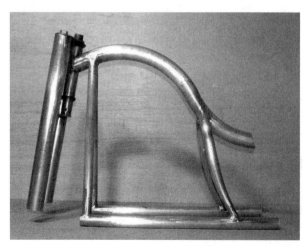

Figure 11-36 *Tack welding the head tube in place*

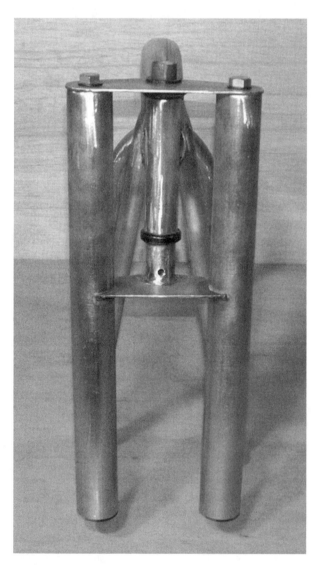

Figure 11-37 *Checking head tube alignment*

welding heat. Figure 11-38 shows the head tube welded solid at both the top and bottom of the joint.

In Figure 11-39, I completed the welding around the entire joint of the head tube, and also installed two gussets made from some quarter-inch steel rod. The gussets were not absolutely necessary, but my head tube was quite long, and I do like to punish all my rides, so it didn't hurt to add this extra strength to the frame. Gussets can be made from virtually any scrap you have lying around, and can greatly enhance the durability and stiffness of a frame.

Your choices for mini bike wheels are unlimited, with hundreds of sizes and types to choose from. I built Sparky from what I had lying around the shop, and the wheel

Figure 11-38 *Weld the sides of the joint last*

Figure 11-39 *Head tube welded and gussets installed*

shown in Figure 11-40 is a 10-inch diameter, 4-inch wide scooter wheel taken from a dead shopping scooter (the type you see in malls and on sidewalks). This wheel represents the most unfriendly wheel possible for use on a project like this, because it has no bearings, no sprocket or pulley mount, and no way to install a brake drum. I choose this route to demonstrate that practically any parts can be hacked to fit. A much better wheel choice can be found by typing "scooter parts" or "mini bike parts" in an Internet search engine. Let's talk a bit about wheels before you start dismembering Granny's scooter!

Wheel diameter

You could use wheels of just about any size for a home-brew mini bike, ranging from those as small as skate blade wheels right up to full size motorcycle wheels, but to get the correct gear ratio, choose carefully to keep complexity and cost to a minimum. Before you settle on a wheel diameter, you should know your motor's full RPM and the desired maximum speed you would like to reach.

To calculate how fast a mini bike will travel, you first need to know how far your vehicle will travel when its drive wheel makes one full rotation. I work this out using the imperial measurement system to keep things simple. To calculate this value (which is the circumference of a circle), multiply the total diameter of the drive wheel by PI (3.1415). So, for Sparky's total wheel diameter of 10 inches, I would multiply 10×3.1415 to get 31.415. This means that Sparky would travel approximately 30 inches if the drive wheel were to make one full revolution. Since my goal for top speed was about

Figure 11-40 *Rear wheel, axle, and bearings*

30 miles per hour, I could now calculate how many maximum revolutions per minute my drive wheel would need to make in order to reach that speed. Since there are 63,360 inches in one mile, a distance of 30 miles would be 1,900,800 inches in total. We know that Sparky's wheel would travel 31.415 inches every revolution, so we just divide 1,900,800 by 31.415 and get 60,506, which is how many wheel revolutions there are in the 30-mile total distance. Now divide that number by 60 minutes (60,506/60) and you get the answer in rotations per minute (RPM), which would work out to 1,008 RPM. So to travel at 30 miles in one hour, Sparky's rear drive wheel must turn around approximately 1,000 times every hour.

The simplified formula for calculating the drive wheel maximum RPM if you know the wheel diameter (WD) in inches, and the desired top speed (TS) in miles per hour, would be 63,360 × TS / (3.1415 × WD) / 60.

Gear ratio

The above formula assumes that the motor is able to directly turn the wheel at the desired RPM (1,008 RPM in my case), but this is almost never a reality since most small DC motors run at 3,000 RPM or more. If I connect the 10-inch diameter wheel directly to the motor at 3,000 RPM, we could calculate the theoretical top speed by using the following formula: 10 (wheel diameter in inches) × 3.14 (PI) × 3,000 (RPM) / 63,360 (inches per mile) × 60 (minutes per hour) = 89.2 miles per hour! Much too fast for my liking, and probably not possible due to wind resistance and friction losses anyhow. Your only option is to reduce the ratio of motor shaft RPM to wheel RPM, using some type of reduction system like a belt or pulley.

Gear reduction is calculated using basic division mathematics. If you have a gear reduction of 1:3, then you divide the motor's top RPM by 3, so the resulting RPM of a motor at 3,000 RPM would be 1,000 RPM through a 1:3 gear reduction system. What do you know, that is just the thing I would need to give Sparky a top speed of 30 miles per hour, assuming a 10-inch drive wheel and a 3,000 RPM top motor speed!

To calculate the actual gear ratio in your transmission system, you will need to divide the diameter of the drive wheel pulley, gear, or sprocket by the diameter of the drive motor pulley, gear, or sprocket. If you had a drive motor pulley of 2 inches, and a wheel pulley of 6 inches, you would have a gear ratio of 1:3, since 6 / 2 = 3. Once you know how to calculate your gear ratio, multiply it by the calculated top speed that would have been produced by your drive motor and wheel if there were no gears, pulleys or chains in the transmission, and then divide it by the gear ratio. So that ridiculous top speed of 89.2 miles per hour would be reduced to 29.7 miles per hour using a 1:3 ratio (89.2 / 3), which is very close to the desired 30 miles per hour top speed I wanted in the first place.

The pulley shown in Figure 11-40 is 6 inches in diameter, so I had to find a 2-inch diameter motor pulley to complete my 1:3 gear reduction. I will discuss the choice of belt or chain drive later on.

Because my wheels did not include any bearings or drive mounting hardware, I was left to install axle bearings and figure out my own method of fastening the drive pulley to the axle. I decided to use a keyed shaft to connect the rear wheel and drive pulley together, and connect this directly to the frame using the flange mount bearings shown in Figure 11-41. I chose this system because it would be easy to implement, and the bearings are easy to find at any yard and garden equipment repair center. The bearings shown here can also be found in dryers and furnace blowers. The inner diameter of the rear wheel was 3/4 inch, so the bearings, pulley, and axle are also this diameter. The plates shown in Figure 11-41 are cut from the same steel used to make the triple-tree plates. I simply traced an area a quarter larger around the bearings to create a nice-looking mounting plate with no

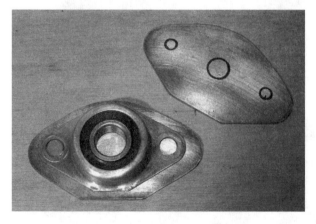

Figure 11-41 *Making the bearing mounting plates*

sharp edges. To make both plates identical, they were held together in the vice and ground at the same time.

If your wheels already have bearings and a sprocket or pulley installed (this is far more common), then your only task will be to fasten the axle to the frame, and you can bypass the bearing plate steps that follow, moving directly to Figure 11-47, where you will see an axle mounting system for wheels with included bearings.

Trace the bolt holes directly through the bearing holes and punch, as shown in Figure 11-42. Since the plan is to install both bearings on the inside of the mounting plates, there is no need to cut a hole through the mounting plates for the axle shaft. This design worked very well and kept the final bike looking neat and clean.

Once both bearing mounting plates are drilled, as shown in Figure 11-43, they will be ready for mounting on the rear of the frame. It is important that both plates be the same shape when the bolt holes are aligned, so check this by placing them together and then grind them if necessary.

The bearing mounting plates are positioned on the frame so that the bearings are facing inwards and that the bearing faces do not extend past the bottom tubes, as shown in Figure 11-44. Basically, the bearing plates will end up directly in the center of the 1-inch EMT tubing that makes up the two bottom tubes. The axle is cut to this exact length and the ends are slightly rounded, so there will be no friction between the ends of the axle and bearing mounting plates if they were to become unsecured for some reason. The bolt heads are rounded and placed on the outside of the frame to further reduce any sharp edges that may come into contact with the rider.

Once the axle has been cut to the proper length, you can install your rear wheel for alignment testing purposes and then tack weld the bearing plates directly to the bottom tubes and rear tubes, as shown in Figure 11-45. I used the wheel to visually check the alignment as viewed from the top of the frame and from the rear. Since both bearing mounting plates have been cut to the exact same size, the wheel should automatically be in perfect alignment, but it doesn't hurt to check, just in case the frame has become distorted from welding. Any slight axle misalignment can be fixed by minor adjustments in the position of either bearing mounting plate.

As shown in Figure 11-46, once you are satisfied with the axle alignment, the bearing mounting plates can be completely welded to the frame on both sides of the joint. It is a good idea to complete the welding while the bearings are held securely to the plates and with the axle (without the wheel) installed. This will help reduce any

Figure 11-42   *Finding the hole centers*

Figure 11-43   *Both bearing plates ready for mounting*

Figure 11-44   *Positioning the bearing mounting plates on the frame*

Figure 11-45 *Tack welding the bearing mounting plates to the frame*

Figure 11-46 *Bearing mounting plates installed*

welding distortion since the plates will be held in the correct place by the axle. Once welded, install the rear wheel and recheck the alignment.

Since my front wheel already had a bearing installed, it was much easier to mount by simply placing a bolt through two "dropouts" welded directly to the forks. This mounting system works for any wheel with an included bearing, and would work perfectly for a rear-drive wheel with sprocket and drum brake as well. As shown in Figure 11-47, the key components that make up this mounting system are a bolt with the same shank diameter as the wheel bearings, two dropouts made from bits of cut

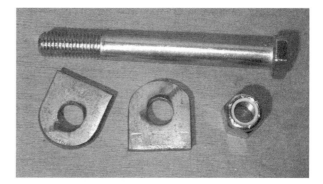

Figure 11-47 *Front wheel mounting parts*

3/32 flat bar or plate, and a locknut. You will also need some type of spacer to keep the wheel from rubbing the frame or fork tubing. This is covered later in this project.

The best place to weld the axle mounting dropouts is directly in the center of whatever tube they will connect to—the front fork tubing, in this case. As shown in Figure 11-48, the bolt is placed through the dropouts and the front wheel with some temporary spacers used to get the wheel aligned between the tubing properly. Use whatever you can find for a spacer right now since you will be replacing it once you know the correct size. Large nuts (shown in Figure 11-48), washers, or tubing cut-offs all work perfectly.

Front wheel alignment is tested the same way as the rear wheel, from the top and from the front. There should be equal distance between the tire and fork tubing, and depending on how the bearings or wheel hub are made, the spacers on each side of the axle may not be the same. In Figure 11-49, my front wheel is perfectly aligned, but the spacers are almost a half inch different. Your wheel may also have a drum brake, so make sure that all the

Figure 11-48 *Front axle installed with a temporary spacer*

Figure 11-49 *Checking front wheel alignment*

brake hardware is installed when you figure out the alignment. You can now tack weld the dropouts in place while held by the front axle.

Once tack welded and aligned, the front fork dropouts can be fully welded to the fork tubing on both sides, as shown in Figure 11-50. On a side note, you may want to make your own custom springer front fork, and all that would take is to hinge the dropouts and add a spring over the top. I did not bother with suspension, since the seat

padding and the pneumatic tires I used seemed to make the ride smooth enough.

Now that you have figured out the correct size for the spacers using washers or nuts, you can cut the real axle spacers from these measurements. As shown in Figure 11-51, a few small pieces of tubing are cut to slide snugly over the axle bolt. The spacers do not have to be an exact fit over the bolt, but they should not be too loose or they will flop all over the place. If you have no luck in finding the correct size tubing to create the spacers, you could use the nuts and washers, but these are quite heavy and may rattle.

Congratulations—you should now have a rolling chassis, something like the one shown in Figure 11-52, ready to take whatever guts you plan to throw into it. Notice the great styling of the flowing frame and those extra chunky front fork legs. The little frame looks indestructible and gives off an old-school, fat-boy look. Of course, you can radically alter the attitude of any motorcycle by simply accessorizing and creating custom fairings later, but we're not there just yet.

Figure 11-51 *Cutting the correct axle spacers*

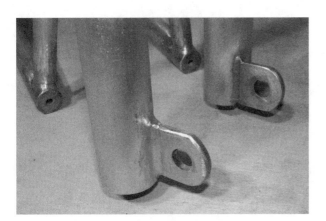

Figure 11-50 *Fork dropouts welded in place*

Figure 11-52 *We have a rolling frame!*

Handlebars are truly a personal choice, and since we based our steering system on bicycle parts, the standard bicycle gooseneck and handlebar can be installed, so your choices are endless. I planned to do some hard riding, so I went with a strong mountain bike handlebar, as shown in Figure 11-53, with the ends of the handlebars cut to a width of 18 inches. I knew the fat-boy look would have been more pronounced with a long chrome set of ape hangers, but because of the extreme acceleration and trail riding I like to do, I doubt they would have held up.

The next step is to get that power plant installed into the frame so you can start burning the rubber off your tires! This is another stage of the build that has no strict set of rules or common standards to follow, so before you cut the two short tubes shown in Figure 11-54, let me explain a few basic rules to follow when mounting your motor.

The motor will probably have several places where a bolt can be installed for mounting, but it is crucial that your main mounting system holds the motor by its face (the end with the drive shaft). The motor face will have at least two large, threaded bolt holes, and may have more than four, so you should try to use as many as you can when planning a mounting scheme. The mounting holes in the side of the motor are only needed to stop the motor from banging around if the front plate flexes due to vibration or extreme torque, and you should never attempt to hold a motor in position by the bolt holes in the sides of the can exclusively, or it will be damaged. The top face is designed to take most of the stress that will be placed on the motor, as well as hold the main motor shaft bearings in place; therefore, it is made from a thick machined plate of steel or aluminum and will become the main mounting face. The motor housing can is nothing more than a rolled piece of sheet metal, and may not even be welded at the seams, so it will offer very little strength, and putting stress on it may cause the permanent magnets to become unglued, or warp the entire motor. If you want to visualize the stress placed on a drive motor, imagine trying to bend the motor shaft in the direction of the drive wheel, as this is essentially what will happen as the energy is transferred from the shaft to the chain or belt.

Knowing this, you will soon see that these two small 3/4 inch conduit tubes, shown in Figure 11-54, will be welded between the bottom tubes just below my motor to allow the installation of a solid mounting plate to hold the motor by its face.

In my design, there is just enough room between the battery box and the rear tire to cram the motor in place,

Figure 11-53   *Handlebars installed*

Figure 11-54   *Motor mounting tubes*

Figure 11-55   *Motor mounting plates installed*

so the two motor mounting tubes are installed directly below the motor so that the mounting plate legs can connect to them. Read ahead to see how and why this is done. Your motor will likely have a very similar mounting system, and this will work for you as well. Figure 11-55 shows the two tubes cut and ground to fit snugly between the bottom tubes just ahead of the rear wheel.

Figure 11-56 shows the motor mounting tubes welded in place. I tested the motor using the mounting plate shown in the next few steps while the rear wheel was installed in the frame, so there was no doubt that these two tubes were in the correct place.

The mounting plate that will hold the motor in place by its face is made from a few bits of 1-inch wide, 3/32 flat bar cut so that they form a face to insert the four mounting bolts and extend to the mounting tubes installed in the last few steps (Figure 11-57). The exact size and length of these plates will depend on your motor face size and drilling, so work directly from it. An easy

way to get a drilling template for your motor is by laying a piece of paper over the motor's face, then rubbing your greasy, dirty, mechanics' fingers over the holes to reveal an exact pattern. You can now punch the plate to drill the appropriate holes.

Figure 11-58 shows the completed and drilled motor mounting plate and the four bolts that will secure the power plant to the plate. If your bolts holes are not drilled exactly, do not try to force the bolts into the motor face as you may strip the aluminum threads. Instead, just drill the offending hole using the next largest size drill bit you have.

The motor mounting plates are shown welded into position in Figure 11-59. There is no magic here—just make sure that the welds are strong and that the motor is

Figure 11-58 *The completed motor mounting plate*

Figure 11-56 *Motor mounting tubes welded in place*

Figure 11-57 *Parts for the motor mounting plate*

Figure 11-59 *We now have the power!*

aligned so the shaft is perpendicular to the bottom tubes. If your motor shaft is misaligned, you may derail the drive chain, or cause the drive belt to prematurely wear out. The center of the motor shaft should also be placed so it is at the center of the drive belt or sprocket on the rear axle. This worked out perfectly in my design, because the motor heat sink sat just over the edge of the bottom tube, out in the open so that air would flow into the fins while riding, much like an air-cooled gas engine. If you have a large motor and a heavy throttle hand, cooling can become very important if you plan to ride for extended periods of time. My original Sparky version had a very small motor taken from a granny trike, and would get seriously hot after 15 minutes of riding, or if there was any wind or hill to deal with.

Figure 11-60 shows the motor, wheels and battery box all nestled together in the frame without any room to spare, exactly as I planned. This photo gives you a better view of the resulting heat sink overhang, which will really help keep the motor cool for just about any length and style of riding. Of course, the battery box can only slide out from the right side of the bike now, but that's OK.

Although the motor mounting plate holds the motor very securely in place by the face so that it can take any amount of torque delivered to the shaft, this does not take into account the vibration the motor may encounter as the vehicle bounces over rough terrain. I like to strap the

motor body to the frame using a hose clamp, just to be safe. If you can't find a hose clamp long enough to make it around the motor and mounting tubes, then join two or more of them together, as I have done in Figure 11-61.

As shown in Figure 11-62, the hose clamp simply wraps around the motor body and the frame so that any vibration will not make the motor jump around or bang on the frame. If you have any distance between the motor body and the frame tubing (I did not), then you should install a stiff piece of rubber, or soft metal block, so that the tightening of the hose clamp doesn't bend the motor mounting face plate. The goal is to stop vibration, not to add any more strength to the mounting system, since it is strong enough already.

The type of rear wheel you plan to use will dictate the style of transmission you will use, which will be either chain or belt. A belt drive will need an idler pulley to maintain belt tension and adequate coverage on the

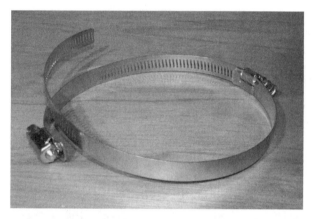

Figure 11-61 *A hose clamp will stop the motor from banging around*

Figure 11-60 *The battery box test fit*

Figure 11-62 *Hose clamp installed*

motor side pulley. A chain drive may or may not need an idler sprocket, depending on your ability to either move the rear wheel back and forth, or simply find the correct length chain. Before you choose a transmission type, if you haven't done so already, let me explain the pros and cons of both chain and belt drive transmissions.

Chain drive

Chain drive is by far the most common method of transmission for motorcycles and scooters, which is why 90 percent of the small wheels you can purchase will have a sprocket mounting flange or a sprocket already installed. Chain drive is considered more efficient than belt drive for the most part, and as long as your chain is not extremely loose, you will probably not need an idler sprocket to pick up the slack. You could even file the motor mounting plate holes into more of a slot shape, so that the motor can be moved back and forth about half an inch to deal with the chain tension. Moving the motor or the rear wheel is a very common way to adjust chain tension, as you know from single-speed bicycles. The main disadvantage of a chain drive is that it is very dirty and extremely noisy. You will realize just how noisy a chain drive can be if you listen to a store bought electric scooter or pocket bike drive past you. The motor is probably 99 percent silent, but that ratchety grinding racket will make the vehicle sound almost as loud as a similar sized gas-powered unit. It almost seems like a disappointment to place a loud grinding transmission on such a clean running, neighborhood friendly vehicle that could be almost 100 percent silent, but the choice is ultimately up to you.

Belt drive

As you can probably tell by now, I personally favor the belt drive solution because it is absolutely silent, doesn't spit grease all over the place, needs no adjustment, and can utilize parts found in an old washing machine or dryer. It is said that belt drive is not as efficient as chain, but to be honest, I have tried them both on the same vehicle on more than one occasion, and noticed absolutely no difference in top speed or overall range, so I disagree based on my own personal experiences. Another

advantage of a belt drive is that it will take some of the shock away from the motor if it is made to accelerate at a dangerous rate, which is what could happen if you decide to not use a motor controller and simply go for the inexpensive contactor throttle. With a chain drive, there is no slip, so you'd better hold on to your lug nuts if you have a simple contactor and a powerful motor, because you will accelerate with as much energy as your batteries can deliver to the motor. With the motor I used, this would result in a runaway motorcycle and a sore butt in a real hurry. With a belt drive, there is some slip under extreme load, and this slip is fully adjustable by adjusting an idler pulley, so you can tailor the transmission for the type of power and motor controller you plan to use. In my design, I have a fairly tight belt for limited slip due to the smooth ramping curve of my motor controller. I can certainly do wheelies and burnouts if I like, but this is by choice, not when I least expect it.

As shown in Figure 11-63, a belt or chain tensioner is simply a flat pulley or sprocket attached to a rod that is pressed into the belt or chain by a spring. The parts were salvaged from a dead washing machine, but these flat idler pulleys can be found at most hardware stores and appliance repair shops. The idler pulley rides in the flat side of the belt, so it does not have a typical V-groove on its surface.

The idler pulley or sprocket (if necessary) will be mounted to the motor mounting plate using a simple bolt, as shown in Figure 11-64. Read ahead a few steps and you will see how these parts come together in order to form the idler system. Again, if you are using a chain, you may not need an idler system, since you could simply slot out the motor mounting holes to allow some degree of chain adjustability. To make slots of the holes, just drill

Figure 11-63 *Belt tensioner parts*

Figure 11-64　*Idler pulley mounting bolt*

Figure 11-66　*Idler pulley installed*

Figure 11-67　*Idler arm mounting sleeve*

the same size holes directly beside the original holes, and then use a small file to create the slot between them.

Figure 11-65 shows the idler pulley arm and pulley mounting bolt which is made by cutting the head off a bolt that will fit through the bearing of the idler pulley, and then welding it to a small bit of flat bar or angle iron. The size and position of this idler system depends on your motor, mounting plate, and belt.

The idler pulley is fastened to the bolt installed in the last step so that it looks like Figure 11-66. The pulley should spin freely on its roller bearings, yet be securely fastened to the arm so that it doesn't rub on the arm.

The idler arm will be fastened to the bolt that is welded to the motor mounting plate. By using a sleeve, it can be positioned up or down, pushing the idler pulley into the underside of the belt. As shown in Figure 11-67, a small piece of tubing or a spacer that fits around the bolt is welded to the other end of the idler arm. This is a

similar idea to that used for the front wheel mounting bolt and spacers.

As shown in Figure 11-68, the idler arm is free to pivot around the bolt on the motor mounting plate, so that the idler pulley can move up or down into the underside of the drive belt. This will allow more of the belt to ride in the small motor-mounted pulley, creating great friction and less slip. The amount of controlled slip will be determined by how hard the idler pulley is pushed up into the belt.

Figure 11-65　*Idler pulley arm and pulley mounting bolt*

Figure 11-68　*Idler arm in position*

To figure out what size of belt you will need, push the idler pulley up about halfway, then run a measuring tape around the outside lip of both pulleys, as shown in Figure 11-69. This measurement will be close enough to purchase the correct length belt, which will probably be rounded to the nearest half inch. Notice the ratio of motor pulley to wheel pulley (1:3). The 2-inch motor pulley and 6-inch wheel pulley gives us that magic 1:3 ratio for an estimated top speed of 30 miles per hour and loads of acceleration. If these were sprockets, they would be about the same diameter, but would be rated in numbers of teeth, rather than dimensions. A sprocket ratio of 15 teeth on the motor to 45 teeth on the wheel would be about right.

The perfect length drive belt will enter and exit the motor pulley at about the same angle, as shown in Figure 11-70. This arrangement puts plenty of belt around that small motor pulley, greatly increasing the friction. If the system had no idler pulley, and the same belt tension, there would be a large amount of slip due to the fact that only a small percentage of the motor pulley would have any belt in the groove. All we need now is a spring to hold that idler in place.

Here is where you can do a little magic to tailor the amount of maximum acceleration your transmission can deliver. If you have a well-behaved motor controller with a sensible acceleration ramp, then this idler spring can be stretched out so that the pulley is quite tight against the belt (it is difficult to push the idler pulley down by hand). This spring is taken from an old mattress box spring, and is quite difficult to pull apart by hand, although I can

Figure 11-70 *Test fitting the drive belt*

muscle it in place with a pair of pliers. As you can see in Figure 11-71, the spring is held under tension by a nut welded to the fork leg so that the spring is pushing the idler pulley into the drive belt. If you want to play around with limited slip, then you could install a small plate with multiple spring mounting holes, or simply use different sized springs. Many washing machines, dryers and furnaces use a system just like this, so keep that in mind as you scavenge for parts.

Besides the connection from the batteries to the motor, the rest of the bike is simple bits and pieces designed to hold the accessories such as seat, lights, fenders, etc. Feel free to go crazy on customization and additions here; there is certainly no strict set of rules and guidelines when it comes to these parts. Figure 11-72 shows the few bits of 1/8-inch scrap flat bar I planned to use to create a sturdy seat mounting platform just over the rear wheel.

I wanted a big phat seat placed directly over the rear wheel as low as possible, so that the rider would sit low and comfortable on the little hog with just enough knee

Figure 11-69 *Measuring the belt size*

Figure 11-71 *Idler tension spring*

Figure 11-72 *Seat mounting plates*

Figure 11-74 *Let there be light*

room for the handlebars. Because of the well-behaved motor controller I planned to install, combined with the front heavy weight added by the batteries, the rider's weight would be near the rear of the bike, with minimal risk of an unplanned wheelie. As shown in Figure 11-73, I cut a bit more from the tail of the main spine tube and installed the seat mounting plates, so that my large seat would be securely held in place by four large woodscrews.

I decided to install all the little stuff before wiring up the motor controller, so the temptation to drive the unfinished, unpainted mini bike would not get me this time. I made fender mounting plates for the front and rear and a headlight mounting tab from various bits of scrap steel and flat bar. Shown in Figure 11-74 are the front fender mounting plate and the headlight tab.

Fender style can really lend a lot to the overall appearance and attitude of your project, so feel free to use your imagination here. A very stylish and strong fender can be made by cutting the sides and top from some thin (1/32) sheet metal using a jigsaw, as shown in

Figure 11-75. I simply traced a pot lid that was slightly larger than the wheels to create the fender sides, then gave them a little flair at the ends to look like those big Harley fenders. These fenders will cover quite a bit of the wheel, and will be strong enough to jump up and down on if sitting on the ground.

There are no fancy tools or mystical black arts used in my garage, buddy, just hard work, basic tools, and a lot of junk! To form the nice round fenders, I just tack weld the side pieces to the top strip, and bend the top strip around the side pieces as I tack weld about every inch or so, as shown in Figure 11-76. The resulting fender is perfectly round, as if rolled by an expensive tool that I don't ever plan to own.

The resulting fender shown in Figure 11-77 is what you get by rolling the top strip along the end pieces as you tack weld. As you can imagine, fenders, fairings, and many other shapes can be made using this simple yet highly effective system.

Figure 11-73 *Seat mounting plates installed*

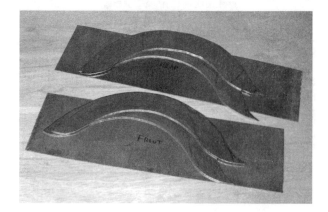

Figure 11-75 *Cutting the fender pieces*

Figure 11-76   *Roll and weld*

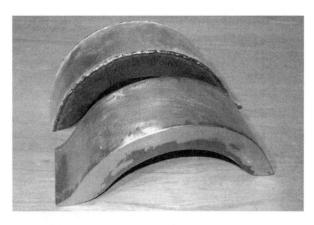

Figure 11-78   *Weld and grind the edges*

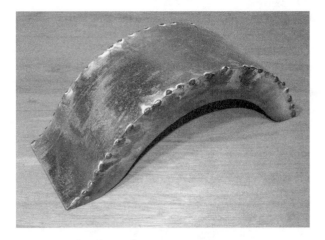

Figure 11-77   *Tack welded fender*

Figure 11-79   *Phat seat requires phat foam*

Figure 11-80   *Foam trimmed and glued to the wood*

Once tack welded together, the entire outside joint can be welded and ground smooth in very little time to reveal the perfectly smooth fender, as shown in the foreground of Figure 11-78. After welding the entire joint, I rough grind the welds with a coarse disc, and then take a sanding disc to clean up the weld until it looks like it was professionally made. Do not weld the inside of the fender joint, or you will warp the fender, pulling the sides together.

I had some 2-inch thick rigid packing foam in my garage, so I thought it would make a perfect seat material to give the seat a real phat look. As shown in Figure 11-79, I first cut the 3/4 plywood to form the seat base, and then traced the area on the foam to rough cut it with a razor knife. This seat is about the same size and shape as a granny trike seat, but feel free to build to suit your style.

Once I glued the seat foam to the base, I took the grinder and traced around the edge of the wood to make the foam follow the exact perimeter of the seat base. As shown in Figure 11-80, the foam is exactly the same shape as the seat, so it is now ready to be covered.

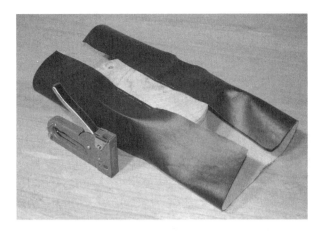

Figure 11-81 *Upholstery work is an art*

Figure 11-81 shows my "poor man's" method for upholstering a seat. I simply take the ultra high-quality "pleather" and stretch it as hard as I can around the seat, using as many staples as possible to hold it in place. If you can muscle the material with enough force, you will get almost all of those wrinkles to flatten out, reducing the "Christmas present" look in your seat. A sewing machine and proper seaming would be a much better alternative, but I am afraid the Atomic Zombie garage has no such equipment!

How many staples are enough? The real question is, how many do you have? I stretch that material until my face turns red, and then blast as many staples as I can into the material to hold it in place. It is pretty amazing how much you can force that material to conform to the seat, even a seat with so many curves, like the one shown in Figure 11-82. When you think you have enough staples, add more.

Figure 11-82 *I keep the staple company in business*

Considering the crudeness of my upholstering methods used, the seat shown in Figure 11-83 actually looks pretty good! Once the material was stretched and stapled, I tapped in some copper thumb tacks with a hammer to give the seat a more authentic look. There are only a few wrinkles near the base of the seat, but I actually think they add to the look.

You now have a place to sit, but your feet need a place to rest, so the foot pegs are next on the list of things to do. The perfect foot pegs can be taken from a freestyle BMX bike—these are light yet strong aluminum tubes that bolt onto the rear axles so riders can stand on them and pull stunts. To fasten them to our frame, simply weld a bolt onto the front of the bottom tube and place the pegs into the bolt. As shown in Figure 11-84, I used a 3/8 bolt and then drilled out the foot pegs to fit over this bolt.

Figure 11-83 *Completed hog-style seat*

Figure 11-84 *Foot peg mounting bolts*

Locknuts were then used. These foot pegs can be found as accessories anywhere that sells bicycles and parts.

As shown in Figure 11-85, the foot pegs should be installed onto the frame on a slight upwards angle, so that they do not strike the ground when your mini bike leans into the corners. Twenty-five degrees should be more than enough angle to allow the mini bike to lean into corners. Because the bike only has a few inches of ground clearance, the foot pegs also work as a kickstand if you simply let the bike rest on them.

The battery box was made to be easy to swap out or remove for charging, so it is held to the frame using this tab and nut system, as shown in Figure 11-86. A small bit of flat bar, or even a washer with one side ground flat, is welded to the battery box near the top and at the bottom, so all you have to do is remove two bolts to pull the battery pack out. The easiest way to get the nut in the correct place over the hole is to install the bolt to the nut,

and then tack weld it to the frame so it ends up in the correct position.

Figure 11-87 shows the bolt holding the battery box securely in place. There is another nut and tab just like this one at the bottom of the battery pack so that it does not rattle around under heavy vibration.

This is the last time during this project that I will ramble on for more than a paragraph, since it is almost time to paint and ride, but since it is time to install some type of motor controller, like the one shown in Figure 11-88, a little bit more information needs to be given.

Simple contactor

The simplest and least expensive method of transferring power from your battery pack to the motor is through a simple switch called a "contactor," which often looks

Figure 11-85 *Foot pegs are installed to allow the bike to lean in turns*

Figure 11-87 *Battery box bolted onto the frame*

Figure 11-86 *Battery box mounting tab*

Figure 11-88 *The motor speed controller*

like a large doorbell that will be mounted on the handlebars so that you can press it with your thumb. This is nothing more than a large-current-capable switch that completes a circuit so that 100 percent of the available horsepower is delivered instantly to the rear wheel. Obviously, a powerful motor and large battery pack will cause jerky, wheelie-prone take-offs, but this can be reduced by using a drive belt with a bit less tension, or by kicking off with one foot so that the bike is rolling a little bit before the motor is engaged. To vary your speed, simply toggle the contactor switch as needed. If you would like to see the many available styles of contactors, just do an Internet search for "ev contactor." You will need to choose a contactor that can handle your drive motor's "stall current," or the battery pack's maximum discharge capacity, since this is the current that will flow through the switch as your motor is asked to accelerate at full capacity from a dead stop.

I have also made contactors from brass bolts, and by installing cables into the solenoid plungers of automotive starter solenoids, but unless you are certain of what you are doing, this approach is not recommended because you could create a contactor that may become stuck in the on position. For the record, Sparky version 1 had a simple contactor made by inserting a brass bolt through the handle of a plastic brake lever, so that the bolt would contact the metal handlebars closing the circuit. This system did work quite well, but was prone to heating up and always threw a huge spark when starting from a dead stop, hence the name Sparky!

Motor speed controller

A motor controller is a far better approach for delivering the battery current to the motor, because you will have full control over the acceleration, top speed, and sometime braking. Motor speed controllers send a pulse-width modulated signal, varying in duty cycle from zero to 100 percent to the motor at a high rate of frequency, so that the motor acts as if you are varying the current. If you had a simple contactor, you could pulse it on and off to try to control your speed. However, the motor speed controller has an onboard computer that does this hundreds of times per second, so the speed control is ultimately smooth. A throttle that has either a lever or conventional twist grip is installed on the

handlebars, and the speed control is so fine that you can make the wheel just barely turn, or rip up the ground with rocket-like acceleration if your motor is capable of doing so. A motor speed controller will also limit the current to the motor, so you will not burn it out if you try to push it too hard, and the acceleration will follow some type of ramp, so an accidental wheelie will not occur if you open that throttle too far from a dead stop.

The downside of a real motor speed controller is of course the price. I highly recommend that you give this option some real consideration, as it will make your riding experience so much more enjoyable, and allow you to keep your machine under complete control, especially in a crowded area where you need precise control of your vehicle. You can also limit the top speed for younger riders, and lock the mini bike when not in use, by adding a key switch if it is not already part of the controller kit. Choosing a controller is much like choosing a simple contactor; you must choose a controller that will handle the rated capacity of your motor at the chosen voltage. Most motor speed controllers do have current limiting, so choosing a lower current controller than required will usually only result in poor performance, but there is a limit, and if your controller is too "wimpy," it will burn out on hills or even when you ride against the wind for too long.

My motor controller is an Alltrax golf cart unit rated for 275 amps continuous, and 400 amps peak for 30 seconds. What this means is that when I am starting from a dead stop, the motor will get 400 amps if my battery pack can deliver it (unlikely), and then the controller will be happy delivering 275 amps to the motor all day or until the batteries give it up. This controller is actually overpowered for my motor and batteries, but I know that there is virtually nothing I can do to blow the controller, and it will climb hills, spin dirt, and outperform most store-bought scooters all day without even getting warm.

When you do a search for "ev parts," or "motor speed controllers," you will find a lot of great information, and there are many different products available to choose from. Like all parts, try to find a reputable company, and ask them for advice when choosing a controller. You really want to avoid an underrated controller for your motor, and if your budget allows, purchase the next step up just to be safe. I plan to build many different electric vehicles with this motor and controller, so I am confident it will hold up to whatever abuse I throw at it.

Regenerative braking

For typical riding, most of my braking comes from the resistance that the controller puts across the motors terminals when I release the throttle. What is happening is that the motor is no longer receiving current; it is being forced to generate it, which puts a load on the motor, effectively slowing it down. If your controller has full-blown regenerative braking (regen), then you will not only get a large amount of smooth, throttle-controlled braking ability, but your batteries will get a little bit of charging as the controller pumps current from the motor back to the batteries. In a large vehicle, this can significantly increase the run time, but in a vehicle like Sparky, this is just an added bonus, especially on a design like mine with no built-in wheel brakes. Obviously, if your controller has no regen, then you'd better have at least one decent wheel brake for safety. Depending on the level of regen your controller has, wheel brakes may not even be necessary, although a bad controller failure could also deplete all your braking ability. A simple motor pulley brake will be presented later as well, just in case you are limited by the wheels and the motor speed controller.

I decided to complete the frame by capping off the extra tubing left over when I cut the bottom tubes rather than cutting them off. I inserted a pair of chrome lug nuts into the ends of the tubes and welded them in place, as shown in Figure 11-89. After painting, I brushed off the paint at the tips of the lug nuts with some steel wool. These small details make a huge difference, if you plan to sit in one place long enough to show off your custom creation. Take this opportunity to check over all the welds, and clean up any rough spots and holes.

Apply a good metal primer after cleaning up the parts with some fine sandpaper. I know it's torture to wait for primer and paint to dry, but if you follow the instructions and do it right, then a simple spray-can job can look just as good as a professional job. I normally prime the parts then let them sit for a day, and then do the spray painting. The spray paint should be left for a few days to cure before you assemble the bike, although I will admit that sometimes my impatience gets the best of me. Fresh

Figure 11-89 *Capping the tubing ends with chrome lug nuts*

paint is easy to scratch and chip for at least a week after it dries, so be careful. Figure 11-90 shows the freshly primed parts taunting me to work faster.

Figure 11-91 shows a fresh-looking, bright orange Sparky assembled again after the paint has cured for a week (I seriously did wait a week). The simple orange with black accents actually turned out quite well, and the little mini bike was starting to resemble a little hog as planned.

Figure 11-92 shows Sparky with all the major components installed, waiting to have the wiring done. Take your time and solder all wires using the proper clips, following the wiring diagram that came with your motor controller. Do not get lazy and leave out the recommended fuses and safety switches just in case you have a crossed wire, or the only smoke show you will be

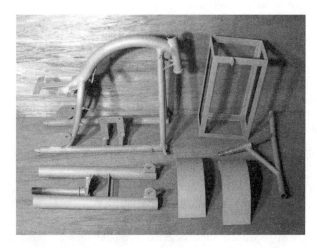

Figure 11-90 *Frame parts primed*

Figure 11-91 *Shiny and new!*

Figure 11-92 *Waiting for a wiring job*

creating is one where all the Metal Oxide Silicon Field Effect Transistors (MOSFETs) in your controller explode. There is no standard wiring diagram I can show you since each motor controller and contactor will have its own wiring map, but I can tell you that you have a 50 percent chance of hooking up your motor in reverse! Most of these small DC motors, mine included, have no polarity and rotation shown on the terminals, so you may end up hitting that throttle and going in reverse. This will not cause any damage, but your first attempt at spinning the drive wheel should be done on blocks before your first test ride.

Most of the wheels available for small scooters and mini bikes will have some type of brake already built into the wheel in the form of drum brakes or band brakes. It is a matter of installing a bicycle brake lever as shown in the manufacturer's instructions, in order to have brakes on one or both wheels. My wheels did not offer any simple method of adding a brake, but since my motor controller had a decent level of braking by simply letting off the throttle, I did not bother trying to install any other brakes. If I had a simple contactor rather than a motor controller, I would certainly need some type of brake, so I devised the system shown in Figure 11-93, in case you have brakeless wheels and a simple contactor.

The brake is just a bicycle pull brake with the pads turned sideways so that they grab hold of the aluminum drive pulley when I engage the brake lever. It may not look like much, but remember that the gear ratio between the motor pulley and the wheel pulley is 1:3, so this brake is effectively grabbing a wheel of 6 inches in diameter. Using only this brake to stop, I can bring the mini bike to a dead stop from full speed in a distance of about 20 feet, which is not too bad.

If you take a sharp knife and cut the brake pads to conform more to the pulley body, then the braking power will be increased even more. Some production scooters and pocket bikes use the same system as this for brakes.

After looking at so many photos of this mini bike being built, you may have forgotten how small Sparky really is! Figure 11-94 brings this into perspective, showing how little real estate he really occupies. Many onlookers chuckle as I unload Sparky from the back of

Figure 11-93 *A sneaky brake, in case you find it necessary*

Figure 11-94 *Yo, look at my bad-ass hog!*

the truck, but their smirks usually turn to slack jaws as I vanish out of sight at speeds normally accomplished by street motorcycles. It's also fun to creep up on a person from behind, and then zip past them since Sparky is virtually silent. The only sounds you usually hear are the sounds that the tires make on the ground and my evil laugh. I can actually sneak up on those Canadian geese shown in the background, but let me assure you, just because you can do something does not make it a good idea! Some people (and geese) don't always appreciate your stealthy electric vehicle!

With the way Sparky is styled, it almost looks like Figure 11-95 has been manipulated somehow to make me look like a giant. It's hard to believe that Sparky can keep up to city traffic, run for well over an hour and completely out-accelerate store-bought scooters or pocket bikes, even the gas ones. Trust me, this little dude has attitude and power!

The nice thing about a motor controller, as compared to a simple contactor, is that younger riders can enjoy the ride as well. Your motor controller can even be strictly controlled by placing a ring on the handlebars that would prevent the throttle lever from engaging to the maximum. If you know a bit about electronics, you could also add a series resistor between the throttle and the motor controller to limit the top speed, by reducing the ability of the throttle's variable resistor to reach the maximum resistance. If you have only a simple contactor, the belt idler spring could be relaxed a bit to limit the overall acceleration, although the mini bike would still eventually reach top speed. Figure 11-96 shows one of our test pilots, Brittany, putting Sparky through its paces over three hours one afternoon.

It doesn't take very long before you get up the courage to want to push Sparky to the limit! With all that power available, you can shoot up any hill, blast

Figure 11-96 *Sparky is well behaved*

Figure 11-95 *Look, that dude is 15-feet tall!*

Figure 11-97 *Silent, fast, and fun*

Figure 11-98 *Still going strong after hours of vigorous riding*

up the dirt, pull wheelies, and even give a buddy a ride, although sitting on the frame tubing sure does hurt!

With all that vigorous riding, Sparky may wear you out before the batteries need a recharge, but when the time comes to plug in, you can expect to be back on the trails within 4 to 6 hours, depending on the size of your battery pack and speed of your charger. If you can't get enough, just have two battery packs so you can cycle them for non-stop fun. Oh, and when you finish your custom electric mini bike, do remember to send us a photo for our gallery, as we love to see what other garage hackers are building!

Project 12: LongRanger Scooter

The LongRanger fills a gap between the electric-motor-assisted mountain bicycle and the fully street-legal electric motorcycle or scooter, allowing you to commute great distances at speeds typical of a standard human-powered bicycle. Because the LongRanger looks and feels much like a standard mountain bike, it will happily live on bike paths and trails normally off-limits to larger motorized vehicles. The LongRanger can be made to travel well over 62 miles (100 kilometers) on a single charge, and will do so at speeds that only a top athlete could keep up for any real distance. Oh, and it does all this without making any pollution or noise whatsoever! Because the LongRanger is built around the battery pack and the type of hub motor you choose, you

can build the base vehicle to suit your needs, altering the top speed, acceleration, weight or range.

It's a good idea to read over the entire plan once before hacking up any bikes or heading to the scrapyard on a search for parts. By reading ahead, you will see why a certain part was chosen, and get some ideas on how to modify the design to suit your needs. With even a slight change in a few of the main frame tubes, you could make a completely different looking chopper, even though the core design has been followed. Don't be afraid to try new things, and remember that hard work is what makes this hobby successful, not spending money on mass-produced products. Long live the real custom!

Regardless of the size and number of batteries you plan to install, you will need a front wheel, suspension fork, and matching head tube from a sturdy mountain bike. To ensure that all the parts fit together properly, it is recommend that you salvage these parts from the same bike, preferably one in decent condition. I actually purchased a brand new mountain bike on sale at the hardware store, and then stripped all of the parts off, leaving what you see in Figure 12-1.

Often, you can purchase a decent quality, full-suspension mountain bike on sale for a hundred bucks, which is not bad considering you end up with a new aluminum front wheel, tire, suspension system, frame, and one rear shock. I know, the quality of an

Figure 12-1 *Choosing a front wheel and suspension fork*

inexpensive department store bike is nowhere near that of a high-end mountain bike purchased from a quality dealer, but at least all the parts are new, and the frame is made of steel, so it can be welded. I ended up using the rear of the bike to make a tadpole trike called the StreetFox, so a hundred bucks was actually a great deal.

Regardless of where you acquire the parts, you will need what is shown in Figure 12-1—a front wheel and suspension fork, a matching head tube with at least a small portion of the frame tubing still intact, two similar rear shocks, and the front brake hardware. Don't worry about the rear triangle of the frame, as it will not be needed.

As you can see in Figure 12-2, the entire front of the mountain bike will be used, which is why it is optimal to work with parts that fit properly together and are in good working order. Stand up the frame with the front tire fully inflated so that the head tube is at the angle it was when the bike was complete, or at an angle typical of a standard bicycle, which is usually somewhere between

Figure 12-2 *Using a guide bar to make a straight cut*

72 and 74 degrees. If your donor frame does not include rear hardware or a rear wheel, like my frame shown in Figure 12-2, you will have to figure out a way to stand up the frame so that the head tube is at the correct angle, since the cutting line is derived this way. As you can see in Figure 12-2, I have placed the rear of the frame on a plastic bucket and a bit of scrap wood, so that the head tube is at roughly the same angle it would have been if the entire bike were still together.

Head tube angle is measured by extending an imaginary line to the ground and then taking the angle of this line with that of the horizontal ground. Rather than messing around this way, simply roll a functional bicycle up next to your supported frame and match up the angles as closely as you can. A few degrees either way is not going to make a huge difference in the completed vehicle. The angle of my head tube, as shown in Figure 12-2, was determined by placing another mountain bike alongside the donor frame for reference, as I propped up the rear of the frame. My head tube angle is approximately 73 degrees.

Also shown in Figure 12-2 is a length of flat bar placed vertically, 4 inches back from the edge of the front tire. This flat bar is used as a guide to mark a line on the top tube and down tube for cutting. Since the battery box will be a perfectly square box riding parallel to the ground, the frame cut lines are to be marked so they are 90 degrees to the ground. To do this, either have a helper use a 90-degree square to check the angle of your guide bar with the ground, or use a level to check this angle as you hold the guide bar against the frame.

Figure 12-3 shows the frame tubing marked for cutting after ensuring that the guide bar was placed 4 inches behind the edge of the front tire, and at 90 degrees to the ground so it was perfectly vertical. A black marker is used to mark the line, which needs to be the same on both sides of the frame.

Once the cut lines have been clearly marked on both frame tubes, use a hacksaw or cut-off wheel in your grinder to cut away the front of the frame from the rest of the frame. I use a cut-off disc (zip disc) for all my metal cutting because it's much faster and more accurate than a hacksaw, but do be careful when hacking up bicycle frames. On the first cut through a complete frame tube, the frame may start to collapse on the grinder disc if under tension, so keep this in mind as you complete any cuts, holding on to the grinder with both hands. To avoid

Figure 12-3 *Marking the frame for cutting*

this, the last half inch of the cut can be made with a hacksaw to avoid any disc snagging that might occur.

Figure 12-4 shows the front of the frame, including a head tube and a few inches of top tube and down tube. To aid welding and avoid filling the garage with horrific, burning paint fumes, the paint was removed around the joint ahead of time using some sandpaper. A sanding disc (flap disc) is also an easy way to remove paint easily.

Although we will discuss the hub motor, controller and batteries in greater detail later, you will need to know the dimensions of the batteries you plan to install on your version of the LongRanger, in order to continue building the frame. If you have not decided on a brand or type of hub motor and batteries yet, then read ahead so you can see how these components will affect the range, weight, and speed of your final design. The obvious goal of the LongRanger is of course long range, so batteries much larger than those considered typical of an

electric-assisted bicycle will be used. As you will soon see, the frame is actually built around the batteries, since they are the key element in the entire design.

Figure 12-5 shows the huge size difference between the standard, hub motor kit batteries and the ones I plan to install in the LongRanger. The small batteries are rated at 7 amp hours each, and the larger ones are rated at 32 amp hours each, so the expected range of the larger batteries is about four and a half times more. The small batteries measure 6" by 2.5" by 4", and the large ones measure 7.75" by 5" by 7". Let's discuss battery size versus expected range.

There are many types of battery chemistries, and each has certain advantages depending on the intended use. Let me make this choice very simple—you should only use sealed lead acid (GEL) batteries for this project. Unlike a typical automotive lead acid battery, a sealed or GEL battery does not have free-flowing liquid, nor does it need maintenance such as "topping up" the fluids, so there is no chance of spilling any corrosive liquid on your vehicle or yourself. Sealed batteries are often labeled as wheelchair or recreational batteries, and the obvious sign that they are sealed is the lack of any type of filling or maintenance valve on the top of the battery. Sealed batteries may cost a little more than the equivalent size wet-cell battery, but the extra cost is worth it in terms of safety and overall battery quality.

The real question is: how far do you want to travel on a single battery charge? Sure, you could go to the battery store and tell them to give you three of the largest GEL batteries they have, but do you really need to commute half way across the country? Also, the larger batteries can get very large and heavy, so you may be very

Figure 12-4 *The front of the frame has been cut*

Figure 12-5 *Comparing battery sizes*

uncomfortable straddling a 12-inch-wide battery box. The batteries shown in Figure 12-5 have never needed a recharge, even after two days of lengthy riding, so as a guide, 32-amp-hour batteries will be more than enough for any single journey you may want to undertake. There are many sizes of GEL batteries in the smaller (under 40 amp hours) range, so your best bet is to look at the estimated range of your hub motor kit with its current batteries and work from there. If the original 10-amp-hour battery pack claims 25 miles (approx. 40 kilometers), then it is a safe guess that a 30-amp-hour battery pack should deliver 120 kilometers of range. Of course, I would divide any manufacturer's range claim by 1.5, as they have been known to exaggerate!

OK, let's sum up all this battery talk. Only purchase GEL batteries, and make sure all of them are the same size and age. A battery larger than 35 amp hours is probably too large, and anything less than 20 amp hours might be too small. Use your original battery's capacity and range as a guide, but take manufacturers' claims with a grain of salt!

Figure 12-6 shows a stack of three GEL batteries separated by blocks of wood so that the connecting terminals do not rub on the underside of the battery above. This battery pack will deliver 36 volts at 32 amp

Figure 12-6 *The mighty 36-volt stack*

hours. Voltage is the sum of all batteries in series, so 3 × 12 volts equals 36 volts. Amp hour rating is the same rating as a single battery in a series connected pack, so since each of these batteries is 32 amp hours, that is the overall amp-hour rating of the pack. Why three batteries? Simple—because the original pack was three batteries and I did not want to overpower the hub motor, or have to search for a different voltage charger. If you add an extra battery to create a 48-volt pack, your hub motor may have no problem with the higher voltage, and it will actually run a bit faster, although you will now have to source a 48-volt battery charger, and make room for one extra battery. Unlike a higher amp-hour rating, with more voltage, range will not increase.

Unless you know what you are doing with batteries, controllers and chargers, the best bet is to simply increase the amp-hour rating of your batteries, and use the same number of batteries and charger that came with the original kit. My charger is only rated for a 1-amp charging rate, which is very small, but it takes care of the 32-amp-hour pack overnight if I do not run it down too far. The typical 36-volt charger is between 2 and 4 amps, so you will be fine.

As shown in Figure 12-6, you must stack up your batteries in such a way that the connection terminals do not rub on the underside of the above battery. Some small pieces of 2 × 4 wood seem to do the job perfectly. Solid rubber hockey pucks may also be a good choice if you have access to them.

Once you have your batteries stacked up and insulated from each other, you can begin making the battery pack rails by using the actual stack of batteries as a guide. As shown in Figure 12-7, four lengths of 1/16-inch thick, 1-inch-wide angle iron have been cut and strapped to the battery pack using an elastic cord. These four lengths of angle iron will form the side rails of the battery pack, and should reach from the base of the lower battery to 1 inch above the top of the upper battery terminal.

Be careful when handling the metal bars around the batteries, especially if the battery terminals are not insulated with a plastic cover. Dropping the metal across the battery terminal will not only cause an extremely loud and startling spark, it could damage the battery or cause a nasty burn. You can safely come in contact with the terminals because the voltage is much too low to deliver a shock, but there is certainly enough amperage there to weld steel!

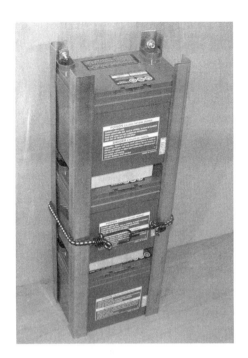

Figure 12-7 *Battery pack side rails*

You should place a few strips of cardboard or thick paper between the angle iron and the batteries for extra clearance, so the batteries are not extremely tight, and this way they can be replaced easily.

The battery pack base is also made by cutting four pieces of the same angle iron in order to form a box, as shown in Figure 12-8. Although I test fit the base on the top of the pack, the measurements will be the same since the rails are the same all the way along the battery pack. The two longer pieces will overlap, as shown in Figure 12-8, and this is not a problem since the ends will all be cut at 45 degrees to form a square.

Figure 12-8 *Cutting the battery pack base*

To make a perfectly square base of the angle iron, each corner needs to be cut at 45 degrees, as shown in Figure 12-9. Mark the ends using a black marker and a 45-degree square by drawing a line from the end corner where the two sides of the angle iron meet. A hacksaw or zip disc is then used to cut off the corners.

Figure 12-10 shows the battery pack base tack welded together for a test fit over the side rails. Make sure the base is square and can slide over the four side rails without forcing them to squeeze the side of the battery. The idea is to make a pack that holds the batteries snuggly, but not so much that you can't slide them in and out of the rails if they need to be removed for whatever reason. If the base seems to be square and fits the rails correctly, weld all four joints on the outside of the box (the side showing in Figure 12-10). The inside of the box is not welded because it would create a raised spot under the batteries, adding stress to the underside of the lower battery, which must carry the weight of all three batteries.

The side rails drop in to the base, as shown in Figure 12-11, and should allow the battery to slide right in with minimal friction. Remove the batteries when tack welding the corner of the joint to avoid burning the thin

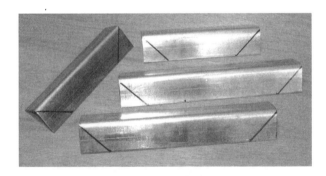

Figure 12-9 *Making the square battery pack base*

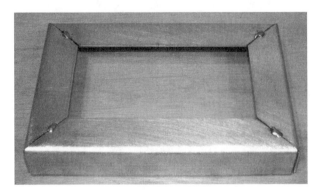

Figure 12-10 *Tack welded battery pack base*

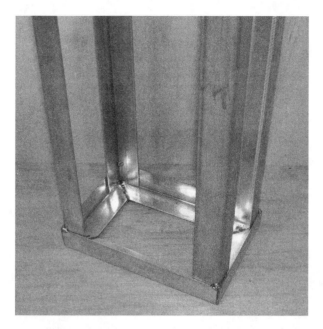

Figure 12-11 *Installing the side rails*

plastic casing of the battery. At this point, only tack weld the side rails in place; this way, the entire box can be squared up once the top has been added.

The top of the battery box is made from 1-inch flat bar, or by cutting a length of angle iron into two pieces by cutting it along the bend. Since the batteries need some way in and out of the battery pack, angle iron could not be used to make the top of the box as it would seal the opening. The top four bars shown in Figure 12-12 are cut so they form a box of exactly the same dimensions as the box on the base of the battery pack. Only tack weld the bars in a few places so you can square up the entire battery pack by forcing it into shape. This is a good time to test fit a battery to make sure it slides into the pack without much resistance.

Once you have your battery pack tack welded together and squared up, you can finish welding all the outer joints to complete the pack. The completed battery pack is shown in Figure 12-13, with all the outer joints fully welded to give great strength to the box. None of the inside joints are welded because this would make it difficult to insert the batteries, and is not necessary for strength. You'd better test fit all three batteries one last time to ensure that heat distortion did not pull the box out of square.

The LongRanger is a battery box with wheels! On a regular diamond frame bicycle, there is little possibility of carrying 60 pounds of large batteries on the frame without dangerously overloading the frame or upsetting the balance. On the LongRanger, carrying the batteries is the purpose of the frame, so the most important part of the frame is a battery box holder. This part of the frame is called the center frame, and it is simply a square frame around the battery pack made of 1.5-inch steel tuning with a wall thickness of 1/16 inch. The four pieces are shown in Figure 12-14.

This tubing can be ordered from any steel supplier, and is very easy to cut and weld. Steel suppliers like to rate tubing wall thickness using a gauge number, so 1/16 wall tubing will be called 16-gauge tubing, and although there is a slight difference between the two, your best bet is to tell them you want a length of 1.5-inch square tubing with a wall thickness as close to .0625 of an inch (1/16) as you can get. If you are a heavy rider (over 300 pounds), then you should ask for the next size after that, which will most likely be 14-gauge tubing, or .078-inch wall tubing.

Figure 12-12 *Adding the top bars*

Figure 12-13 *Battery pack welding completed*

Figure 12-14 *Tubing for the center frame*

How much tubing you will need is completely dependent on the perimeter of your newly created battery pack. Simply measure the perimeter of the battery pack around its longest dimension and then add a few inches to spare when you source your tubing. The rear swingarm will also be made from the same tubing, so add 3 feet to the battery box perimeter and you will easily have more than enough tubing.

To make a perimeter frame around the battery pack, you need to create a frame out of the 1.5-inch tubing that will have the same inside perimeter as the outside of the battery pack. Add a quarter inch to the perimeter of the center frame, so the battery pack can have a little room to slide in and out of the main frame. With this design, you can have more than one battery pack for quick change if you do not want to wait overnight to charge the pack. This design also makes it very easy to remove the entire battery pack for use on another vehicle or for maintenance if needed. Figure 12-15 shows the four lengths of 1.5-inch square tubing cut at 45-degree angles,

so that the resulting frame has an inside dimension approximately a quarter inch larger than that of the battery pack. The battery pack can drop right into the center frame.

Using a 90-degree square, set up the center frame on a flat surface and then tack weld the four corners together, as shown in Figure 12-16. With a tack weld on each corner, recheck the alignment with the square and adjust if necessary. The frame will be easy to manipulate into shape with only a single tack weld at each corner. When you are satisfied with the alignment, check one last time by dropping the battery pack into the frame, and then add a few more tack welds around the frame to keep it from moving.

When you have checked the center frame alignment using a 90-degree square and you are certain that the battery pack will fit inside, fully weld each outer corner joint, and then check the alignment one last time. The final welding is done on each side of the 45-degree angle joints to complete the center frame welding, as shown in Figure 12-17.

Welding the inside joint of the center frame should be done using the smallest weld bead you can possibly make, in order to avoid warping the frame or creating a high spot that will interfere with the installation of the battery pack. Figure 12-18 shows the small weld bead

Figure 12-16 *Squaring up the center frame*

Figure 12-15 *Cutting the center frame tubes*

Figure 12-17 *Center frame welding completed*

Figure 12-18   *Cleaning the center frame welds*

used on the inside of the center frame joint, as well as the side welds, which have been ground flush with the square tubing. Grinding the side welds flush with the tubing is an optional step, simply done for appearance only. This grinding is first done with a heavy grinding disc to take down the excess weld material, and then it is finished using a sanding disc. Avoid taking out any of the original tubing metal or too much of the welded area or you will weaken the joint. The final ground welds should have slight bumps, not valleys.

As you can see in Figure 12-19, the center frame is aptly named because it will join the front and rear of the bike together. To figure out where the front of the frame will join the center frame, place the center frame on level ground so that it has at least 6 inches of ground clearance between the ground and the underside of the tubing. If you plan to take on some serious off-road terrain, you

Figure 12-19   *Setting up the frame*

may want to increase the ground clearance an inch or two, but for normal riding, 6 inches should probably be enough. With the center frame stood up, as shown in Figure 12-19, the front of the frame will simply find its own place, due to the fact that it is already cut at the perfect angle to meet the front for the center frame's vertical tubing. Because we cut the front of the bike away from the frame at the correct head tube angle, you can be assured that the final head tube angle will be correct, as long as the center frame is parallel to the ground.

With the center frame propped up to an appropriate height from the ground, place the front of the bike up against the square tubing, and have a helper hold it in the correct position by looking straight on at the head tube with the front wheel between his or her legs. The head tube must be parallel with the center frame tubing, so that the wheel is perfectly in line with the rest of the frame. A solid tack weld placed at the very top and bottom of the tubing should hold it in place. You may want to throw a cover over the front tire and forks to avoid welding spatter damaging the rubber parts.

After you apply the tack welds to the front of the frame and the center frame, visually inspect the front wheel and head tube for alignment with the center frame. The head tube and vertical square tubing should be as parallel as possible so your front wheel tracks properly. If everything looks good, carefully remove the front forks so you do not disturb the tack welds, and then place the center frame on your work bench for welding. You can now recheck the alignment visually, and make slight adjustments with your trusty hammer. If the head tube looks obviously misaligned, break one of the tack welds and redo it. Figure 12-20 shows a larger weld made on the top and bottom of the joint after checking alignment with the forks removed.

To weld the front of the frame to the center frame, begin with the top and bottom joint on each tube, as shown in Figure 12-21. Starting here will keep side-to-side distortion to a minimum when you weld the side of the joints. It should still be possible to bang the head tube with a hammer to make slight adjustments, and it's never too late to cut the welds and fix obvious errors. A good welder fixes problems as soon as possible.

Figure 12-22 shows the completed welds between the center frame and the front of the frame. The sides of the joint are done after the top and bottom to avoid distortion, and they are done only a little at a time, switching sides

Figure 12-20   *Head tube welded to the center frame*

Figure 12-22   *Front frame completely welded*

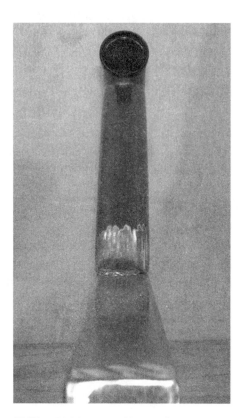

Figure 12-21   *Weld the top and bottom first*

after an inch-long weld. The head tube should remain in good alignment using this welding technique, and this can be checked by reinserting the front forks and wheel for a final visual inspection. This weld should only be cleaned by removing spatter or bumps, but do not grind the weld metal or it may weaken the joint.

Until a few years ago, the most common way to add clean motive power to a small vehicle, such as a moped, scooter, or bicycle, was to drop in a chain, belt, or friction-driven electric motor. This system was OK, but due to the minimal availability of quality small motors, often the resulting drive system was inefficient, noisy, and lacked proper speed control. You could spend a larger amount of money to get a quality motor speed controller, a more efficient motor, and larger batteries, but by the time you add all these costs together, the final price tag could easily be $1,500 or more, and the motor may be a little on the large side for a bicycle-type project. There have been numerous, inexpensive, electric add-on kits for bicycles popping up on the market, but often they are inefficient and cheaply made, with barely enough power to get you over a slight hill or moving into a headwind, and range is often a serious problem. Enter the hub motor!

What type of hub motor should you purchase? Obviously, there are hundreds of them on the market now with prices varying from approximately $200 to over $1,000. Like all products, you usually get what you pay for, but there are several factors that will influence the price of a hub motor kit, such as internal construction, output power, rim quality, and the type of batteries and charger that come with the unit. My hub motor, shown in Figure 12-23, is a 450-watt brushless unit built into a 20-inch wheel. It comes with the twist throttle, controller, charger, wiring, and the batteries shown in Figure 12-5. The price on my hub motor kit is approximately $500, and is considered to be a higher quality unit due to the brushless construction. A brushless hub motor has no wear parts, and will never require maintenance, whereas a brushed motor will require changing the internal brushes about once a year if well used. Brushless motors

Figure 12-23 *A complete hub motor kit*

that looks too good to be true almost always is. Avoid the junk and you will have a quality electric vehicle.

The rear wheel on the LongRanger is actually a front hub motor, since there are no sprockets on the bike. The rear wheel will be placed behind the center frame at the same 4-inch distance from the rear tube as the front wheel is away from the front tube. It does not matter if you have a 20-inch or 26-inch rear wheel, as the same clearance is needed. Figure 12-24 shows the placement of the rear wheel as the bike is positioned level to the ground.

The rear suspension swingarm is made from the same 1.5-inch square tubing as the center frame, which is why I stated that you should have a few feet extra. The two longer tubes are cut to a length appropriate for the chosen rear wheel and tire size, and the small tube is cut to a standard front hub length. The three tubes are shown cut in Figure 12-25, based on the drawing in the next figure.

If you are using a front hub motor (one without a sprocket), then it will fit on a standard front bicycle fork,

are also more efficient, so if you can afford the extra cost, choose a brushless design.

As for motor power, there will be plenty of data on the manufacturer's website regarding speed and range. Again, range is usually inflated, but often top speed is rated very accurately. Due to top speed regulations, almost all hub motors will top out around 22 mph (35 km/h), which is dependent on the rim size. A smaller rim will slightly reduce your top speed, but will increase your hill climbing ability. For this reason, my LongRanger has been fitted with a 20-inch rim to allow better hill climbing power and off-road use. My top speed peaks at around 17 mph (28 km/h). The wattage rating of the hub motor is what will determine the amount of hill climbing power and acceleration you can expect, with 450 watts being at the lower end of the scale and 1000 watts at the upper end of the scale. My 450-watt motor can climb a fairly steep hill, and take on any headwind, but I also have the added torque due to the smaller 20-inch wheel. As a general guide, shoot for a 450–550-watt motor for a 20-inch wheel, and a 600–750-watt motor for a 26-inch wheel. If you really want serious acceleration and hill flattening power, a motor of 750 watts or more will be extreme, although at the cost of reduced range.

My recommendation is to try a bicycle fitted with a hub motor kit, and see how it runs without any pedaling input. You can really tailor the power characteristic of the LongRanger from "running all day long at bicycle speeds" to "taking off like a rocket and pulling wheelies," depending on your motor choice. And as always, a deal

Figure 12-24 *Positioning the rear wheel*

Figure 12-25 *Cutting the swingarm tubing*

which has dropouts placed 4 inches apart. For this reason, the smaller swingarm tube that joins the two longer tubes is cut to a length of 7 inches using 45-degree cuts at each end, so its inner width is 4 inches, as shown in Figure 12-26. The two longer tubes are cut to a length that is equal to the distance from the axle of the wheel to the edge of the tire, plus 1.5 inches for the width of the smaller tube and 1.5 inches for tire clearance. For this reason, you need to have the tire you plan to use installed on the rim and fully inflated when you make these measurements. The length of my two longer swingarm tubes (for my 20-inch wheel) was 11.5 inches, but I recommend that you get this measurement based on your rim and tire. If you are planning on using a massive rear tire, such as a 16-inch motorcycle tire mounted to a 20-inch bicycle rim (yes, this works), then the swingarm may need to be wider than the standard 7 inches.

With the measurements derived from Figure 12-26, the three swingarm tubes are cut, as shown in Figure 12-27.

The small tube has a 45-degree cut at both ends, and the longer tubes have a 45-degree cut at only one end. Place the three tubes on a flat surface and use a bit of scrap steel to keep the open ends in place as you tack weld the assembly together. As shown in Figure 12-27, the 4-inch inside width is held by the smaller swingarm tube and the bit of scrap angle iron welded across the open end. Once tack welded, pick up the swingarm and visually inspect it for alignment. Both legs should be perfectly parallel to each other, and the ends should be square.

With the swingarm tack welded and checked for alignment, you can complete the welding by starting with the outer corner welds and then the top and bottom welds. The inside of the joint welding is done last, and with a small bead to minimize the inward pulling of the swingarm legs from head distortion. Clean up any welding spatter or cold spots, but do not grind the welds flush on any joint of the swingarm as it needs to remain as strong as possible. The completed swingarm is shown in Figure 12-28 after completing the welding around the entire joint. When the welding is complete, you can grind off the tack welds holding the scrap tube at the open end of the swingarm, and don't be alarmed if the legs pop inward about a quarter inch, as this is to be expected due to the shrinkage caused by the inner welds.

The rear dropouts which hold the rear wheel to the swingarm are made from two pieces of 3/16-inch thick angle iron cut to 1.5 inches in length. At 1.5 inches, the dropouts will have plenty of room to carry the wheel axle and will cover up the open ends of the swingarm tubing. Figure 12-29 shows the 1.5-inch wide angle iron marked for cutting. Since you will need two more 1.5-inch long

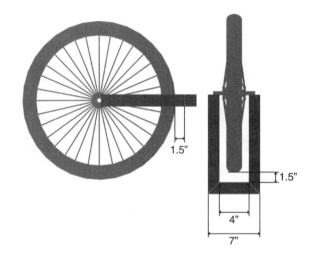

Figure 12-26 *Swingarm clearance*

Figure 12-27 *Tack welding the swingarm tubes*

Figure 12-28 *Finishing the swingarm welds*

Figure 12-29 *Cutting the rear dropouts*

Figure 12-31 *Drilling the axle holes*

pieces of angle iron later to make the swingarm mounts, cut four identical pieces now from the same length of angle iron.

To find the center of the angle iron, draw an X across the corners, as shown in Figure 12-30, then use a center punch to mark the hole, which will be in the center. The two pieces in Figure 12-30 will be used to form the dropouts, but you can mark the other two pieces the same way, as they will also need to have a hole drilled into the center of one side.

Drill a hole through the dropouts where you marked the center point (see Figure 12-31). Start with a small drill bit and work your way up to a hole that is about 1/16 of an inch larger than the diameter of the axle on your hub motor. The axle size of various hub motors may vary, so it is best to measure it directly or by checking the axle fit in the hole, as you progressively drill it out with a slightly larger drill bit. My hub motor axle fits nicely into a half-inch hole. Also shown in Figure 12-31 is a marker

line transferred from the edge of each hole to the end of the face, which will be cut using a hacksaw or zip disc to form the dropout slots.

The lines shown in Figure 12-32 are cut out to make the dropout slots, as shown in Figure 12-32. Now your rear wheel axle can slide onto the dropouts, just like the rear wheel on a standard bicycle or motorcycle. Try to make the slots as uniform as possible, and no wider than the original hole. A hacksaw or zip disc can be used to cut the slots slightly inside the line, and a flat file can be used to perfect them.

The two dropouts are mounted to the swingarm, as shown in Figure 12-33. Here I have tack welded them to the outer edge of the swingarm tubing so I could test fit the wheel, making sure it sits straight in the swingarm tubing. If your swingarm was put together perfectly straight, then the dropouts will also be perfectly aligned by lining them up squarely with the ends of the tubing.

Figure 12-34 shows a perfectly aligned rear wheel as viewed down the length of the swingarm. If you are

Figure 12-30 *Finding the center*

Figure 12-32 *Completed dropouts*

Figure 12-33 *Mounting the dropouts to the swingarm*

Figure 12-35 *Completely welded dropouts*

Figure 12-34 *Checking alignment*

careful, you can place a cloth over the hub motor and make a few more tack welds to the dropouts once you are happy with the wheel alignment, using the axle as a guide to hold them in place. Once you have several tack welds on the aligned dropouts, remove the wheel for final welding.

The dropouts are welded all the way around the open joint, as shown in Figure 12-35, which also caps off the open end of the swingarm tubing. This is another weld that should only be cleaned up slightly in order to maintain full strength. When the welding is complete, recheck the wheel alignment, which can be fine tuned if necessary by slightly adjusting the axle as you tighten up the axle nuts.

The rear swingarm will be allowed to pivot against the lower part of the center frame so it can act as a suspension system, held in place by a pair of common mountain bike spring shocks. For the smoothest

suspension, a tube that can carry a pair of brass bushings (pictured in Figure 12-36 on the right) and a half-inch-diameter bolt should be used. You could get away without the brass bushing and just use the bolt, but there would be a bit more friction and possibly some squeaking as the suspension moved up and down due to the metal on metal grinding. It is always good practice to insert brass or plastic bushing in any moving part pivoting on a bolt or rod. The tube shown in Figure 12-36 (on the right) is 1.5 inches wide, just like the frame tubing, and has a center diameter that will accept the brass bushings, which have a center diameter of half an inch for the bolt. The two pieces of 1.5-inch angle iron, also shown in Figure 12-36, are the same as the ones used to make the rear swingarm dropouts, and have been center marked for drilling the half-inch bolt hole.

The tube that carries the swingarm pivot bolt and brass bushing will be installed at the lower rear of the center frame, as shown in Figure 12-37. To keep the pivot tube lined up with the underside of the frame, both the frame and tube are placed on a flat surface; another square tube, in my case. Tack weld the pivot tube in place while it is

Figure 12-36 *Swingarm pivot hardware*

Figure 12-37 *Mounting the pivot tube*

Figure 12-39 *Connecting the swingarm*

resting on the flat surface, but do this with the bushings removed or they will be damaged by the heat. Most brass bushings have been impregnated with oil, so overheating will cause the oil to burn out of the brass.

The pivot tube is to be welded all the way around its perimeter, as shown in Figure 12-38. Start with the side welds, making sure the tube is level with the underside of the frame by using the flat surface as an alignment guide. The weld along the top and bottom is completed last. When the part has cooled, you can tap in the brass bushings, as shown in Figure 12-38.

Because we are using a flat surface as an alignment guide, it is very easy to get the swingarm and mounting brackets installed by placing everything on the flat surface for tack welding, as shown in Figure 12-39. The two angle iron brackets are bolted onto the pivot tube, and then the swingarm is pressed up against their face so it is in the center of the frame. From here, you can place a few tack welds on the swingarm tubing and brackets,

holding them in place. To get the swingarm centered on the frame, make sure the distance between the ends of the brackets and swingarm tubes is equal, or make the center of the swingarm and pivot tube an alignment guide.

Once you have a few good tack welds holding the swingarm to the frame by the brackets and pivot tube, move the assembly around so you can visually inspect it for alignment. Placing the rear wheel back in the dropouts is a good quality-control check, as it lets you see the alignment between the rear center frame tube and the wheel. The rear wheel should be perfectly in line with all the other vertical tubing. When you are happy with the alignment, complete the welding around the three outer edges of the bracket, as shown in Figure 12-40. The inside weld is not done because it is not necessary and may cause the plates to collapse together, making it impossible to slide over the bushings.

Figure 12-38 *Pivot tube welded in place*

Figure 12-40 *Swingarm brackets welded*

You are very close to having a rolling chassis! At this point, the rear wheel will pivot freely, allowing the bike to drop to the ground, so you will need your ground clearance prop one last time to hold up the frame, as shown in Figure 12-41. The goal is to determine where to mount the two rear suspension spring shocks, so that the rear swingarm has the most effective travel while keeping the center frame level to the ground. Both tires should be fully inflated, and all front fork hardware should be completely installed, so you have an accurate idea of how the bike should be standing when at rest.

I recommend that you use two adult-sized mountain bicycle coil shocks to create the suspension on the LongRanger. This system works well with my 60 pounds of batteries and riders weighing up to 200 pounds. If you weigh more than that, or have installed some monster batteries, you may want to use heavier spring shocks, like the ones found on small minibikes or mopeds. These shocks can be purchased at many go-cart or recreational shops, and will have a bit more travel and compression strength than those found on a mountain bike. If you think you may need larger spring shocks, tell the parts counter clerk that you are looking for two identical rear spring shocks that would be good for a small- to medium-sized off-road minibike.

Regardless of the suspension spring type you choose, the simple tabs shown in Figure 12-42 will probably be the easiest way to connect the springs to the swingarm tubing. These tabs are made from 3/16-inch-thick, 1-inch-wide flat bar, cut to a length of 1.5 inches. The holes in the tabs are drilled to match those on your suspension springs, and the ends are then rounded off to remove any dangerous sharp edges.

Figure 12-42 *Shock mounting tabs*

When you have the two mounting tabs cut and drilled, bolt one of them onto the base of one of your suspension springs, and then do a test fitting, as shown in Figure 12-43. The tabs should be mounted between the tire and the inside of the swingarm, where the space has been left between the tire and tubing. When the spring is held in place, there should be no interference between any part of the spring and the tire or frame tubing. Avoid mounting the tabs any closer to the end of the swingarm (towards the frame), or you will reduce the effectiveness of the suspension springs by adding mechanical advantage to the swingarm, which is acting like a lever. Basically, the closer the tabs are to the pivot point, the less effective the springs will become. If you are using non-bicycle suspension springs, some testing may be in order before finalizing the welds.

Figure 12-44 shows the suspension spring tabs tack welded to the swingarm just ahead of the edge of the small tube. The two springs have been tested to fit

Figure 12-43 *Determining the shock position*

Figure 12-41 *Getting ready to install the rear shocks*

Figure 12-44 *Suspension tabs tack welded to the swingarm*

Figure 12-46 *Suspension completed*

without any rubbing on the frame or tire, so the tabs can be completely welded in place now. If you are not sure how your springs will perform, leave the tack welds for now and move along until your springs are installed so you can sit on the frame for testing. The suspension should only compress a slight bit when you sit on the bike, not bottom out to the max. You might have noticed that mountain bike springs are somewhat adjustable—tightening the top nut clockwise to compress the spring will add more starting tension to the shock.

When the two shock mounting tabs are welded to the swingarm, you can bolt the shocks in place and use them as a guide to mark the area on the frame where the top mounting bolts will need to be welded. Figure 12-45 shows the perfect place on the frame where my two mountain bike shocks need to be mounted for maximum travel. When choosing bolts for the shocks, use high-grade, new bolts with fine threads and locknuts.

Figure 12-46 shows the two rear suspension springs ready for action after welding the top bolts to the frame.

Figure 12-45 *Marking the shock mounting bolts*

The two shocks have been adjusted by turning the top nuts down against the springs, until the swingarm only dropped slightly as I stood on the frame where the batteries will normally be living. If your swingarm flops to the ground easily under your weight, then you may have bad springs, or really need to crank down those adjusting nuts. If you are not sure how the shocks will perform, wait until your pre-painting test ride to give them an actual performance test. You can always find better shocks and remount the tabs and bolts with little effort.

Now that your LongRanger is rolling around the garage on two wheels, it's just a matter of bolting all the parts on so you can whiz around the streets all day on a single battery charge. OK, there is a bit more work to be done, but the rest of the job will seem easy. The battery pack is designed to drop into the center frame, making it easy to remove for swapping out, or making the bike lighter for easy transportation. To secure the battery pack into the main frame, a few tabs cut from some thin 1-inch flat bar are made, like the ones shown in Figure 12-47. These tabs are approximately 1.5 inches long and have quarter-inch bolt holes drilled through them. The actual size of these tabs is not highly critical, as you will soon see.

The battery pack mounting tabs are welded to the top and bottom of the battery pack in such a way that they hold the battery pack into the frame, centered along the 1.5-inch square tubing. Shown in Figure 12-48 is one of the top mounting tabs welded to the flat bar that creates the top of the battery pack box. To figure out where to weld these tabs, have a helper hold the battery pack in the center frame, then mark the position of the tabs as you

Figure 12-47 *Battery box mounting hardware*

Figure 12-48 *Installing the battery pack tabs*

hold them against the frame. All tabs are installed on one side of the frame so you can slide out the battery pack; which side is your choice.

Figure 12-49 shows the battery pack installed by fastening the four tabs to the frame using machine screws. The screw holes are determined after the tabs are

Figure 12-49 *The top battery pack tabs*

welded to the battery pack by making a mark through the holes in the tabs with a black marker. Center punch and drill the appropriate sized holes in the square tubing for whatever screws you plan to use.

Your electric bike is almost completed. Figure 12-50 shows the rolling chassis with removable battery pack installed in the frame. The next step will give you a place to sit on the bike.

Seating is usually a personal touch, and often builders will opt for a ready-to-install seat. A perfect ready to mount seat would be one from an old moped or a small off-road motocross bike. You could even use some of the frame parts and the seat from the original suspension mountain bike you used for parts, but since I have no love of "wedgie" style bicycle seats, I normally make my own seats from scratch. Another thing to think about is how high, or how far back, you want the seat, so before you start cutting any tubing, read ahead a bit to see if you like how my seat was made. Because of the strong square tube center frame and ample seating room above the rear wheel, you have total freedom to make your own type of seating. Figure 12-51 shows a few bits of ¾-inch electrical conduit that I have dug from my scrap bucket and welded together to form a pair of seat rails.

As a general rule, I never leave open tubing, so the ends of the 3/4 round tubes have been capped off by welding a pair of washers around the ends of the tubes (Figure 12-52). Capped tubing looks so much more professional once painted.

I wanted my seat a bit on the low side so riders of all ages could reach the ground on my LongRanger. The seat

Figure 12-50 *Battery pack fully installed*

Figure 12-51 *Creating a seat frame*

Figure 12-52 *Capping the ends of the tubes*

Figure 12-53 *A motocross-style seat mount*

Figure 12-54 *Seat support tube*

Figure 12-55 *Completed seat mounting hardware*

rails are curved so that the final seat height would be a few inches lower than the top of the center frame tube. If you are not sure how high to make your seat, check to see how high you would have the seat on a regular bike. Again, if you decided to use an actual bicycle seat, you could incorporate the frame's seat tube and clamp to create a fully adjustable seat, just like a regular bike. Figure 12-53 shows the curved seat mounting rails I plan to use, which will hold my home-brew plywood and foam seat in the bike.

The curved seat mounting rails are welded to the back of the center frame tube, as shown in Figure 12-54, so the seat will be several inches lower than the top of the frame. Once the top part of the seat rails is welded to the frame, a pair of support tubes (one shown in Figure 12-54) is added for strength. These support tubes are made from some scrap half-inch round tube, the same tubing used on the seat stays of a typical bicycle frame.

Figure 12-55 shows the completed seat mounting hardware after adding the two small support tubes which give the system great strength. Always triangulate small tubing for strength, or it will not hold up to any real weight. This also applies if you chopped up a bicycle frame to use the original seat post hardware to create a fully adjustable seat. Have a look way back at Figure 12-1, and you can see how easy it would be to

Figure 12-56   *Foot peg hardware*

Figure 12-57   *Foot peg mounting bolts*

simply cut up the seat mounting hardware from the original frame and weld it right to your new frame. The large gusset on the original frame would become your triangulation.

Now that your butt has a place to call home, you need to find a place for your feet to rest, since they won't be doing any work on this vehicle! A simple foot peg system can be made from a pair of aluminum BMX "trick pegs" and a few half-inch bolts. The BMX foot peg holes need to be drilled or filled out for the half-inch bolts, and the bolts should be quality high-grade to avoid bending if you plan to take your LongRanger over some rough off-road terrain. Although I did not create the LongRanger for off-road use, it can certainly take some abuse, but if off-road and trail riding are your ultimate goals, then BMX foot pegs may not be suitable. BMX pegs are fixed to the frame, so if you are blasting down a narrow trail with stumps all over the place, you may end up catching a peg on an obstacle, which could damage the bike or your body. If you plan to take it to the "back woods" often, then consider a pair of folding pegs, like the type on an off-road minibike or motocross bike. These can be purchased at many recreational sports stores or bike part dealers.

I found the most comfortable place to rest my feet was the center of the frame, as shown in Figure 12-57. This is another personal preference that has much room for alteration, so you may want to finish your seat and simply sit on the bike to find your optimal foot peg position. When you do weld the foot peg mounting bolts to the frame, keep in mind that the battery pack box needs to slide in and out of the frame, and that your foot pegs might need an extra nut or two to make them a bit

wider to account for the width of the batteries. Figure 12-57 shows the foot peg mounting bolts welded to the frame, as well as a single nut used to space the pegs out a bit wider.

Because we used a fully functional front end to make the LongRanger, front brakes are already accounted for and will operate just like they did on the original mountain bike. However, rear brakes are not included unless your hub motor has regenerative braking, which is not all that common, so you will have to add them. Since the LongRanger can weigh 80 pounds by itself, it is highly recommended that you have brakes on both wheels for safety. A very easy way to add rear brakes is by cutting the brake studs from an old bicycle frame, or by removing the entire front brake assembly from a suspension mountain bike fork, to transplant onto the rear swingarm. Figure 12-58 shows a typical mountain bike front suspension fork, which includes a plate that holds all the front brake hardware. This entire plate will be removed to keep the brake hardware intact.

Because the entire brake assembly was cut from the mountain bike suspension fork, it can simply be installed

Figure 12-58   *Salvaging brakes for the swingarm*

on the underside of the rear swingarm in one piece. To align the brake, squeeze the lever onto the rear wheel and tack weld the assembly directly to the swingarm tubing. This method also works if you are installing individual brake studs cut from the rear of a bicycle frame. Side-pull brakes taken from very cheap bicycles should be avoided because they offer minimal braking force, and this LongRanger design is a fairly heavy vehicle.

As shown in Figure 12-60, a solid bead of weld on each side of the brake assembly will secure it to the underside of the rear swingarm. Now your LongRanger has two quality brakes that will allow it to stop as good as any good quality bicycle. Tape is placed over the brake mounting studs while welding to avoid damage.

Congratulations, as you can now get out your paint mask to finish your new electric vehicle! Figure 12-61 shows my frame after a good coat of primer has been applied, using my usual technique of hanging the frame outdoors on a clothesline while I paint it using typical primer in spray cans. Spray cans will never yield a paint

Figure 12-61 *Priming the frame*

job as good as a factory job, but if you take your time and follow the recommended application instructions, you can achieve a very good paint job for only a few dollars. All fresh coats of paint or primer should be left to cure for at least a day before you handle the parts or try to reassemble the bike. Yes, I know this is a painful process that can try your patience—garage hackers are often guilty of wet paint assembly!

After a day of doing nothing, the freshly painted frame is ready for reassembly. I had already test ridden the LongRanger before primer was applied, so I knew it was just a matter of bolting all the parts on properly and getting the cables and wires installed. As a rule, I do not keep used brake cables unless I know their history, so the front cable would obviously be fine, and the rear brake cable was replaced. Figure 12-62 shows the freshly

Figure 12-59 *Mounting the rear brakes*

Figure 12-60 *Rear brake assembly installed*

Figure 12-62 *The painted frame*

painted LongRanger, looking like a cross between a mountain bike and a small 1970s Japanese motorcycle. I chose tall BMX handlebars for my bike, but use whatever style feels most comfortable to you, of course.

I did not have a motocross seat in my vast parts collection, so rather than coughing up fifty bucks for a new one, I decided to fashion my own crude seat from some scrap wood, a bit of foam, and some vinyl material. Figure 12-63 shows a few pieces of plywood cut to conform to the seat rails, and held together with a strip of 1-inch flat bar and a few small woodscrews.

To create a nice soft seat, I glued down some medium stiffness, 2-inch-thick foam, as shown in Figure 12-64. Seat padding should not be made from extremely soft foam, like the stuff found in couch cushions, or it will not be very effective. Decent seat foam can be found at many material outlets, or it can be salvaged from furniture packing.

Figure 12-65 shows my "poor man's" method for upholstering a seat. I simply take the ultra high-quality "pleather" and stretch it as hard as I can around the seat, using as many staples as possible to hold it in place. A sewing machine and proper seaming would be a much better alternative.

The completed seat is shown in Figure 12-66, after stretching and stapling the material around the foam and wooden base. The reflector at the rear of the seat keeps me safe at night and also hides the ugly seam in the material. The seat is held to the curved tubing by four woodscrews fastened from the underside of the seat. It's quite comfortable!

Brakes should not be overlooked on this vehicle as it is heavy and capable of some decent speeds. Shown in Figure 12-67 is the rear swingarm brake getting a new cable and proper adjustment. Brake pads should be very close to the rims when idle, but not rubbing so that any amount of lever movement will result in braking force.

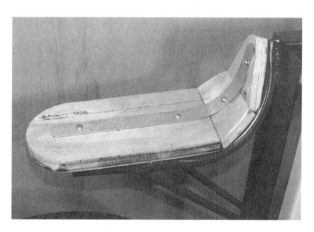

Figure 12-63 *Making a simple seat base*

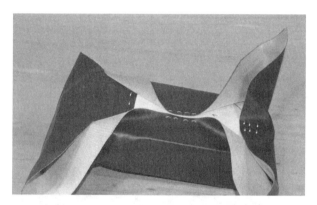

Figure 12-65 *A simple upholstery job*

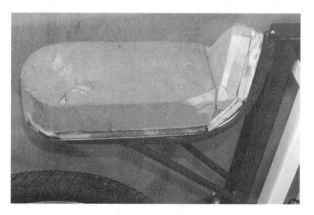

Figure 12-64 *Padding the seat*

Figure 12-66 *Completed seat*

Figure 12-67 *Setting up the brakes*

Do not be tempted to take your first test run without brakes on this vehicle.

Cables and wires require a bit of planning and some creative thinking. Tie wraps are good for holding down cables, but do not force any brake cables into tight bends or your brakes may fail to release properly, or have excessive friction due to the extra friction induced into the tightly bent cables. The handlebars should also be able to turn their full range without creating any tension on the cables, so keep this in mind when wrapping up all these wires and cables. Figure 12-68 shows the brake cables neatly tied in such a way that they work perfectly from any steering angle.

You should also check to see which way your hub motor is going to spin when you engage the throttle before you tie down all those wires. My motor had no indication of rotation, so once I knew which way it needed to turn, I marked it right on the motor shell. The

motor controller box, fuses and all other included kit hardware should be installed according to your kit instructions. If you have extra wire, simply wrap it up and neatly tuck it under the seat out of the way. Make sure that the wiring has no exposed conductors, and that no wire or cable is rubbing on any moving part of the bike, such as the wheels or swingarm. Figure 12-69 shows the ready-to-run LongRanger, all wired and cabled up. I also included a front and rear fender to keep water from spraying around, and it gave the bike a more serious look.

If you installed your hub motor kit properly, you should get a "green light" when you turn on the ignition key or switch. You should take your first ride with fully charged batteries and head around the block a few times in your neighborhood, so you can make sure your brakes are in good working order before hitting any heavy traffic areas. Depending on the type of hub motor you have, you may be able to twist the throttle and get moving from a dead stop, or you may have to give a kick off the foot pegs to get rolling first. My hub motor requires a rolling speed of 2.5 mph (4 km/h) before it will engage, which may seem annoying at first, but actually makes the bike much safer when parked. Because the motor does not engage from a standstill, kids can't do any harm if they come up and unexpectedly crank the throttle as they yell, "Cool bike!"

I always like to put my machines through their paces, sometimes riding in ways that would void the warranty of lesser machines. Figure 12-71 shows the tail end of a jump-style landing as I took the LongRanger on various tests along trails at the campsite. I often encountered

Figure 12-68 *The art of cable wrapping*

Figure 12-69 *Adding the motor controller*

Figure 12-70 *Let's ride!*

Figure 12-72 *Adding a trailer?*

Figure 12-71 *Does this void the warranty?*

Figure 12-73 *Running on solar energy*

people riding those department store electric scooters and I passed them as if they were standing still, especially if they attempted to climb any hill. The LongRanger is practically silent while operating, as if powered by the wind alone. The fact that the LongRanger operates as quiet as a whisper makes it perfect for camp or quiet neighborhoods, where those horrifically loud, gas-powered scooters are seriously annoying.

Even the small electric kick scooters seemed to make an incredible amount of noise compared to the LongRanger.

After climbing some pretty steep hills without any trouble or drag at all, I really wanted to see what the limit was, so I found a rope and a willing test pilot with a kick scooter to add some extra drag to my ride (Figure 12-72). Pulling the scooter and pilot was no problem for the LongRanger, even up a moderate hill, which was extremely impressive considering that a department store electric scooter with a motor rated at 500 watts would choke on lesser hills. Having a quality-made and highly

efficient hub motor really made a difference, and I could tow the scooter up all hills that I tried, except for a very steep hill that even some gas-powered vehicles have issues climbing. A 750-watt hub motor would have probably snapped the tow rope, or caused me to pull a wheelie up the hill.

I think the LongRanger is a perfect way to commute long distances without making any pollution, noise or environmental impacts. You might not need a license or insurance to ride a small electric bicycle like this in many parts of the world. In communities where there are no existing electric bicycle laws, you should be able to blend right in with typical bicycle traffic. By adding a solar charger to your system, you could have a maintenance-free form of transportation that will cost you just pennies to run, and you can ride proud knowing that you built it yourself. I hope you enjoy riding your new electric vehicle, and urge others to build their own version of this non-polluting alternative transportation. Leave the gas-powered vehicles at home!

The Silent Speedster is a fast and sporty scooter that operates without emitting noise or carbon pollution. For only a few pennies, you can zip along on roads, bike paths, and off-road trails, often making the trip in less time than it would take in a motorized vehicle, since you can avoid gridlock or bumper-to-bumper traffic jams. The Silent Speedster is perfect for the road, as well as the trail, and it allows you to enjoy nature without disturbing the silence. Many inexpensive electric scooters have chain-driven drive wheels or gear reduction motors, so they make quite a bit of noise, or suffer from inefficient power transfer, but the Silent Speedster features a quiet and efficient hub motor design. The Silent Speedster looks much like a typical bicycle, and includes all the standard safety equipment, so you will be able to live happily on the public roadways and bike trails without disturbing the peace. Here is your chance to make a difference, and leave the gas guzzler parked in the driveway whenever you can.

The Silent Speedster can be built to travel well over 30 miles (50 kilometers) on a single charge, and will do so at speeds equal to or greater than the posted city limit. Because the Silent Speedster is built around the battery pack and the type of hub motor you choose, you can build the base vehicle to suit your needs, altering the top speed, acceleration, weight or range.

It's a good idea to read over the entire plan completely before hacking up any bikes, or heading to the scrapyard on a search for parts. By reading ahead, you will see why

a certain part was chosen, and get some ideas on how to modify the design to suit your needs. Maybe you want to build a two-seater version of the Speedster, or create a grocery carrying box. With a little input from your own drawing board, you could easily modify this project to suit your own needs.

What type of hub motor should you purchase? Obviously, there are hundreds of them on the market now with prices varying from approximately $200 to over $1,000. Like all products, you usually get what you pay for, but there are several factors that will influence the price of a hub motor kit, such as internal construction, output power, rim quality, and the type of batteries and charger that come with the unit. My hub motor kit shown in Figure 13-1 is a 450-watt brushless unit built into a 20-inch wheel. It comes with the twist throttle, controller, charger, wiring, and the three batteries shown in Figure 13-1 (smaller pack). The price on my hub motor kit is approximately $500, and is considered to be a higher quality unit due to the brushless construction. A brushless hub motor has no wear parts, and doesn't require maintenance, whereas a brushed motor will require changing the internal brushes about once a year if well used. Brushless motors are also more efficient, so if you can afford the extra cost, choose a brushless design.

Figure 13-1 *Choosing a hub motor kit for your vehicle*

As for motor power, there will be plenty of data on the manufacturer's website regarding speed and range. Again, range is usually inflated, but often top speed is rated very accurately. Due to top speed regulations, most hub motors will top out at around 22 mph (35 km/h), which is dependent on the rim size and the number of batteries in series. A smaller rim will slightly reduce your top speed, but will increase your hill climbing ability. For this reason, my Silent Speedster has been fitted with a 20-inch rim to allow better hill climbing power and off-road use. To get an extra speed boost, I added another battery to the pack to increase my voltage from 36 volts up to 48 volts, adding 25 percent to my top speed. My top speed now peaks at around 28 mph (45 km/h). The wattage rating of the hub motor will determine the amount of hill climbing power and acceleration you can expect, with 450 watts being at the lower end of the scale and 1,000 watts at the upper end of the scale. My 450-watt motor can climb a fairly steep hill, and take on most headwinds with ease, but I also added torque due to the smaller 20-inch wheel. As a general guide, shoot for a 450 to 550-watt motor for a 20-inch wheel, and a 600 to 750-watt motor for a 26-inch wheel. If you really want serious acceleration and hill flattening power, a motor of 750 watts or more will be extreme, giving you a top speed beyond the city speed limit with an extra battery, although at the cost of reduced range.

Check with the manufacturer of the hub motor controller concerning how much voltage it can handle. Typically, the kit will come standard with three 12-volt batteries, giving an overall voltage of 36 volts. This is how the top speed is determined. My controller had no problems with an extra battery and gave me a nice speed boost at 48 volts. There are many controllers that will allow you to run up to 72 volts or more, and this type of performance will be approaching that of a small motorcycle, with top speeds of 43 mph (73 km/h), and tire spinning acceleration. At 40 mph, you may not go unnoticed on the city streets, so check your local laws regarding electric scooters.

My recommendation is to try a bicycle fitted with a hub motor kit, and see how it runs without any pedaling input. If you plan to add the extra battery, then expect at least 20 percent more top speed, with a noticeable difference in acceleration. You can really tailor the power characteristic of the Silent Speedster from "running all day long at bicycle speeds" to "taking off like a rocket and pulling wheelies," depending on your motor and battery choice. And as always, a deal that looks too good to be true almost always is. Avoid the junk and you will have a quality electric vehicle.

Since the plan is to add at least one extra battery to the pack, a 20-inch wheel is your best bet, as top speed will certainly be fine due to the increase given by the extra battery. A 20-inch wheel also gives you the added benefit of higher torque for speedy acceleration and good hill flattening power. A 26-inch wheel will also work, but you may find that the scooter reaches speeds that are a little too fast for comfort using an extra battery and the tradeoff will be hill climbing ability and overall range. Most sellers of hub motor kits have the option of 20- inch or 26-inch wheels, so I recommend a 20-inch wheel for this project. You will also need a matching front wheel, as shown in Figure 13-2.

Choose a good quality tire with a heavy sidewall, minimal tread, and a PSI rating of at least 75. A knobby tire is only useful on rugged terrain, and heavy tread on mainly smooth surfaces will only rob you of power, reducing your overall range. The wheel should also have at least 36 spokes and an aluminum rim so that your brakes work properly. Chrome rims are very low quality, and should never be used on anything except for choppers and show bikes, as the rim will not allow pad brakes to grip properly.

If you are following the plan details, then you will need a solid 20-inch BMX frame to butcher for parts. Figure 13-3 shows an old steel BMX frame that was pulled from the city dump for use on this project. You will need the complete rear triangle, front forks, head tube, seat post, and handlebars, so make sure these parts are not damaged. Since we will be hacking up the frame and welding new tubing to it, an all-steel construction is

Figure 13-2 *Finding a matching front wheel and tire*

Figure 13-3 *The "donor" frame*

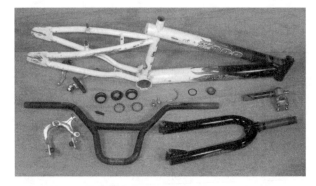

Figure 13-4 *Taking everything apart for inspection*

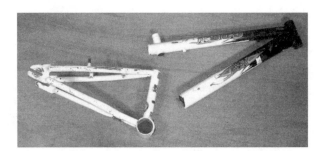

Figure 13-5 *Cutting away the rear triangle*

important, so avoid aluminum and chromoly frames, as these will require special welding equipment and skills to work with. Often, you can purchase a perfect donor BMX bike from the department store for under $200, and it will have everything you need, including two good tires and the front wheel, so consider this as you price out parts you do not have in your junk pile.

Take the donor bicycle completely apart, clean out any grease and loose dirt from the bearings and cups, and then check all the parts for excessive wear or damage. You will not need the top tube or down tube from the BMX frame, so do not worry if it is damaged. The front forks and rear triangle should be in good shape, and cantilever brake studs on either the rear triangle or front forks will be beneficial since they are more efficient than caliper-style brakes. Figure 13-4 shows my donor bike taken completely apart for inspection. Also notice the cantilever brake studs on the rear of the frame, something that can really help stop a fast moving vehicle that carries a 20-pound hub motor and 30 pounds of batteries.

Your first cuts will be made at the bottom bracket and down tube joint, and at the top tube and seat tube joint, as shown in Figure 13-5. Your goal is to liberate the rear triangle in one piece from the rest of the frame. Notice my BMX donor frame had an odd top tube that extended past the seat tube, which is why it was cut along the top of the seat tube rather than at the top tube. Either way, you should end up with a complete rear triangle with the seat tube cut just above the point where the seat stays are welded.

If you are using a cut-off disc in your grinder to make the cuts, then beware when cutting either the top tube or down tube, as the frame will tend to collapse on your disc as you get through the last bit of metal, which could cause your grinder to stall, or run out of your hand. If you have no way to secure the frame while cutting, then leave a sliver of metal during each cut rather than going all the way through the tubing, and you can simply bend the rear triangle back and forth to release it after the cuts are made. If using a hacksaw, don't worry about it and simply hack away!

Once you have the rear triangle cut from the rest of the frame, grind off any excess material left over from the cutting process so the new tubing will form a good joint for welding. Be careful not to take off too much material that you make the tubing thinner, especially around the seat tube. A little lumpiness at the cut area is not a problem, and will be hidden once the new tubing is installed. I usually start with a coarse grinding disc to remove most of the metal, and then finish up with a zip disc or even a sanding disc to avoid digging into the good tubing. Figure 13-6 shows the rear triangle after cleaning up the cut area in front of the bottom bracket. Your rear triangle may also need cleaning up around the seat tube if

Figure 13-6 *The cut and cleaned rear triangle*

the top tube was originally welded directly to the seat tube.

You will also need to cut the head tube from the original bicycle frame, as shown in Figure 13-7, leaving as little stumpage as you can to reduce the amount of grinding necessary. The length of the head tube is not critical, and this will vary from frame to frame, depending on the size of frame used. Some head tubes also have varying inside diameters, so make sure you have a set of compatible bearing cups, as well as the 20-inch front forks that fit your particular head tube. The easiest way to ensure the front forks and all head tube hardware are compatible is to take them from the same donor bicycle like I did; this way, you are not stuck with incompatible parts.

Clean up your head tube by grinding away any excess tubing material and paint, being careful not to dig into the tubing walls. You don't have to remove all the paint, but it is best to keep the area to be welded as clean as possible, since this will aid in the welding process and keep the horrible stink of burning paint fumes to a minimum. The two holes in the head tube can also be

welded over at this point, although it may not be necessary as the frame tubing will probably cover at least one of them later on. Figure 13-8 shows the fully ground head tube ready to begin a new life.

Although we will discuss the hub motor, controller, and batteries in greater detail later, you will need to know the dimensions of the batteries you plan to install in your version of the Silent Speedster, in order to continue building the frame. If you have not decided on a brand or type of hub motor and batteries yet, then read ahead so you can see how these components will affect the range, weight, and speed of your final design. The goal of the Silent Speedster is a good mix of range and top speed, so the batteries that originally came with the hub motor kit will be replaced with larger batteries for extended range, and one extra battery will be added for a little more top speed. As you will soon see, the frame is actually built around the batteries, since they are the key element in the entire design.

Figure 13-9 shows the large size difference between the standard hub motor kit batteries and the ones I plan to install in the Silent Speedster. The small batteries are rated at 7 amp hours each, and the larger ones are rated at 12 amp hours each, so the expected range of the larger batteries is about two times more than the original pack. The small batteries measure 6" by 2.5" by 4", and the new ones measure 6" by 3.75" by 4". Let's discuss battery size versus expected range.

There are many types of battery chemistries, and each has certain advantages depending on the intended use. Let me make this choice very simple—you should only use sealed lead acid (GEL) batteries for this project. Unlike a typical automotive lead acid battery, a sealed or GEL battery does not have free-flowing liquid, nor does

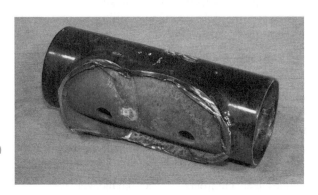

Figure 13-7 *Cutting away the head tube*

Figure 13-8 *Cleaning up the head tube*

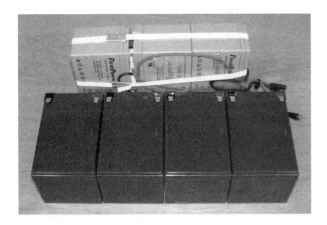

Figure 13-9   *Choosing your battery pack*

it need maintenance such as "topping up" the fluids, so there is no chance of spilling any corrosive liquid on your vehicle or yourself. GEL batteries can also be installed in any orientation, unlike wet batteries which must always sit straight up. The ability to mount a GEL battery in any orientation gives you a lot more room to arrange the pack to fit your needs. Sealed batteries are often labeled as wheelchair or recreational batteries, and the obvious sign that they are sealed is the lack of any type of filling or maintenance valve on the top of the battery. Sealed batteries may cost a little more than the equivalent size, wet-cell battery, but the extra cost is worth it in terms of safety and overall battery quality.

The batteries shown in Figure 13-9 have given my Silent Speedster a range of about 30 miles (50 km), and this is more than enough for a fun day of riding. There are many sizes of GEL batteries in the smaller (under 12-amp-hour) range, so your best bet is to look at the estimated range of your hub motor kit with its current batteries and work from there. If the original 6-amp-hour battery pack claims 15 miles (24 km), then it is a safe guess that a 12-amp-hour battery pack should deliver 30 miles (50 km) of range. Of course, I would divide any manufacturer's range claim by 1.5, as they have been known to exaggerate!

As stated earlier in this chapter, only purchase GEL batteries, and make sure that all of them are the same size and age. A battery larger than 20 amp hours is probably too large for this project, and anything less than 10 amp hours might be too small. Adding more batteries in series increases your top speed, but overall range is the product of the amp-hour rating of a single battery. If you want to go faster, add more volts, and if you want to go further,

add more amps! Use your original battery's capacity and range as a guide, but take manufacturers' claims with a grain of salt!

We will start building this vehicle by creating the battery pack box before cutting any tubing for the frame, since the frame is actually built around the battery box. As shown in Figure 13-10, place your three or more batteries side by side, and then measure the entire length of the pack. The overall length of my battery pack is 15.5 inches, and the width is 6 inches. These measurements will be used to create the angle iron battery box base. Round up to the nearest quarter inch to be safe, as a slightly larger battery box is a lot better than one that will not fit your batteries. Now is also a good time to tape up the battery terminals to avoid an accidental short while you are working around the batteries.

The battery box is made from some 1-inch angle iron with a thickness of 1/16 inch. To make a perfectly square base of the angle iron, each corner needs to be cut at 45 degrees, as shown in Figure 13-11. Mark the ends using a black marker and a 45-degree square, by drawing a line from the end corner where the two sides of the angle iron meet. A hacksaw or zip disc is then used to cut off the corners. Remember that the inside dimensions of the battery box base need to be slightly larger than the size of the battery pack as measured in the last step, or the batteries will not fit into the frame. Test the box by

Figure 13-10   *Tack welded battery pack base*

Figure 13-11   *Cutting the battery box base pieces*

setting up the angle iron pieces around the battery pack to make sure they are not too small.

Start by welding the battery box base corners, as shown in Figure 13-12, so you can easily square up the base if necessary. With all four corners tack welded together, you can test fit your batteries one last time.

Use a 90-degree square to ensure that the battery box base is perfectly rectangular and then place it on a flat surface for welding. Add a small tack weld on each 45-degree joint to secure the frame for one last battery test fitting. Figure 13-13 shows the battery box frame ready for tack welding as it sits on a flat surface.

Once you have the battery frame tack welded around the corners and on each 45-degree joint, drop in the batteries one last time before you complete the joints. You should have a little space between the inside walls of the frame and the batteries, but not so much that they will move around. If your batteries do not seem to fit into the frame, then you need to cut a few tack welds and try again. Figure 13-14 shows the batteries snuggly fitting into their new home.

To complete the welding of the battery box base, start by welding the 45-degree joints on the underside of the frame, as shown in Figure 13-15, and then finish by

Figure 13-12 *Welding the box corners first*

Figure 13-13 *Battery pack base tack welded*

Figure 13-14 *Test fitting the batteries in the frame*

Figure 13-15 *Battery box base completely welded*

welding the outer corners. Do not weld the inside of the joints where the batteries need to fit, or you may find that they will no longer fit properly into the frame.

The batteries will be secured to the battery box frame by a "seatbelt" made from a length of 1-inch flat bar with a 1/16-inch thickness. This battery strap will keep the batteries from bouncing around or falling out of the frame if the scooter is overturned or laid down for transporting. The length of the strap is the same as the distance around the loaded battery pack from one side to the other, as will be shown in the next few steps. Figure 13-16 shows the battery strap cut to the length needed to hold my batteries into the base frame.

Unless you are a wizard with making deadly accurate measurements and bending flat bar, a slotted hole will probably be the best idea to ensure that the battery strap holds the batteries down snugly. The first quarter-inch hole is drilled in the battery strap so that it will meet the side of the battery box half way up the angle iron, and

Figure 13-16 *Squaring up the center frame*

then a line is extended past the edge of each hole to the bottom of the strap, as shown in Figure 13-17. The slot can then be cut away by tracing along the line with your grinder disc or by sawing it out with a hacksaw. The battery strap will now have a half inch of adjustment to help ensure that it holds the batteries firmly to the base frame.

The battery strap is bent to fit around the battery pack so that the slotted hole ends up on each side of the pack, as shown in Figure 13-18. The strap does not need to be perfect now, since the slotted hole will allow you to adjust the tension of the strap around the battery pack, ensuring that the batteries cannot move or vibrate. The thin flat bar is easily bent by placing the area to be bent in a vice and then working it with a hammer.

The slotted hole should end up on the angle iron frame, as shown in Figure 13-19. A nut will then be welded to the angle iron frame so the strap can be secured and adjusted by installing a bolt on each side. With your batteries installed in the frame, place the bent strap in place and then mark the center of the slotted hole so you know where to weld the nut.

Figure 13-19   *Placement of the slotted hole*

Figure 13-20 shows how the 1/4 bolt holds the battery strap in place. The 1/4 nut is welded directly to the angle iron frame under the slotted hole, and then the bolt is installed to hold it in place. You will likely need to cut the bolt down to the correct length, as a standard bolt will be too long, so this is another job for your cut-off disc.

The two main tubes that make up the Silent Speedster frame are called the "front boom tube" and the "rear boom tube." The rear boom tube will carry the batteries and join directly to the bottom bracket of the rear triangle, and the front boom tube will join the head tube to the rear boom tube. Both boom tubes are made from 1.5-inch mild steel square tubing with a 1/16-inch wall thickness. This tubing can be ordered from any steel supplier and is very easy to cut and weld. Steel suppliers like to rate tubing wall thickness using a gauge number, so 1/16 wall tubing will be called 16-gauge tubing, and although there is a slight difference between the two, your best bet is to tell them you want a length of 1.5-inch square tubing with a wall thickness as close to .0625 of

Figure 13-17   *Making the slotted hole*

Figure 13-18   *Bending the strap to fit the battery pack*

Figure 13-20   *Battery strap secured to the frame*

an inch (1/16) as you can get. Round tubing should be avoided for this project, as it will be difficult to mount the battery box properly and cut the required angles.

Figure 13-21 shows the two boom tubes after cutting. The short tube is the rear boom tube and the longer one is the front boom tube. The length of the rear boom tube is based on the overall length of your battery box as measured from end to end (with battery strap included), and the front boom is initially cut to a length of 24 inches, and will be shortened later based on your head tube and front fork position. If you are using 26-inch wheels instead of 20-inch wheels, then make the initial front boom at least 4 inches longer. The next drawing shows how each tube will be cut.

Figure 13-22 shows how each of the two main boom tubes needs to be cut. The rear boom tube (shorter tube) runs parallel to the ground, and carries the battery box, so it is cut to a length of the battery box as measured from the top. The front boom tube (longer tube) is cut to an initial length of 24 inches just to be safe, and will be cut down later when fitting the head tube. If you are using 26-inch wheels instead of 20-inch wheels, then make the front boom length 28 inches to compensate. One end of each tube is cut to an angle of 32 degrees, so the resulting angle between the tubes will be 116 degrees when they are welded together. The calculation used is $(90 - 32) \times 2$.

The two boom tubes will be welded together at the ends with the 32-degree angle cut, so that the resulting boom will create a 116-degree angle between the two tubes. If you cut the two ends to 32 degrees, the angle between the two tubes will be very close to 116 degrees when you check it after tack welding the two tubes together, as shown in Figure 13-23. A few degrees off will not have much effect on the final design, but do try to get it as close as you can, and make the tack weld with the tubes lying on a flat surface so they are perfectly in line. Two tack welds on each corner of the joint, as shown in Figure 13-23, will secure the tubing for a final angle check before you finish the welding.

Once you have the two boom tubes tack welded together and the angle is close to 116 degrees between them, you can complete the welding by starting with the top and bottom of the joint, as shown in Figure 13-24. This order of welding helps prevent side-to-side distortion of the tubing from welding heat, which could make the boom twisted. If you were to start welding on the side of the joint, the two tubes would not be perfectly in line, and could be out enough to show a visible flaw in our final design. A good welder always thinks ahead, and tries to work around heat distortion as much as possible.

Figure 13-23 *Tack welding the boom tubes*

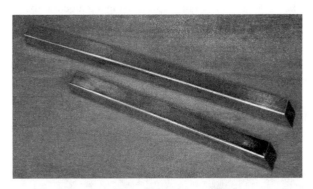

Figure 13-21 *Cutting the two frame boom tubes*

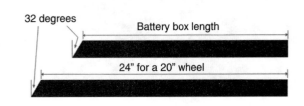

Figure 13-22 *Front and rear boom tube lengths*

Figure 13-24 *Welding the boom tubes together*

The sides of the boom joint can be completed once the top and bottom of the joint have been fully welded. Use the same size of bead on each side so distortion is not prominent on either side. Figure 13-25 shows the entire boom joint fully welded and inspected for pinholes.

If you trust your welding, then you can go ahead and clean up the joint flush with the tubing wall, as shown in Figure 13-26. Avoid taking out too much metal or you will weaken the joint. A small hill over the weld is much better than a valley, and this process will not weaken the joint if the welding was done properly. A poor weld may not have proper penetration, so grinding the area flush may actually weaken the joint. If you see holes or flux in the joint after grinding the weld metal flush with the tubing, then your welds may have been done too cold. The welds shown in Figure 13-26 have been ground clean using a sanding disc (flap disc) and an angle grinder.

The head tube will be welded to the end of the front boom (the longer tube), and the rear boom (shorter tube) will be welded to the bottom bracket on the original BMX rear triangle. To mate properly with the bottom bracket, the tubing must have a round fish-mouth cut made in it, roughly the same diameter as the bottom bracket shell. As shown in Figure 13-27, a round object with the same diameter as the bottom bracket is used as a guide to trace out a line to be cut from the rear boom tube. Keep in mind that the cuts are made on the sides of the rear boom tube, not the top and bottom where the cuts for the head tube will be made.

The joint between the rear boom and the bottom bracket should look like the one shown in Figure 13-28. Keeping the gap to a minimum facilitates the welding process and reduces warping as the weld metal cools.

Once you have the fish-mouth cut made in the end of the rear boom, it is time to join the rear triangle to the boom. The ground clearance needs to be set so that there are 4 inches between the underside of the boom and the ground. Install a wheel into the rear triangle and mock up the frame, as shown in Figure 13-29. This is done so that you know at what angle the rear triangle needs to be welded to achieve the 4-inch ground clearance. This process must be done with one of your wheels and a fully inflated tire, in order to make sure that the final angle is

Figure 13-25   *Side boom joint completed*

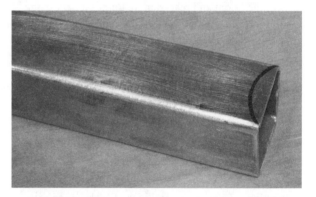

Figure 13-27   *Making a cut for the bottom bracket*

Figure 13-26   *Cleaning up the boom welds*

Figure 13-28   *Checking the bottom bracket joint*

Figure 13-29   *Setting the ground clearance*

Figure 13-31   *Welding the lower boom joint*

the same as this one. Normally, I do not like to weld a frame in place like this, but it does make the process a lot easier to simply drop a tack weld onto the bottom bracket and boom joint right where it sits. Make sure the boom tubing is in line with the rear wheel and rear triangle before you make your first tack weld, and then recheck this alignment.

The large tack weld shown in Figure 13-30 was done as the frame was standing in place, as shown in the previous photo. By making a large tack weld, you can carefully handle the frame once the rear wheel is removed, which will make it much easier to visually inspect the boom and rear triangle for alignment from several different angles. It should be possible to force the rear triangle side to side a little bit if necessary, in order to make it perfectly in line with the boom tubing.

If you are satisfied that the boom tubing and rear triangle are in line, then you can add another large tack weld on the underside of the bottom bracket and rear boom joint to hold the two pieces together for one last visual alignment check. When you are sure that things are aligned, complete the weld on the underside of the bottom bracket, as shown in Figure 13-31. Try to make

one continuous weld across the joint to minimize any side-to-side distortion.

The top of the bottom bracket and boom joint can be done once the underside has been completed. The sides of the joint are done last, and should be done in the same manner to ensure that one side does not flex any further than the other. If you think the alignment of the rear triangle and boom is off on one side, then weld the joint on the same side you wish to pull the alignment. It is a welders' trick to use heat distortion to one's advantage. Figure 13-32 shows the completed bottom bracket and rear boom joint, welded all the way around.

The frame may feel extremely sturdy after you complete the welding between the bottom bracket and the rear boom, but it is not strong enough to be used as it is. The square tubing boom could easily support up to 250 pounds like it is, but the bottom bracket shell would have to endure more stress than it could handle, and would simply fold up over the first hard bump. For this reason, the addition of a smaller top tube will give the frame incredible strength, as well as offer protection and support for the top of your battery pack. This top tube is

Figure 13-30   *Boom and bottom bracket tack welded*

Figure 13-32   *Boom and bottom bracket welding completed*

just a length of thin walled (1/16) 1-inch square or round tubing that joins the top of the seat tube to the front boom, as shown in Figure 13-33. The length of this tube is completely dependent on the distance from one tube to the other, so just measure across the distance and cut a tube a few inches longer so you can work backwards. The top tube runs perfectly parallel to the rear boom tube.

Figure 13-34 shows how the top tube will meet the seat tube using another fish-mouth cut to keep gap spacing to a minimum. This is why I recommended that you start with a top tube a little longer than necessary, as it is easy to grind a little bit out of the joint at a time and work backwards until the tube fits perfectly. It is all too easy to make a tube a quarter inch too short, which will be a nightmare to weld in place due to the massive gap in the joint.

The top tube runs parallel to the rear boom, so whatever height is needed at the seat tube end should be transferred along the entire length of the top tube right up to the joint at the front boom. Once you have the fish-mouth cut at the seat tube end of the top tube, you can tack weld it in place or hold it there, while you mark the point where the top tube needs to connect with the front boom. I made a mark for the center of the top tube, as shown in Figure 13-35.

Figure 13-33   *Installing the top tube*

Figure 13-34   *Top tube and seat tube joint*

Figure 13-35   *Keeping the top tube parallel to the rear boom*

Once you have the front boom mark made, weld the top tube in place, as shown in Figure 13-36, and your frame will be extremely solid and able to support a great deal of weight. Although frame flex is not much of a worry now, it is always good practice to start welding the two sides of a joint that will cause the least amount of flex. For this joint, it is the top and bottom again, just as it was for the rest of the frame tubing.

Although I ended up welding a seat post directly to the top of the seat tube, it still made sense to cap off the open end of the tubing, as it no longer has a clamp and will not need to have a seat post inserted into it ever again. You have tons of room to create your own seating system for the Silent Speedster, and you may not even use the seat tube as a support, so capping off the open end is a good idea to make the final product look a lot more professional. A simple method of capping off an open tube is to weld in a washer and then fill in its hole.

A little work with a sanding disc does wonders for the once-open seat tube, as shown in Figure 13-38. Now dirt and moisture can't enter the bottom bracket if you decide not to use the seat post mounting system shown later in the plan.

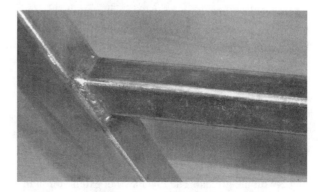

Figure 13-36   *Welding the top tube to the front boom*

Figure 13-37 *Capping the top of the seat tube*

Figure 13-39 *Setting the 70-degree head tube angle*

Figure 13-38 *Seat tube cap completed*

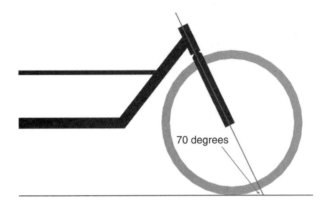

70 degrees

Figure 13-40 *70-degree head tube angle*

Once you join the head tube to the rest of the frame, you will be able to drop in your batteries and hit the streets! OK, there is a bit more work to do, but once the head tube is installed, you will have a rolling chassis, and that is a significant milestone in any vehicle build. The head tube needs to be installed so that the angle formed between it and the ground is approximately 70 degrees, as described in the following drawing. Note that 70 degrees is a good head tube angle for a vehicle of this speed and wheelbase, but if you know what you are doing, then you may want to alter this angle slightly for a more relaxed, or more responsive, ride. You will need to mock up the frame one last time, keeping that 4-inch ground clearance and making sure the rear boom is parallel to the ground, as shown in Figure 13-39. The front wheel and head tube are then held in place at 70 degrees so you can draw a line on the top of the front boom for cutting. You should leave 2 inches of clearance between the edge of the front tire and the front boom for fender placement.

Shown in Figure 13-40 is the imaginary line that extends through the head tube and makes the 70-degree angle with the ground. The line may not extend through the front forks as they probably have some rake, so just

remember that it is the head tube angle that counts. To set up this angle, I tape a long, straight stick to an adjustable square and then place it on the ground once it is set at 70 degrees. The long stick will help you set the proper head tube angle.

If you can't seem to get your front wheel to stay in place at the required 70-degree angle, then have a helper trace the cut line as you hold the head tube against the side of the front boom, as shown in Figure 13-41. Remember that the fish-mouth cut will take up about half

Figure 13-41 *Marking the line to cut the boom*

an inch, so add this to your cut line or simply trace the line about a half inch past the head tube as it is held in place. Your goal is to mount the head tube at 70 degrees, and end up with 2 inches of clearance between the front boom and the front tire.

The fish-mouth cut on the end of the front boom is made just like the one for the bottom bracket earlier, using a similar diameter tube or object as a guide when tracing the line. A half-inch deep cut should be all that is needed to cradle the head tube into the 1.5-inch square tubing for a perfect fitting joint. Figure 13-42 shows the boom tube after grinding out the half-inch deep head tube fish-mouth cuts on the top and bottom of the tubing.

After making the fish-mouth cuts, prop up the frame again and drop in the head tube to make sure it is close to the 70-degree angle. A degree in either direction will not make a huge difference, but try to get it as close as possible, grinding a little bit out at a time until the angle is very close. Figure 13-43 shows the perfectly snug joint

between the head tube and the top of the front boom, as the head tube rests at the desired 70-degree angle.

When you have the head tube joint up to par, make a single tack weld on the top of the joint, as shown in Figure 13-44. I did this with the frame set up on the ground, so I could examine the head tube alignment from all angles, ensuring that the head tube is in line with the rest of the frame. The head tube should be parallel with the seat tube and rear wheel. You can work the head tube side to side a slight bit if it is necessary, and then add another tack weld on the bottom of the joint to hold it securely in place. Due to the fish-mouth cut, the 70-degree angle should already be good, but it doesn't hurt to check it one last time.

With a good tack weld on the top and bottom of the head tube joint, the frame can now support itself as long as you do not put any weight on it. This is your last chance to check the head tube for alignment visually, and make the needed adjustments by tapping it or forcing it by hand. Look at your frame from all angles, making sure both wheels are parallel when the front wheel is pointing straight ahead. If everything looks good, carefully remove the front forks and rear wheel so you can complete the frame welding on your workbench.

As you have probably guessed, start by welding the top and bottom of the head tube joint and then move on to complete both sides. I like to drop in an old pair of head tube cups, as shown in Figure 13-46, so that you keep the head tube as round as possible, as the heat from welding tries to distort it. Avoid any grinding of this weld, as it is very strength-critical and should be as strong as possible.

Now you have a rolling frame that supports weight without any problem, so drop in a set of handlebars and you can safely handle the bike and sit on the frame.

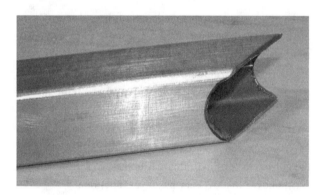

Figure 13-42 *Cutting the head tube fish-mouth*

Figure 13-43 *Checking the head tube angle*

Figure 13-44 *Tack welding the head tube in place*

Figure 13-45 *Head tube alignment final check*

Figure 13-46 *Completing the head tube welding*

Figure 13-47 *Where shall I sit?*

There are so many seating options that can make this scooter into a multitude of different vehicles, each with a different look and feel. You could go for the stand-up scooter look and simply bolt on a pair of BMX foot pegs to the side of the frame, or how about a sideboard-style plank? For the sit-down scooter, you could add a nice fishing boat seat, or make your own seat system, complete with shock absorber and luggage compartment. I decided to use the simplest bicycle-style seating method, since it recycled some of the leftover bicycle parts, and would cost almost nothing.

No matter what type of seat you choose, you have to decide where your butt is going to live on the scooter by trial and error. I found that the bucket placed near the rear of the frame shown in Figure 13-47 was just perfect for my size and sitting style, so I used that as a guide. When choosing a sitting placement, keep in mind that the foot pegs will be placed 6 inches from the ground, and that you should be able to reach the handlebars without stretching. Other than that, feel free to use your imagination and make any kind of seat you like.

Because of the angle of the rear triangle, I knew that a seat post would never be able to function the way it used to, so the top of the seat tube was capped off. Of course,

this doesn't mean you can't weld a seat post to the top of the seat tube at whatever angle is necessary, and that is exactly what I intended to do. I found a really ugly 1970s exercise bike seat, a few seat stays from an old bicycle frame, and the original seat post that came with the donor BMX bike. The fat seat will be replaced with a nicer one, and the curve in the BMX seat post was perfect for placing that seat exactly where I wanted it, as per the bucket test earlier.

The two support tubes made out of the old seat stays are absolutely necessary in order to make your seat post rigid, especially when it is placed at such a laidback angle. A gusset would have also worked, but I wanted to use up some of the scrap frame tubing I had lying in a pile in the corner of the garage. When you are making your seat mounts, be aware of your rear brake hardware, making sure you are not placing any part in the way of the correct operation of the rear brakes. Figure 13-49 shows how I made a very basic yet functional bicycle-style seat mount for my Silent Speedster.

Now that you can sit and steer your scooter, it's time to add the muscle: the battery pack. Because of the simple

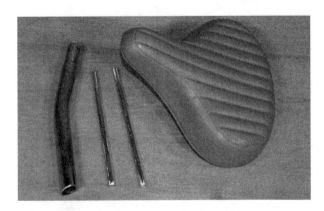

Figure 13-48 *Recycling some old bicycle parts*

Figure 13-49 *Installing the seat post*

square tube frame, the battery pack can sit right on top of the rear boom, and only needs four small tabs to secure it in place. This system lets you have more than one battery pack for a quick change if you plan on riding your scooter non-stop, and do not want to wait overnight for the battery pack to recharge. The battery box tabs are shown in Figure 13-50, and are made from some scrap 1-inch flat bar cut into four 1-inch long segments with a hole drilled in the center. Read ahead a bit to see how this simple battery mounting system works.

The two lower battery pack tabs are welded to the underside of the battery box frame, as shown in Figure 13-51. The tabs are offset from the center of the battery box frame so the battery pack ends up centered in the frame when installed. Since the tabs press up against nuts that have been welded to the side of the rear boom tubing, take this into account when figuring out how far to offset the battery pack tabs. The offset from center will be 3/4 of an inch plus the thickness of the nut. Three quarters of an inch is half the width of the 1.5-inch frame tubing.

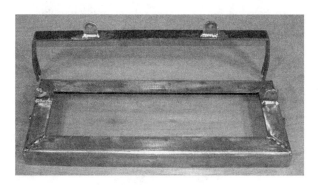

Figure 13-50 *Installing the battery pack tabs*

Figure 13-51 *Lower battery pack tabs installed*

Figure 13-52 shows how the tabs hold the battery pack to the rear frame boom. Under each tab is a nut welded to the frame, just like it was done earlier when you made the battery box strap. Now that the lower battery tabs are installed and the battery pack is held in place, you can figure out the placement of the upper battery tabs.

The upper battery box mounting tabs work just like the lower tabs, but another small length of flat bar will come down from the top tube to place the nut in the correct position. As shown in Figure 13-53, a marker is used to

Figure 13-52 *Battery tab nuts and bolts in place*

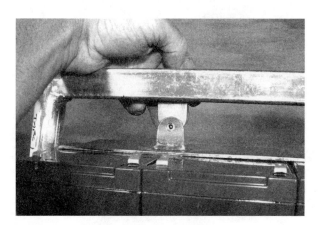

Figure 13-53 *Installing the upper battery box tabs*

mark the point where the nut needs to be welded to the flat bar that will hold it in place. Also, keep in mind that the battery pack needs to slide out of one side of the frame, so all nuts should face the same side of the frame.

Figure 13-54 shows the upper battery mounting tabs installed and securely holding the battery pack in place. Now the entire battery pack can be removed by taking out the four bolts that hold it in place.

The rear of the upper battery tabs is shown in Figure 13-55. The other side is the side that the battery pack slides out of.

Foot pegs can be placed practically anywhere on the scooter that you like, but as a rule, try to keep them between 5 and 6 inches off the ground, so you can hit the corners at full speed without having a foot peg strike the ground. If you plan to do a lot of off-road riding, then consider foldable pegs, as these are much safer when obstacles such as trees and stumps may be thrown in the mix. A very simple foot peg system using commonly available bicycle parts can be made using some BMX stunt pegs, a 4-inch length of 1/16 wall tubing, and a high-grade bolt with a diameter of half an inch. These

Figure 13-56 *Foot peg hardware*

parts are shown in Figure 13-56. Drill the 1/2 hole in the BMX foot pegs, as they are pre-drilled and threaded for bike axles.

Figure 13-57 shows how the foot peg hardware comes together, using the long bolt to press both foot pegs against the $^3/_3$-inch tube that will then be welded onto the scooter frame. A decent grade bolt is a must, as a soft steel bolt could bend if you put too much weight on a single peg.

Where you choose to install the foot pegs is up to you, but don't place them lower than 5 inches from the ground, or you may not be able to navigate corners without slowing down. At 6 inches from the ground, you will be able to lean your scooter much farther than you will ever have the nerve to do, unless motorcycle racing is your hobby. I mounted the foot peg tube on the inside of the front boom, as shown in Figure 13-58. Because the front boom is at an angle, there is no interference with the battery pack, and my feet end up in a nice relaxed position, while still maintaining a 6-inch foot peg to ground clearance.

The tube that holds the foot pegs and bolt is welded directly to the frame, as shown in Figure 13-59. In addition to making sure you do not place the pegs lower than 5 inches to the ground, the only other thing you need to do is ensure that the foot pegs are level, or

Figure 13-54 *Battery pack securely mounted*

Figure 13-55 *Upper battery tabs rear view*

Figure 13-57 *Foot peg hardware setup*

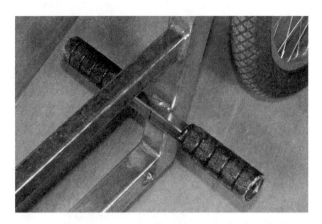

Figure 13-58 *Foot pegs installed onto the frame*

Figure 13-59 *Foot peg mounting tube welded to the frame*

perpendicular to the frame. Also, never choose the battery box as a foot peg mounting place.

The last thing you need to do before painting the frame is to cover up the unused open bottom bracket. This step is actually optional, but honestly, after all that work, why not finish the job, and cover that open hole which obviously looks out of place? By covering the open bottom bracket shell, you hide the last trace of that home-built look, and cover up any places that moisture or critters might want to take up residence. Any scrap sheet metal can be used to cover the holes, and I simply rough cut them out with the cut-off wheel from some scrap sheet metal, as shown in Figure 13-60.

Once the two thin metal discs are cut out, they are tack welded to the bottom bracket shell, as shown in Figure 13-61.

Weld the two bottom bracket caps all the way around, as shown in Figure 13-62. Due to the thin sheet metal used, the welder can be turned down a great deal. At first, the weld looks ugly and rough, but that will soon be fixed.

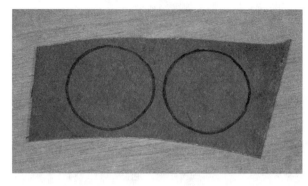

Figure 13-60 *Making some bottom bracket caps*

Figure 13-61 *Capping off the open bottom bracket*

Figure 13-62 *Welding the caps in place*

After a little work with the sanding disc, the bottom bracket caps look perfect, hiding the last telltale sign that the scooter frame might be home built. It's the little details that get noticed the most, especially by fellow builders.

My hub motor kit came with a motor controller that straps to the frame. I added the two small tabs, shown in Figure 13-64, to keep the controller from wandering around over time due to bumps and vibration. Thinking ahead to small details like headlight mounting, fenders,

Figure 13-63 *Grinding the caps clean*

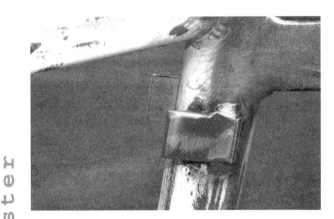

Figure 13-64 *Motor controller mounting tabs*

and small accessories helps avoid the need to touch up paint when you realize it later.

Waiting for paint and primer to dry is a boring experience, but don't lose patience and try to get your trike on the road before the paint or primer has cured. Often, primer needs a full day to dry, and paint will not be ready for handling for at least two days. If you read deep into the application directions, you might even see that proper paint curing time is a week or more, so the paint will be easily scratched if you are not careful. I always use department store spray cans, and usually the results are very good if I can avoid the urge to assemble a wet-paint frame. Figure 13-65 shows the frame freshly primed with the smaller parts painted and curing.

Paint also has a curing time that should be considered. Often, you can handle the frame carefully after a day or two of drying, but the paint will chip and scratch easily until the curing period has passed, which is often more than a month! I have been guilty of assembling a project with tacky paint, but I do try not to rough handle any part until at least a week has passed. Figure 13-66 shows the

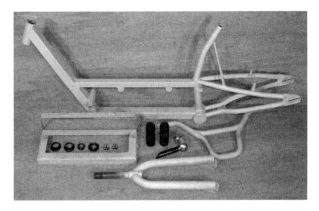

Figure 13-65 *Primed and curing*

results of a few days of hard yet fun work in the garage: a freshly painted scooter chassis ready for wiring and cabling. Notice the ugly retro seat has been replaced by a modern gel seat.

Take your time with the electrical system, minding your positives and negatives. Always use proper terminals, and always solder every connection. Crimp-on connectors are for clowns! Seriously, crimped connectors are failure prone and induce losses or electrical sparks into your system as they work loose over time. Your battery pack should have a connector that can be easily unplugged for your charger, and all wiring should be the proper gauge. If you are not certain what size of wire to use, then look at the original wiring that came with your hub motor kit, and use larger wire than that. A fuse between your battery pack and the connector is also a good idea, in case you short the connector or fry the controller. If you have no idea whatsoever on how to wire four batteries in series, then ask a friend—it's no big deal, and we all have to start someplace. Figure 13-67 shows my series wired battery pack using soldered

Figure 13-66 *The basic rolling chassis painted*

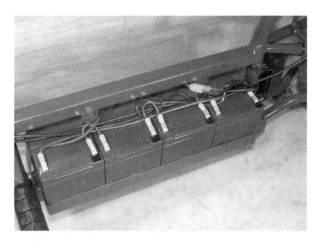

Figure 13-67   *Neatly wiring the electrical parts*

spade connectors and an automotive connector plug.

The right-hand side of your handlebars will probably be quite busy, including a throttle, brake lever, and a warning bell to get slow pokes out of your way. Figure 13-68 shows my right-hand control cluster, with all the goodies in close proximity. In my area, the law requires a horn on all electric bicycles, and although I was tempted to add a large car horn, I decided to be nice to my battery pack and install a silly bell instead. "Ding!" Look out, high speed scooter coming your way!

The rear brakes are of the higher quality cantilever style, as shown in Figure 13-69. These brakes squeeze directly against the rim for efficient grip unlike the standard pull brakes, which work more like pliers. No matter which brake system you have, always use new cables, and make sure you test them intensively before hitting the street. An electric vehicle like the Silent Speedster is a lot heavier than a bicycle, and can outrun

Figure 13-69   *Rear brake detail*

many cyclists, so you certainly want good brakes. Never force a cable into a tight bend, and make sure no cable rubs on any moving part, especially when you turn the front wheel from side to side.

My front brake is a side-pull, caliper-style brake (Figure 13-70), which does work, but not nearly as good as the type used on the rear of the scooter. Each brake lever has an adjustment screw, and it should be screwed all the way in when you initially set up the brakes. If the brakes don't grab hard enough, slowly turn the lever adjustment screw counterclockwise to stiffen up the brakes. A tiny bit of brake rubbing is OK, as long as you can spin the wheel freely.

Figure 13-71 shows the Silent Speedster ready for action. I have added a headlight, flashing rear light, and will also be adding two fenders and a rear view mirror. I will probably also make a simply battery box cover, but for now, I enjoy showing off the internals when I get

Figure 13-68   *Throttle, brake, and bell*

Figure 13-70   *Front brake detail*

Figure 13-71 *Ready to ride!*

amazing that the machine makes no pollution or noise whatsoever. Only the sound made by the tires hitting the road can be heard, and the motor will happily run for hours without generating any noticeable heat. The handling characteristics are amazing, and I felt completely at home in the thick of traffic, with a clear view of the road ahead. Just before Figure 13-73 was taken, a deer was standing only a few feet from the parked Silent Speedster, and did not even notice when I rode up. You can really enjoy the silence of the outdoors without any noise pollution on an electric vehicle.

I took the Silent Speedster up to a few large hills and was totally impressed at the hill flattening power delivered by the hub motor. I already knew this hub motor could tackle large hills, but the added battery really gave it a nice boost, and there was almost no noticeable speed loss on a very large railroad overpass that I went up and down several times. When I made it to the top of the hill shown in Figure 13-74, I was still managing top speed, even against the very strong wind that day. On the way down a steep hill, the hub motor acts like a governor, only allowing a gliding speed of a few percent more then the top speed of the motor, so you rarely have to brake on long steep hills.

Cornering is a real blast on the Silent Speedster because of the liberal foot peg to road clearance. You can hit the corners at full speed, as long as the road is clear and you have the nerve to lean that far over. I was racing in a small clean parking lot at full speed for quite a while, and had a very good lean going into the corners at high speed. The turning circle is also very good, so U-turns in the middle of the street only take up one lane. In

talking with fellow builders and electric vehicle enthusiasts, so the batteries are going to remain in full view for the time being. Accessories really change the look of a scooter, so get down to your bike shop and see what you can find. If you are good with composite construction, the addition of a fairing would really put your scooter over the top.

I am really happy with the way the Silent Speedster has turned out. The finished product in Figure 13-72 shows nice clean lines and solid construction. Oh, and I must thank "TheKid007," a fellow Atomic Zombie Krew member and builder, for coming up with the name Silent Speedster. I was so busy in the garage that I did not even have time to think up a good name.

Considering how fast the scooter takes off, and the fact that I can almost keep up with city traffic, it is truly

Figure 13-72 *The Silent Speedster calling for a pilot*

Figure 13-73 *Sneaking up on nature*

Figure 13-74 *Hill climbing is no problem*

Figure 13-76 *Nothing but open road ahead!*

Figure 13-75 *Speeding along on battery power*

Figure 13-75, I am turning around at the top of an overpass for yet another run up and down the hill to test the ability of the hub motor.

I think the Silent Speedster is a perfect way to commute distances without making any pollution, noise or environmental impacts. You probably do not need a license or insurance to ride a small electric bicycle like this in many parts of the world, and in communities where there are no existing electric bicycle laws, you should be able to blend right in with typical bicycle traffic. By adding a solar charger to your system, you could have a maintenance-free form of transportation that will cost you just pennies to run, and you can ride proud knowing that you built it yourself. I hope you enjoy riding your new electric vehicle, and urge others to build their own version of this non-polluting alternative transportation. Leave the gas-powered vehicles at home and let the mighty electron move you!

Project 14: Kids' Electric Trike

Figure 14-1 *The basic parts needed for the trike*

Here is a fun, little, electric trike for the "young 'uns" that can be built in a single evening from a pair of children's bicycles, a few bits of scrap tubing, and an old stand-up electric scooter. Actually, this project is so simple that you can make it out of practically any battery-powered DC motor and any scrap bicycle parts you may have lying around the shop. Since speed is certainly not the goal here, even an old cordless drill will make a fine power source for this vehicle, allowing your kids to run around the yard for as long as the batteries hold out. You can purchase one of those small, plastic, electric kids' cars at just about any toy store, but the all-plastic construction and low-wattage motor may not keep up with the kids' demanding driving habits. If your yard or park is not perfectly smooth and hill free, the department store vehicle may wear out in a hurry, or simply fail to traverse the terrain. By using a pair of kids' bikes for the wheels and forks, this electric trike becomes a high-quality vehicle, capable of driving on gravel, up hills, through the mud, and even on the grass.

Since kids' bikes come in many sizes, with wheels measuring 10 inches, 12 inches, 14 inches, and 16 inches, you should not have any problem scrounging up the parts for this project by visiting the local dump or hitting the yard sales around the neighborhood. The parts are really not all that critical, but you will want the two front wheels and two front forks to match, at least for size and shape, since they will be placed side by side on the trike.

Figure 14-1 shows the basic parts that will be needed in order to put the kids' trike together. You will need

three bicycle wheels (two of which should match), three front forks (two of which should match), one head tube and bearing set to match one of the front forks, and some type of DC motor that can power one of the wheels. I chose an old stand-up electric scooter motor as the power source because it could be placed against one of the trike wheels to make the vehicle move, thus requiring no transmission, gears or chains. These little stand-up scooters are also plentiful at yard sales and scrap piles, since it is often the frame that bends before the motor fails. As shown in Figure 14-2, the entire rear end of the electric scooter is cut from the frame, leaving only the motor and the small drive wheel.

Figure 14-2 also shows what will become the front of the trike: a front fork and matching head tube with all of the included bearing hardware. This head tube actually came from the electric scooter, and it just happened to fit the kids' bicycle fork perfectly. With all the bearing

Figure 14-2 *The front fork and head tube*

hardware installed, the fork will spin freely and without friction, which is a sure sign that all the hardware is properly matched. If you take your front forks and head tube from the same bicycle, then you will be certain that all hardware matches.

To make the kids' trike, you need two wheels and forks for the rear and one front fork, wheel and head tube. For this reason, the two rear wheels and forks should match, and the front wheel can be whatever size you want. On my trike, I decided to use three matching wheels since they were easy to locate, and a front fork that was slightly larger than the two at the rear. You will also notice that all the wheels are actually front wheels, since there is no drive chain needed. You could use a rear wheel if you cannot locate three front wheels, but a little force will be needed in order to widen the fork legs to take the axle.

Most cheap stand-up electric scooters have a 100 to 300-watt motor connected to a small rubber wheel, like the wheel you would find on a road skate (Figure 14-4). These motors will typically move the scooter at a speed of 15 to 20 miles per hour when using a good 24-volt power source. For use on a kids' trike, I decided to use only one large 12-volt battery, reducing the top speed to about 10 miles per hour, but extending the run time by many hours. Since the motor already turns the small rubber drive wheel, all you need to do is let it rub against one of the trike wheels, and you will have a transmission system. Whatever speed the scooter would be capable of will be the top speed of your trike, since there is no gear reduction. As long as the scooter drive wheel is turning the bicycle wheel by contact at the edge of each wheel, there is a 1:1 gear ratio, so wheel size makes absolutely no difference to the final speed of the vehicle.

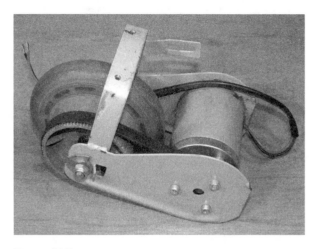

Figure 14-4 *Scooter motor and drive wheel*

Cut the stem from each of the rear forks just above the crown, as shown in Figure 14-5. The cut area will be the point at which you will attach your new frame tubing, as this is the strongest area of the fork. If for some reason your forks are not exactly the same, try to cut each stem so that the distance from the axle to the cut area will match on both forks.

Since I am building this kids' trike from parts I had lying around the garage, I simply placed the two rear forks on the ground, and found a pair of 1-inch-diameter square tubes that would create a triangular frame, as shown in Figure 14-6. I think the square tubing came from an old table I hacked up, but since the trike is only going to hold the weight of a child, you don't need to use heavy tubing. As for the size of the frame, simply have your young pilot sit on a chair, and make some basic measurements of how much room they will need in order to sit comfortably between three wheels. I went for a wheelbase about the same as the original bicycle, and a

Figure 14-3 *Front and rear trike wheels*

Figure 14-5 *Cutting the stem from the two rear forks*

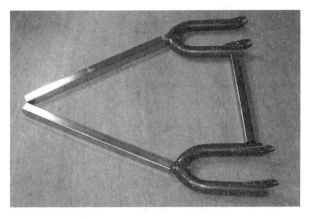

Figure 14-6 *Laying out the basic frame*

Figure 14-8 *Welding the rear frame tube to the forks*

width of about the same as the wheelbase, creating an equilateral triangle footprint.

The basic triangular frame needs three tubes to hold it together, one at the rear to join the two rear forks, and two more tubes at the front to form a triangle between the fork crown area and the head tube. Shown in Figure 14-7 is the tube that will join together the two rear forks, creating the frame to whatever width you like. The fish-mouth cut will make welding a breeze, and the tube should be placed about half an inch from the end of the fork leg tube so there is room to make a weld.

With the parts sitting on a flat surface, tack weld the rear frame tube to the two rear forks, as shown in Figure 14-8, so that the fork legs are at 90 degrees to the rear frame tube. Because both forks are sitting on the same flat surface, vertical wheel alignment will be perfect, and as long as you get both forks running parallel, horizontal wheel alignment will also be perfect. If the two rear wheels are badly out of alignment, there will be scrubbing on the road, which will decrease

battery run time, so try to get the wheels running as true as possible.

Once you have completely welded the rear frame tube, join the two rear forks tighter so that you can continue the frame layout, adding the two front frame tubes that will form the triangle, as shown in Figure 14-9. The two front frame tubes will determine the wheelbase (length) of your trike, which should be fairly close to the original bicycle wheelbase, so that your pilot can reach the handlebars comfortably. Also shown in Figure 14-9 is the front head tube, which should be set at an angle to match the angle at which it was installed on the original bicycle frame. The joint between the head tube and the two front frame tubes needs to be ground out to conform to the head tube, but it is fine to simply rough cut the tubes for now, as you will want to install the wheels in order to figure out the correct head tube angle.

If you welded the head tube at 90 degrees to the front frame tubing, then there is a good chance that you will have the correct head tube angle, as can be seen in Figure 14-10. To make sure, I installed all three wheels to ensure that the head tube angle was almost the same as it was on the original bicycle before hacking it to bits. Since this is a slow-moving kids' trike, head tube angle is not a real concern, so just make your best guess. If it looks right, then, dude, it probably is right! Now you can

Figure 14-7 *Joining the two rear forks together*

Figure 14-9 *Continuing the frame layout*

Figure 14-10 *Figuring out the correct head tube angle.*

finish all the frame welding to secure the main frame triangle to the head tube and rear forks.

The biggest complaint when it comes to kids and their electric vehicles is how fast the battery goes dead. On the original stand-up scooter, you could barely go around the block once before the battery was weak enough to slow down the scooter. By using a much larger battery and half the voltage, the new trike will run almost all day on a single charge, and allow the vehicle to get up hills that would normally exceed the smaller battery packs' immediate discharge capacity. Sure, the large battery is quite heavy, but that simply adds stability to the trike, so it won't tip over as three kids try to stand on it as they roll over the curb. Another very important thing to note about the battery shown in Figure 14-11 is that it is a Gel

cell, or non-spillable battery, which is a must for a kids' vehicle, which could see all kinds of abuse you never even thought of. Unlike a typical lead acid battery, this battery will not leak any dangerous chemicals, so don't use any other type of battery.

The battery will live between the two rear forks, as shown in Figure 14-12, supported by a few bits of flat bar which have been welded to a base made of angle iron. The battery box base should place the battery at least 4 inches from the ground, so it doesn't bang on small objects which may roll under the trike as your kids drive it through hostile territory.

The drive motor can live just about anywhere on the trike, on any one of the three wheels, but it stays neatly out of sight when placed under one of the rear forks, as shown in Figure 14-13. Use whatever means is necessary in order to place the scooter's drive wheel up against the

Figure 14-12 *A simple angle iron battery box*

Figure 14-11 *A large, 32-amp-hour, 12-volt battery*

Figure 14-13 *Installing the motor*

trike's wheel, so that there is a decent amount of friction between the two wheels. Inflate the tire to the proper pressure before you do this, and try to keep both the drive wheel and trike wheel in line, so that useless friction is avoided. There should be enough friction between the two wheels that it takes a bit of force to make them slip, but not so much that they are difficult to turn by hand. The little scooter motor is not all that powerful, so a little slipping off the start is better than so much friction that your motor overheats and burns out. A nice smooth tire, or at least minimal tread at the point of contact, is also a good thing.

Figure 14-14 shows the drive motor and all three wheels installed on the basic trike frame. If I hold up the drive wheel and drop the motor wires across the battery, the drive wheel hums along, and seems to power the trike with a reasonable amount of force when it is on the ground. Take note of which way the wheel turns according to wire polarity, so you don't end up with a backwards moving vehicle after finalizing the wiring. Yes, the scooter wheel must turn in the opposite direction of the trike's rear wheel, or counterclockwise, to move forward.

A very simple footrest can be made with a piece of bent tubing, or a pair of conduit elbows welded together, as shown in Figure 14-15. Again, I try to use up whatever bits of scrap tubing are lying around the garage, so feel free to experiment with whatever footrest designs you think will work.

The seat will be placed over the battery, helping protect the wiring, and keeping meddling hands out of the electrical bits. Using some 3/3 tubing, or whatever scraps you have in the junk pile, make a basic seat frame, like the one shown in Figure 14-16, that can be welded to

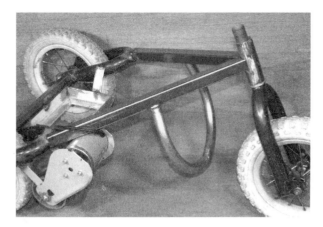

Figure 14-15   *A place to rest those lazy feet*

Figure 14-16   *Making a seat frame*

the frame, allowing the seat to cover the battery yet also allowing for the removal of the battery. An old kitchen chair with metal legs is a good source of metal for the seat frame as well as the seat cushion.

There is nothing critical about the seat frame, other than allowing a comfortable sitting position and easy removal of the battery for charging or swapping. The simple seat frame shown in Figure 14-17 allows the battery to drop into the battery box from the front of the trike once the seat cover has been removed.

Once you finish up all the welds and paint your trike, your young pilot will be able to hit the trails on a long-running, environmentally friendly vehicle. Of course, you have to do a bit of electrical wiring first in order to transfer electrical power from the battery to the motor, but since it is a simple matter of adding a switch between the two, there is no need for a circuit diagram. If your scooter came with a throttle switch like mine did, then simply install the electrical system exactly the way

Figure 14-14   *Drive motor installed*

Figure 14-17 *Installing the seat frame*

Figure 14-18 *Ready to ride!*

Figure 14-19 *Rollin', rollin', rollin'*

it was on the scooter. If you have no throttle switch, then find a contact switch that can handle your motor's power (10 to 30 amps typically), and you are ready to roll. I also added a front pull brake to the completed trike shown in Figure 14-18, although it was probably not really necessary due to the limited top speed, and the fact that it stops moving within a few feet once the throttle is off.

The completed kids' electric trike runs for many fun hours on a single battery charge, and can take the abuse that a young pilot can inflict without any problems. Top speed is limited to a kid-safe level due to the lower voltage, and even if the trike stalls, the motor will not burn out because there is a limited amount of slip in the friction drive. Figure 14-19 shows the brave test pilot, Dylan Lange, putting the trike through its paces. Oh, and yes, the trike can also move a fully grown kid around, as I have found out!

Conclusion

Well, there you have it, folks! We took a pile of recycled bicycle parts, some scrap tubing, a few basic tools, and carved out an entire fleet of unique and functional vehicles. This hobby is something that anyone can enjoy without spending a lot of money or needing an engineering degree. If you can imagine a new kind of human-powered or electric vehicle, then chances are you will be able to build it with a little effort. Builders like you have picked up that welding torch and made bicycles that fly, float, and even reach speeds beyond the posted highway limits. Once you get the basic skills and get past that first build, you will be hooked forever, always modifying or improving your cool inventions. Inventing mechanical devices is an art much like painting.

Your workbench is the canvas, the welder is the brush, and that pile of junk in the corner of your garage is a wonderful palette of paint. Don't be afraid to set your imagination free and alter the designs presented here to suit your own needs. Mix and match projects, change dimensions, or simply use the plan as a general guide to build something altogether different—a little improvization is what makes a true garage hacker shine.

Thanks for your support, and we hope to see your work displayed in our forum and online gallery. If you would like to connect with other fellow garage hackers to share ideas, discuss your own projects, or seek help on some of the vehicles presented here, then log on to www.atomiczombie.com and say Hi.

Well, we're off to the garage now where half a dozen unfinished mechanical monsters are waiting for me to help them come alive. There will be sparks, smoke, grinding, hammering, and a lot of sweat, but eventually that garage door will creak open once again, and who knows what might roll out this time!

Cheers and happy building!

Brad and Kat
The AtomicZombie Krew

Index

H

I

K

L

M